S0-ANO-399

NITROGEN IN DESERT ECOSYSTEMS

US/IBP SYNTHESIS SERIES

This volume is a contribution to the International Biological Program. The United States effort was sponsored by the National Academy of Sciences through the National Committee for the IBP. The lead federal agency in providing support for IBP has been the National Science Foundation.

Views expressed in this volume do not necessarily represent those of the National Academy of Sciences or the National Science Foundation.

Volume

1 MAN IN THE ANDES: A Multidisciplinary Study of High-Altitude Quechua/*Paul T. Baker and Michael A. Little*

2 CHILE-CALIFORNIA MEDITERRANEAN SCRUB ATLAS: A Comparative Analysis/*Norman J. W. Thrower and David E. Bradbury*

3 CONVERGENT EVOLUTION IN WARM DESSERTS: An Examination of Strategies and Patterns in Deserts of Argentina and the United States/*Gordon H. Orians and Otto T. Solbrig*

4 MESQUITE: Its Biology in Two Desert Scrub Ecosystems/*B. B. Simpson*

5 CONVERGENT EVOLUTION IN CHILE AND CALIFORNIA: Mediterranean Climate Ecosystems/*Harold A. Mooney*

6 CREOSOTE BUSH: Biology and Chemistry of *Larrea* in New World Deserts/*T. J. Mabry, J. H. Hunziker, and D. R. DiFeo, Jr.*

7 BIG BIOLOGY: The US/IBP/*W. Frank Blair*

8 ESKIMOS OF NORTHWESTERN ALASKA: A Biological Perspective/*Paul L. Jamison, Stephen L. Zegura, and Frederick A. Milan*

9 NITROGEN IN DESERT ECOSYSTEMS/*N. E. West and John Skujiņš*

Additional volumes in preparation

US / IBP SYNTHESIS SERIES 9

NITROGEN IN DESERT ECOSYSTEMS

Edited by

N. E. West
John Skujiņš

Utah State University

Dowden, Hutchinson & Ross, Inc.
Stroudsburg Pennsylvania

Copyright © 1978 by **The Institute of Ecology**
Library of Congress Catalog Card Number: 78-17672
ISBN: 0-87933-333-2

All rights reserved. No part of this book may be reproduced or transmitted in any form or by any means—graphic, electronic, or mechanical, including photocopying, recording, taping, or information storage and retrieval systems—without written permission of the publisher.

80 79 78 1 2 3 4 5
Manufactured in the United States of America.

Library of Congress Cataloging in Publication Data

Main entry under title:
Nitrogen in desert ecosystems.
(US/IBP synthesis series ; 9)
Bibliography: p.
Includes index.
1. Desert ecology. 2. Nitrogen cycle. I. West, Neil E. II. Skijiņš, J. III. Series.
QH541.5. D4N57 574.5'265 78-17672
ISBN 0-87933-333-2

Distributed world wide by Academic Press,
a subsidiary of Harcourt Brace Jovanovich,
Publishers.

FOREWORD

This book is one of a series of volumes reporting results of research by U.S. scientists participating in the International Biological Program (IBP). As one of the fifty-eight nations taking part in the IBP during the period of July 1967 to June 1974, the United States organized a number of large, multidisciplinary studies pertinent to the central IBP theme of "the biological basis of productivity and human welfare."

These multidisciplinary studies (Integrated Research Programs) directed toward an understanding of the structure and function of major ecological or human systems have been a distinctive feature of the U.S. participation in the IBP. Many of the detailed investigations that represent individual contributions to the overall objectives of each Integrated Research Program have been published in the journal literature. The main purpose of this series of books is to accomplish a synthesis of the many contributions for each principal program and thus answer the larger questions pertinent to the structure and function of the major systems that have been studied.

Publications Committee: US/IBP
Gabriel Lasker
Robert B. Platt
Frederick E. Smith
W. Frank Blair, Chairman

PREFACE

Desert biology has long been dominated by attention to the aridity factor. There is still little doubt that this feature is the major cause of uniqueness in desert ecosystems. However, other factors can become limiting to biological production during those times when moisture is adequate. Foremost among these secondary limiting factors are nutrient cycling processes.

Because of little or no leaching, it generally has been assumed that desert soils must be nutrient rich. Scientists in Australia, a continent notorious for its nutrient limitations, began to notice that if two wetter than average years occurred consecutively in their arid areas, plant—and thus animal—productivity showed a decline during the second year. This provided the first hint that nutrients in desert soils might be secondary limiting factors. A series of experiments proved that this was indeed the case. Experimental work with grasses from the center of Australia showed that marked increases in plant yields occurred with the addition of water only when nutrients were also added. Nitrogen was shown to be the major deficient element. Similar phenomena were thought likely to apply in North American deserts.

Part of the research launched by the US/IBP Desert Biome was aimed at unraveling the nitrogen cycle in several types of North American deserts. This was done in the hope that increased understanding of the nitrogen cycle in semiarid to arid ecosystems would provide further essential knowledge of their behavior in the natural state and upon human impact. This understanding could then lead to a possible enhancement of productivity of some desert ecosystems.

As compared to other ecosystems, the scientific research on nutrient cycling in arid lands had been negligible until this decade. We can foresee many years of intensive research work which will only scratch the surface of the wealth of information that could, and should, be obtained for complete understanding of the nitrogen cycle in desert and semidesert areas.

This volume may be considered an initial report on progress toward improving our understanding of the role of one chemical element in the structure, function, and control of desert ecosystems. In the following chapters, the authors review the previous state of knowledge on the nitrogen cycle in these contexts and add their own new expansion of data and conclusions. The result is, we hope, a comprehensive overview of the topic.

Although uncultivated deserts are not presently productive by current agronomic standards, they do cheaply produce food and fiber for many people with little energy or material input. Our efforts have, we believe, given some leads on how to better maintain or expand the production to meet human needs on these lands for a world growing more hungry, especially for nitrogen-derived protein.

We wish to thank the reviewers listed below for their many helpful comments, but we take full responsibility for this volume as authors and editors alike.

Dr. James L. Charley
University of New England
Australia

Dr. F. E. Clark
USDA Science and Education
Administration

Dr. David Coleman
Colorado State University

Dr. C. C. Delwiche
University of California, Davis

Dr. Y. Dommergues
Centre de Pedologie Biologie
France

Dr. W. L. Hunt
Colorado State University

Dr. H. F. Mayland
USDA Science and Education
Administration

Dr. D. Pramer
Rutgers State University
New Jersey

Dr. Robert L. Todd
University of Georgia

Dr. John Waid
LaTrobe University, Australia

Dr. Jack B. Waide
Clemson University
Clemson, South Carolina

Dr. R. G. Woodmansee
Colorado State University

This work received its major financial support from the U.S. National Science Foundation under Grants GB15886, GB32139X, GB41288X and BMS7402671 A04 to Utah State University for which D. W. Goodall and F. H. Wagner were project leaders. We wish to thank both for their constant encouragement.

N. E. West
J. J. Skujiņš

CONTENTS

LIST OF CONTRIBUTORS

Peter L. Comanor
Associate Professor of Biology, Department of Biology, University of Nevada, Reno, Nevada

Paul J. Eberhardt
Assistant Professor of Soil Science, Department of Plant and Soil Science, Tennessee Technological University, Cookeville, Tennessee

Raymond B. Farnsworth
Professor of Agronomy, Department of Agronomy and Horticulture, Brigham Young University, Provo, Utah

Joel E. Fletcher
Professor Emeritus of Civil and Environmental Engineering, Utah Water Research Laboratory, Utah State University, Logan, Utah

Clayton S. Gist
Ecologist, Oak Ridge Associated Universities, Oak Ridge, Tennessee

Richard B. Hunter
Post-graduate Research Scholar, Environmental Biology Division, Laboratory of Nuclear Medicine and Radiation Biology, University of California, Los Angeles

David W. James
Professor of Soil Science, Department of Soil Science and Biometeorology, Utah State University, Logan, Utah

Jerome J. Jurinak
Professor of Soil Science and Head, Department of Soil Science and Biometeorology, Utah State University, Logan, Utah

Gale E. Kleinkopf
Associate Research Plant Physiologist, Department of Plant and Soil Sci-

ences, University of Idaho Research and Extension Center, Kimberly, Idaho

James O. Klemmedson
Professor of Range Management, School of Renewable Natural Resources, University of Arizona, Tucson, Arizona

Brian P. Klubek
Research Associate, Department of Soil Science, North Carolina State University, Raleigh, North Carolina

Mac McKee
Research Associate, Policy Research Program, Utah State University, Logan, Utah

Robert T. O'Brien
Professor of Biology, Department of Biology, New Mexico State University, Las Cruces, New Mexico

Donald B. Porcella
Professor of Environmental Engineering and Associate Director, Utah Water Research Laboratory, Utah State University, Logan, Utah

Evan M. Romney
Research Soil Scientist, Environmental Biology Division, Laboratory of Nuclear Medicine and Radiation Biology, University of California, Los Angeles, Los Angeles, California

Robert R. Rychert
Assistant Professor of Biology, Department of Biology, Boise State University, Boise, Idaho

Pasquale R. Sferra
Professor of Biology, Department of Biology, College of Mount St. Joseph, Mount St. Joseph, Ohio

John J. Skujiņš
Professor of Biology and Soil Science, Departments of Biology and Soil Science and Biometeorology, Utah State University, Logan, Utah

Darwin L. Sorenson
Research Microbiologist, Utah Water Research Laboratory, Utah State University, Logan, Utah

Sadek M. Soufi
Professor of Plant Biochemistry, Department of Biochemistry, American University, Beirut, Lebanon

Eugene E. Staffeldt, deceased

Patricia Trujillo y Fulgham
Research Technician, Department of Animal and Range Science, New Mexico State University, Las Cruces, New Mexico

Thomas C. Tucker
Professor of Soil Chemistry, Department of Agricultural Chemistry and Soils, University of Arizona, Tucson, Arizona

Arthur Wallace
Professor of Plant Nutrition and Research Plant Physiologist, Environmental Biology Division, Laboratory of Nuclear Medicine and Radiation Biology, University of California, Los Angeles, Los Angeles, California

Neil E. West
Professor of Range Ecology, Department of Range Science, Utah State University, Logan, Utah

Robert L. Westerman
Associate Professor of Soil Science, Department of Agronomy and Soils, Oklahoma State University, Stillwater, Oklahoma

NITROGEN IN DESERT ECOSYSTEMS

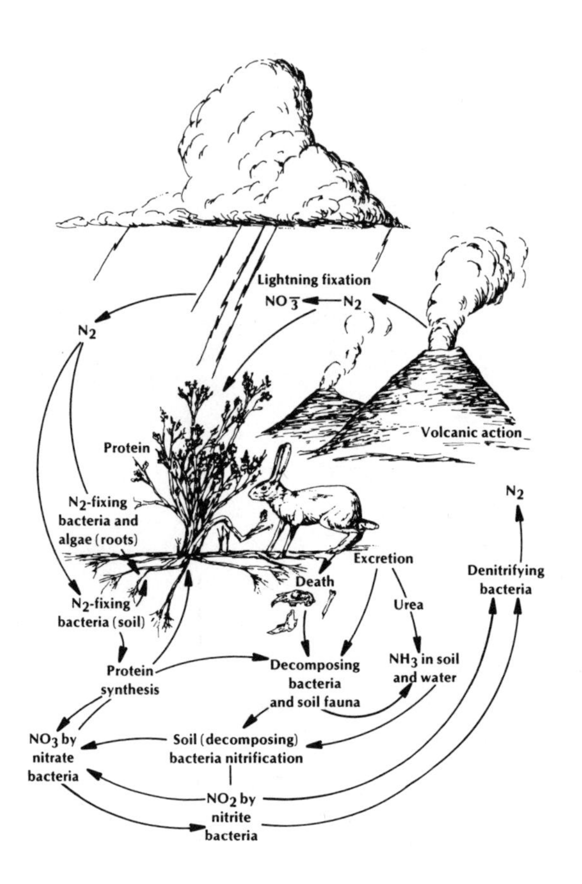

Lightning fixation
NO3 N2
N2
Volcanic action
Protein
N2
N2-fixing
bacteria and
algae (roots)
Excretion
Denitrifying
bacteria
Death
Urea
N2-fixing
bacteria (soil)
NH3 in soil
and water
Protein
synthesis
Decomposing
bacteria
and soil fauna
NO3 by
nitrate
bacteria
Soil (decomposing)
bacteria nitrification
NO2 by
nitrite
bacteria

1

STRUCTURAL DISTRIBUTION OF NITROGEN IN DESERT ECOSYSTEMS

*N. E. WEST and J. O. KLEMMEDSON**

INTRODUCTION

Much of the past effort in ecology has been expended in describing the distribution of organisms in relation to their environment. Data collection has often been taken only at one point in time and space, giving an essentially static view of ecosystem structure. The goal of an emerging ecosystem science is to add the dynamic aspects so that our understanding will include how ecosystems function and control mechanisms operate. However, the study of these time-based patterns is usually started from a descriptive point of reference. By knowing the spatial distribution of organisms, and the qualities and degree of various environmental influences, we can more adequately prepare to sample over a relevant time span.

Descriptive data on nitrogen distribution can give clues to the relative importance of various ecosystem compartments in the total cycle. With only a few assumptions, first approximations of nutrient fluxes can be made (Charley 1972; West, in press b). From this kind of rudimentary understanding, priorities can be set for initiating research on the various processes and for more directly assessing the dynamics of the system.

The approach suggested above will be followed in the development of this volume. In this chapter we will discuss how nitrogen is structurally distributed in desert ecosystems; that is, how nitrogen is located in both the biotic and the abiotic environments. Values given will be standing crops, peak quantities or yearly averages. Relative importance of various compartments and flow rates will be inferred. Authors of subsequent chapters will review the current understanding of temporal patterns and processes contributing to the nitrogen cycle. Occasional later reference will be made to probable control mechanisms operative in these ecosystems. Control of ecosystems is understood poorly at present, however, and necessarily will be built on a more comprehensive theory of ecosystem structure and function.

---

*The junior author's major contribution to the manuscript was made during his tenure as a Charles Bullard Forest Research Fellow at Harvard University, Harvard Forest, Petersham, Massachusetts.

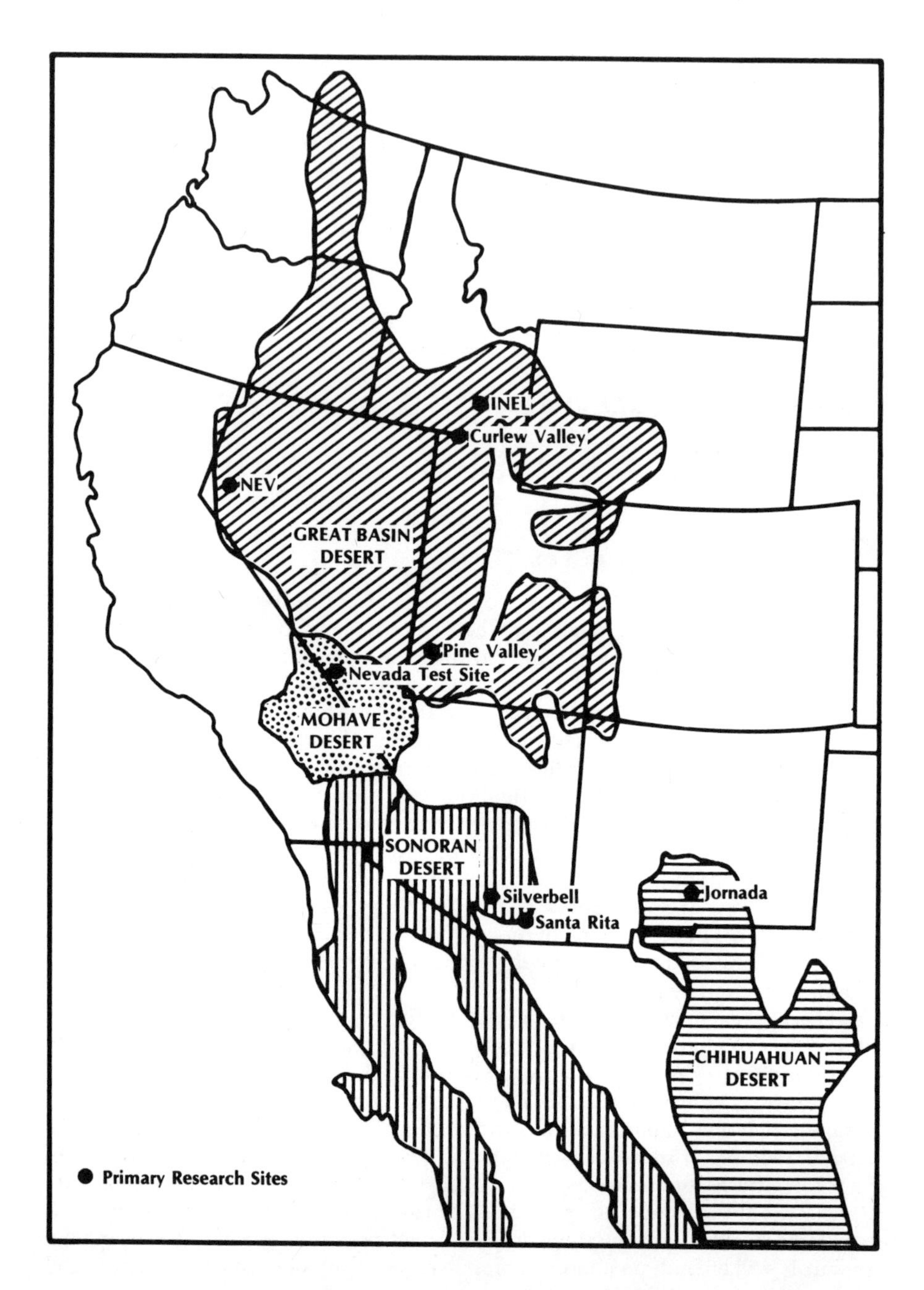

Figure 1-1 *Map of the North American regional deserts [after Shreve 1942]. Locations of US/IBP Desert Biome and other research sites mentioned in this and subsequent chapters are designated.*

COMPARTMENTS

An ecosystem is a complex array of biotic and abiotic compartments together with all the biological-environmental interactions which permit it to exist as a functioning whole. To make some order out of this complexity, ecologists have divided the biotic part of ecosystems into functional groups based on their role in energy flow. These trophic compartments include producers, consumers and decomposers. Abiotic compartments include the inorganic and nonliving organic parts of the environment (mainly soil) and those physical factors and forces associated with the atmosphere (i.e., carbon dioxide, precipitation, wind). We will primarily discuss the distribution of nitrogen in each of these compartments with examples from North American desert ecosystems.

A distinction which we and succeeding authors will frequently use pertains to the different regional deserts of North America (Fig. 1-1). Except for the mountains, the area within these desert boundaries generally receives less than 30 cm of total annual precipitation. Some daily temperature extremes in summer exceed 25 C for all the areas. The duration of heat and drought, however, varies considerably between sites as the climatic diagrams of Figure 1-2 show. The Sonoran and Chihuahuan deserts are warm-winter deserts showing affinities to the subtropics, whereas the Great Basin Desert is classified as a cold-winter desert because of temperate zone placement. The Mohave Desert is transitional between the "warm" and "cool" deserts (McGinnies et al. 1970; Walter 1971, 1973). Considerable differences in ecological responses will be noted in these different synecological contexts.

Producers

Only photosynthetic and chemosynthetic organisms are independent and have the mechanisms to fix energy and incorporate vital elements into biomass de novo. All other organisms depend either directly or indirectly upon these producer organisms for their energy and most of their nutrients. Some producer organisms have the inherent capability to take up atmospheric nitrogen and to metabolize it into protein. Blue-green algae belong to this group of free-living nitrogen-fixers that are especially important in desert ecosystems. The blue-green algae are apparently the major nitrogen-fixing entities of a variety of microorganisms which comprise cryptogamic soil crusts found throughout American deserts (Cameron 1963; Fletcher and Martin 1948; Mayland et al. 1966; Shields et al. 1957). Blue-green phycobionts and free-living blue-green algae account for nitrogen fixation of lichen crusts (Shields et al. 1957; Snyder and Wullstein 1973b), and *Azotobacter*-like organisms are associated with many desert cryptogams (Snyder and Wullstein 1973b). Although the blue-green algae are apparently most effective in nitrogen fixation as phycobionts of lichen-crusts (Shields 1957; Shields et al. 1957), their extensiveness and effectiveness vary considerably in different desert regions (Mayland et al. 1966; Rogers et al. 1966; Lynn and Cameron 1973).

In most ecosystems, symbiotic nitrogen-fixing bacteria associated with legumes produce most of the fixed nitrogen. Legumes, however, are generally minor components of American cold-winter deserts; hence, symbiotic fixation of nitrogen involving higher plants has not been considered important heretofore. There is some

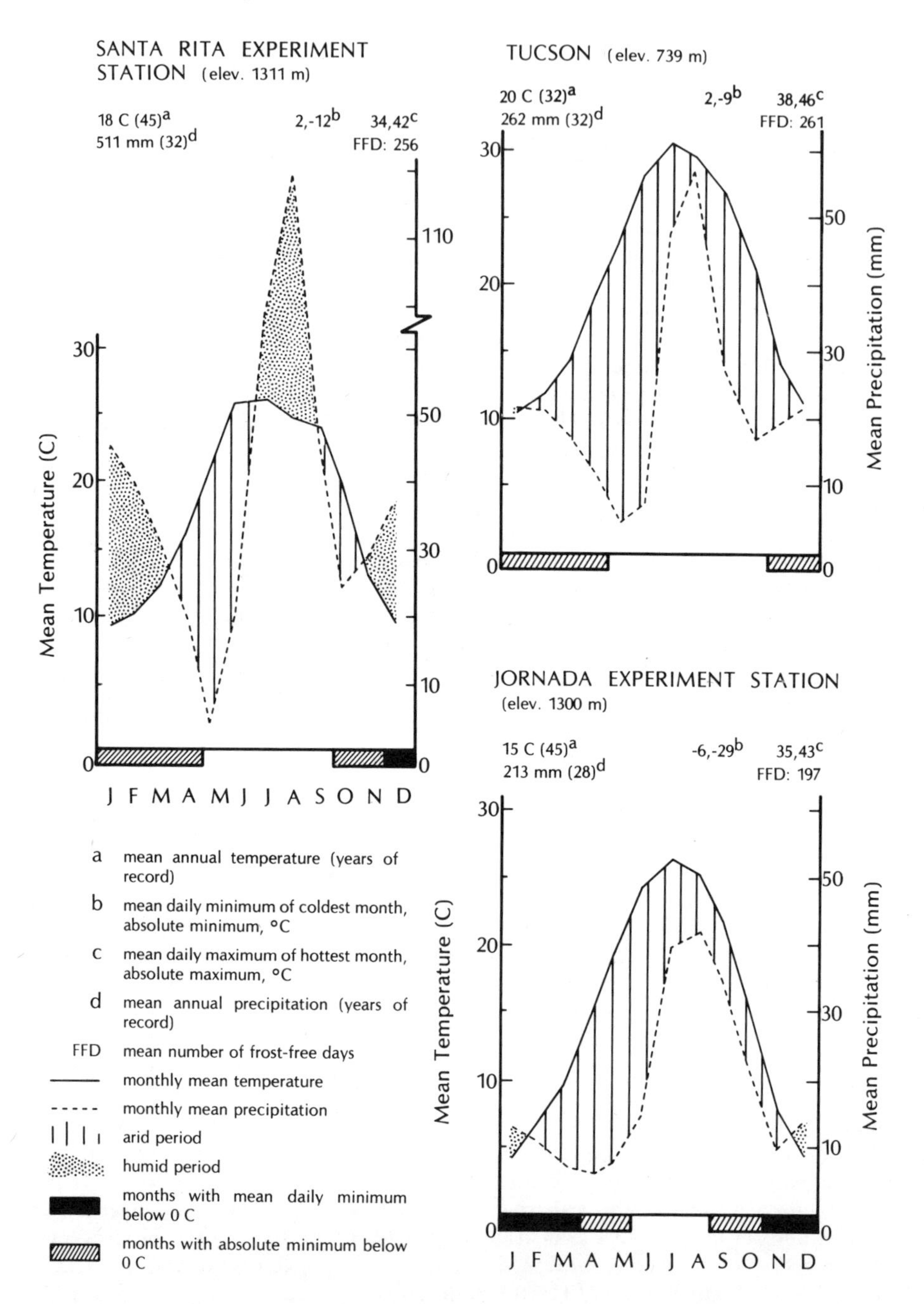

Figure 1-2 *Climatic diagrams, following Walter [1963], for representative stations in the North American desert regions. Representative stations are as follows: Sonoran Desert region, Santa Rita Experiment Station and*

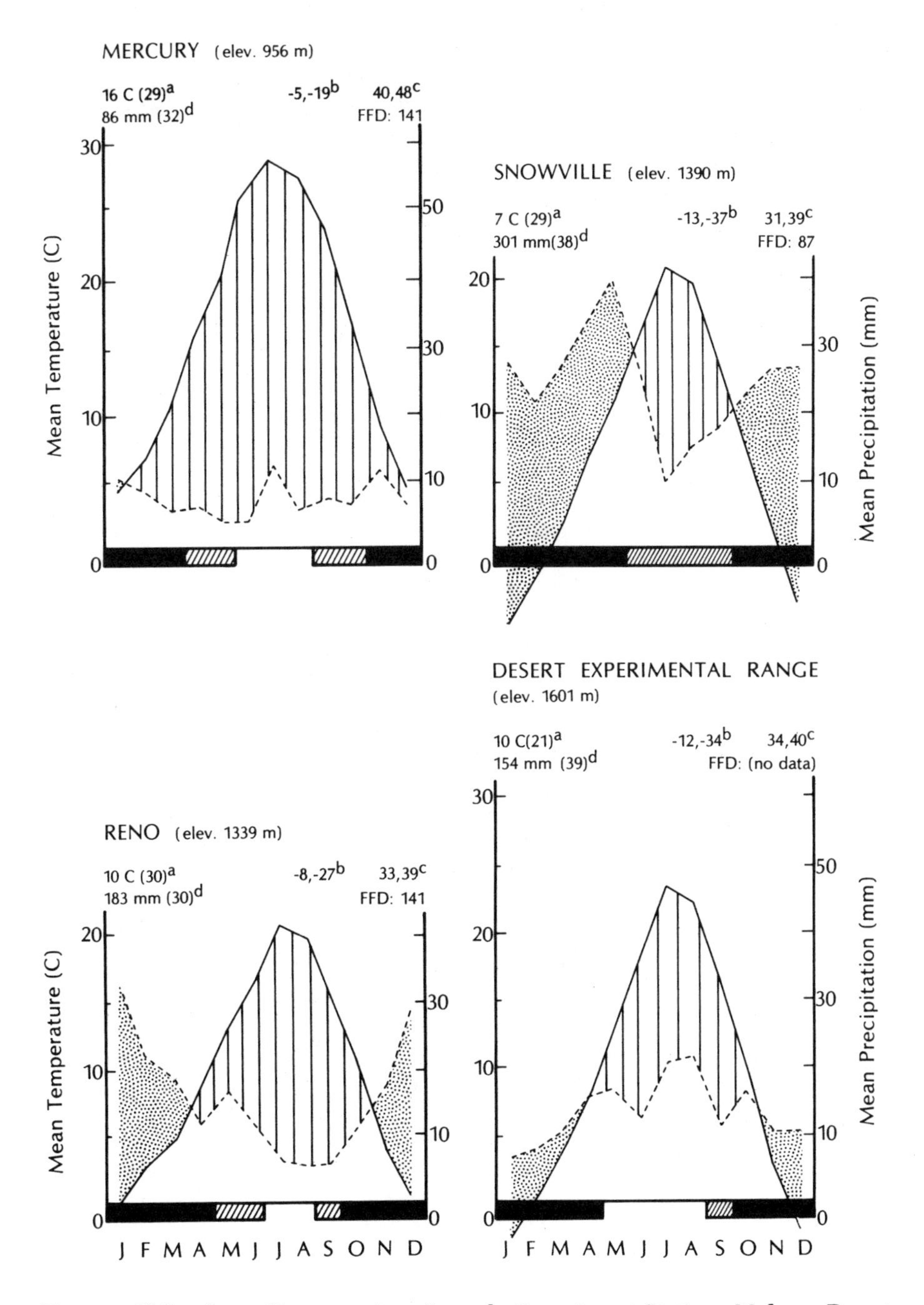

Tucson; Chihuahuan Desert region, Jornada Experiment Station; Mohave Desert region, Mercury [Nevada Test Site]; Great Basin Desert region, Snowville [near Curlew Valley], Reno and Desert Experimental Range [Pine Valley].

reason to change this attitude. Farnsworth et al. (Chapter 2 of this volume) discuss the growing evidence that rhizospheral or nodulating organisms are found on such nonleguminous desert genera as *Artemisia*, *Coleogyne*, *Hymenoclea*, *Tetradymia*, *Krameria* and *Opuntia*.

Our warm-winter deserts have many species of leguminous shrubs, but there is little evidence that these desert legumes function effectively in symbiotic nitrogen fixation. Although the potential for nodulation among legumes has often been considered universal (Nutman 1965), environmental conditions and absence of the proper endophyte frequently prevent nodule formation in legumes in arid regions (Beadle and Tchan 1955; Allen and Allen 1961; Bond 1967, 1971; Vincent 1974). In two recent studies (Garcia-Moya and McKell 1970; Klemmedson and Barth 1975) plants of *Acacia*, *Cassia*, *Cercidium* and *Prosopis* collected from the Mohave, Sonoran and Chihuahuan deserts were devoid of root nodules. These shrubs were found to function in roles similar to nonleguminous shrubs in the accumulation and redistribution of nitrogen and other elements within their ecosystems (Fireman and Hayward 1952; Charley and West 1975). There is the chance, however, that nodulation might be of short duration and so far missed in sampling.

The distribution of nitrogen in the biomass of ecosystems at a number of Desert Biome sites in the United States is portrayed in Table 1-1. These data, although based on somewhat different sampling schemes, disclose large differences in the shoot:root distribution of nitrogen. The high amounts of nitrogen shown for these shrub systems reflect relatively high concentrations of nitrogen in all plant parts. Klemmedson and Barth (1975) report nitrogen percentages in *Prosopis juliflora* and *Cercidium floridum* near 5 percent for leaves and flowers, dropping to around 1 percent for dead wood. These values are somewhat higher than those recorded for leguminous and nonleguminous shrubs in the Mohave Desert (Garcia-Moya and McKell 1970), but lower than the maxima reported by Rodin and Bazilevich (1967).

Consumers

This compartment includes animals that consume producer organisms (primary consumers) or those that prey on the primary consumers (secondary consumers). As in other ecosystems, desert consumers account directly for very little of the biomass or energy flow of the system. In reality, they don't "consume" plant material, or at least not very much of it; they ingest and metabolize a very small percentage of the producer biomass, removing only a fraction of its energy and nutrient content for growth and body maintenance. The remaining biomass is left in the ecosystem for utilization by organisms further along the food chain and for eventual transfer back to inorganic form where it once again becomes available to producer organisms. Thus, by modifying vegetation and litter, consumers play an important role as accelerators of nutrient cycling and as ecosystem controllers. In the process of foraging, animals comminute plant material and make it more readily decomposable. Some consumers, including termites, millipedes and centipedes, have enzyme systems, often in association with symbiotic gut microbes, which appear essential to the primary degradation of resistant materials. The coprophagous organisms may be included here, even though some of them may be classified either as decomposers or as consumers. The term detritivore is sometimes used to better describe these organisms from a functional point of view.

Table 1-1 *Distribution of Nitrogen in the Standing Crop Biomass of Vegetation at Several Shrub-Dominated Desert Biome Study Locations* [g/m²]

	Great Basin		Mohave	Sonoran		Chihuahuan
	Sagebrush	Salt desert	Wolfberry	Paloverde	Mesquite	Creosotebush
Plant part						
Above-ground			18.5			
Leaves	1.7	1.3		1.6	5.0	4.2
Stems	3.8	2.7		62.7	49.8	17.2
Reproductive parts	.2	.1		0.5	0.2	1.1
Litter	3.5	2.8		5.5	7.8	14.9
Below-ground						
Roots	15.7	12.1	8.5	3.9	9.4	141.8
Total	25.1	19.0	27.0	74.2	72.2	179.2
Major species and their percentage of total community standing crop	*Artemisia tridentata* 87%	*Atriplex confertifolia* 98%	*Lycium andersonii* 39% *Ambrosia dumosa* 30% *Krameria parvifolia* 21%	*Cercidium floridum* 93%	*Prosopis juliflora* 96%	*Larrea tridentata* 62%

Location	Idaho Nat. Eng. Lab. (INEL)	Curlew Valley, Utah	ERDA Nevada Test Site, Nevada	Santa Rita Exp. Range, Arizona	Santa Rita Exp. Range, Arizona	Jornada bajada, Jornada Exp. Range, New Mexico
Date	Jul 1968	Jul-Aug 1968, 1969, 1970; Average of 3 yr	15 Mar 1967	Average of 24 shrub systems collected seasonally from May 1971 to Sep 1973	Average of 34 shrub systems collected seasonally from Feb 1972 to Sep 1973	1 Apr 1972
Source of information	West (1972)	Bjerregaard (1971)	Wallace and Romney (1972a)	Klemmedson and Barth (1975)	Klemmedson and Barth (1975)	Biomass data from Whitford et al. (1973) multiplied by average N values in Spector (1956)

Like plants, animals also have a role in distribution of nutrients, but because of their mobility (particularly large consumers) the distributional role of animals contrasts with that of plants. Whereas plants tend to concentrate nutrients more or less in one location, animals both concentrate and disperse nutrients. Animals tend to disperse nutrients widely to the extent of their wanderings, primarily via excretory products but also by virtue of other activities. Their nesting, burrowing and roosting habits are likely to concentrate nutrients. Thus, activities of consumer organisms on the one hand reinforce and on the other counteract the nutrient redistribution activities of plants.

Table 1-2 gives some rough estimates of the amounts of nitrogen in animal biomass at several Desert Biome study locations. The importance of individual organism groups and associated pathways with respect to nitrogen transfer is debatable for any desert region because of gaps in our knowledge and probable high errors in arriving at these estimates. For example, because the nitrogen in feces and urine is so susceptible to losses from volatilization and denitrification, the importance of this pathway for redistribution and transfer of nitrogen to the soil system is uncertain (Rixon 1969; Gist and Sferra, Chapter 10 of this volume). The recent discovery of nitrogen fixation in the gut of termites (Benemann 1973; Breznak et al. 1973) is an example of new relationships that may be found to complicate the picture while adding to total understanding of desert nitrogen budgets.

Decomposers

In mesic environments (those with more favorable moisture and temperature than deserts) standing crops and productivities are higher, and decomposer organisms must be more abundant and active to mineralize the volume of biomass accumulated. By contrast, smaller populations of these organisms exist in deserts primarily because of lower primary and secondary productivity. Populations of these organisms fluctuate markedly in both size and activity in response to greater variation in rainfall and other seasonal conditions favorable for decomposition. Abiotic processes of decomposition (oxidation) appear relatively important in nutrient cycling of desert ecosystems (West, in press b), particularly with respect to nitrogen. In turn, many biotic representatives of more mesic environments, such as *Azotobacter* and *Clostridium*, both of which fix nitrogen, are almost totally lacking under the dry, well-aerated and low-carbon conditions of the desert (Tchan and Beadle 1955).

Low carbon:nitrogen ratios, of both desert soil and vegetation (Beadle and Tchan 1955; Charley 1972; Klemmedson and Barth 1975), contribute to rapid mineralization of organic matter when moisture and temperature are favorable. Hence, the microbial energy source which is normally scant in the desert disappears rapidly, leading to the virtual starvation of microbial life until the next crop of litter falls. Standing crop inventories of microbial biomass have not been included with the tabular data of this chapter. Such data are scarce and usually fail to portray the dynamic nature of the decomposer populations or the magnitude of energy flow through the group. Because of their need for carbon, most of the microorganisms are located in or on the plant detritus and in the rhizosphere.

Table 1-2 *Distribution of Nitrogen in the Consumer Standing Crop Biomass at Several Desert Biome Study Locations* [g/m^2]

Group	Great Basin	Mohave	Sonoran	Chihuahuan
Lagomorphs	0.04	0.0004	0.0018	0.53
Rodents	0.001	0.0004	0.004	0.012
Reptiles	0.0001	0.013	Not available	0.005
Livestock	0.02	--	0.057	--
Insects	0.210	0.014	0.013	0.031
Site(s)	Southern validation site, Curlew Valley, Utah	Rock Valley, ERDA Nevada Test Site, Nevada	Southwestern Research Sta., Portal, and Santa Rita Exp. Range, Arizona	Jornada bajada, New Mexico
Source of information	Biomass from Balph et al. (1973) multiplied by ave. N values for nearest taxa in Spector (1956)	Biomass from Turner et al. (1973) multiplied by ave. N values for nearest taxa in Spector (1956)	Biomasses from the following: Lagomorphs from Hungerford et al. (1972, 1973) Rodents, biomass from Chew and Chew (1970), dry wt and % N in Spector (1956) Livestock calculated from an average grazing capacity of 2.5 AUY/section Insects, biomass of 5 major species of termites only from Nutting et al. (1973); biomass values multiplied by ave.N values	Biomass from Whitford et al. (1973) multiplied by ave. N values for nearest taxa in Spector (1956)
Cropping date	Aug 1972	10 Apr 1972	Various	Various

Soils

Although desert soils are generally low in total nitrogen (Table 1-3) because of low moisture and high temperatures (Jenny 1930), they still contain the bulk of the nitrogen in the ecosystem. Figures ranging from 70 to 98 percent of ecosystem nitrogen, depending on kind and amount of vegetation, have been obtained for the Sonoran and Great Basin deserts (Klemmedson and Barth 1975; Bjerregaard 1971). This is not an unusually high proportion of ecosystem nitrogen and is comparable to grassland and forest ecosystems (Welch and Klemmedson 1975; Rodin and Bazilevich 1967). Soil nitrogen is present in a complex array of inorganic and organic forms (Harmsen and Kolenbrander 1965; Bremner 1965a), much of it adsorbed to the clay fraction in relatively unavailable forms or present as resistant organic substances (e.g., humic and fulvic acids) of variable stability (Paul 1970). Hence, the pool of nitrogen that is readily available for plant growth is difficult to assess. Although high levels of nitrate can be expected in arid soils (Delwiche 1956; Nishita and Haug 1973), losses due to leaching are seldom high. On the contrary, inorganic nitrogen may be subject to heavy denitrificaiton and some volatilization losses to the atmosphere (Klubek et al., Chapter 8 of this volume; Westerman and Tucker, Chapter 7 of this volume).

Vertical and horizontal patterns of soil nitrogen are largely due to the influence of vegetation and are probably demonstrable on a universal basis throughout the world's deserts. Typical vertical patterns are shown in Figure 1-3, where nitrogen is shown to be concentrated in the upper part of the profiles. This pattern can be expected to be most pronounced where vegetation has high shoot:root ratios and

Table 1-3 *Total Nitrogen within the Active Part of the Soil Profile [where roots occur, but minus sieved roots] at Several Desert Biome Study Sites [g/m² surface area of soil column]*

Form	Great Basin		Mohave	Sonoran	
	Sagebrush	Salt desert	Wolfberry	Mesquite	Paloverde
Total N	520	831	105	262	245
Profile depth	90 cm	90 cm	45 cm	60	60
Plant community dominants	*Artemisia tridentata*	*Atriplex confertifolia*	*Lycium andersonii*	*Prosopis juliflora*	*Cercidium floridum*
Site	Idaho Nat. Eng. Lab.	Curlew Valley, Utah	Mercury Valley, ERDA, Nevada Test Site, Nevada	Santa Rita Exp. Range, Arizona	Santa Rita Exp. Range, Arizona
Source	West (1972)	Bjerregaard (1971)	Wallace and Romney (1972a)	Klemmedson and Barth (1975)	Klemmedson and Barth (1975)

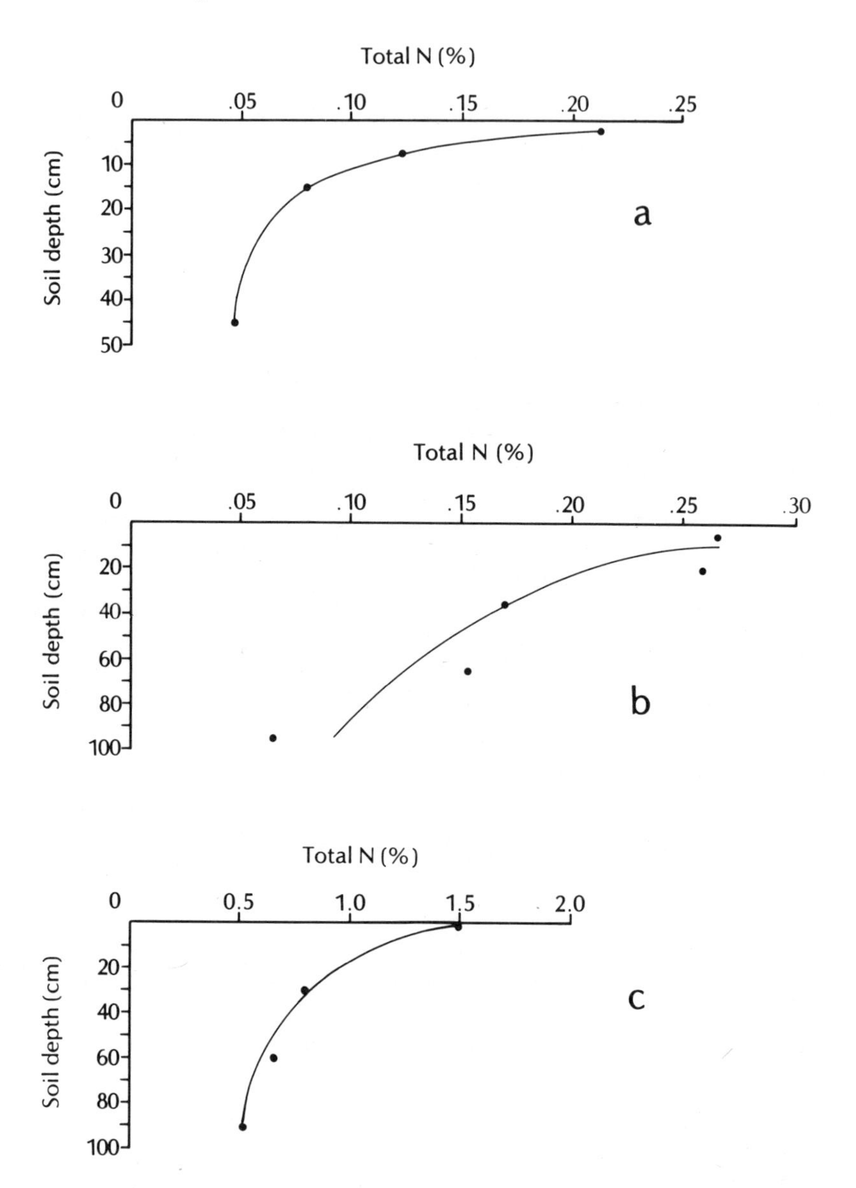

Figure 1-3 *Vertical distribution of percent total nitrogen in some generalized soil profiles at three North American desert sites. a = big sagebrush* [Artemisia tridentata] *site on Pahute Mesa, Nevada Test Site* [*Wallace and Romney 1972a*]; *b = mixed desert shrub and succulent site in Sonoita sandy loam, Santa Rita Experimental Range, Arizona* [*Tucker and Westerman, unpubl. data*]; *c = shadscale* [Atriplex confertifolia] *site in Thiokol series, xerollic calciorthid, Curlew Valley, Utah* [*Bjerregaard 1971*].

shallow root systems and less pronounced in ecosystems where a large proportion of the biomass is concentrated in the root system. Presumably, shrub-dominated ecosystems of the Great Basin and Chihuahuan deserts (Tables 1-1 and 1-3) should illustrate the latter phenomena best (Fig. 1-3). Horizontal patterns, although detectable under mesic conditions (Zinke 1962; Zinke and Crocker 1962; Charreau 1974), are striking in desert situations where the scattered occurrence of vegetation results in "islands of fertility" (Garcia-Moya and McKell 1970) or mosaics of nitrogen accumulation (Nishita and Haug 1973; Charley and McGarity 1964; Charley and West 1975) and nitrogen availability (Tiedemann and Klemmedson 1973b) coinciding with the pattern of vegetation. Presumably the distinctiveness of these mosaics is a function of longevity and the scattered vegetal pattern of deserts. Figure 1-3 portrays nitrogen depth functions for some average profiles; considerable anisotropy is evident in the soil under or between shrubs. Figure 1-4 displays the differences in total nitrogen under and between creosotebushes (*Larrea tridentata*) in southern Nevada. Figure 1-5 shows that these kinds of patterns have both horizontal and vertical components. These patterns are chiefly the result of plants absorbing nutrients through their root systems and redepositing it on the desert floor as mulch. Decomposer activity is enhanced by moderated temperature and increased infiltration and retention of soil moisture in the shade of desert shrubs (Tiedemann and Klemmedson 1973a). Animals are attracted to the "islands" for cover and food supply and in turn affect nitrogen distribution. This structural pattern should be recognized in all studies of the nitrogen cycle in desert ecosystems.

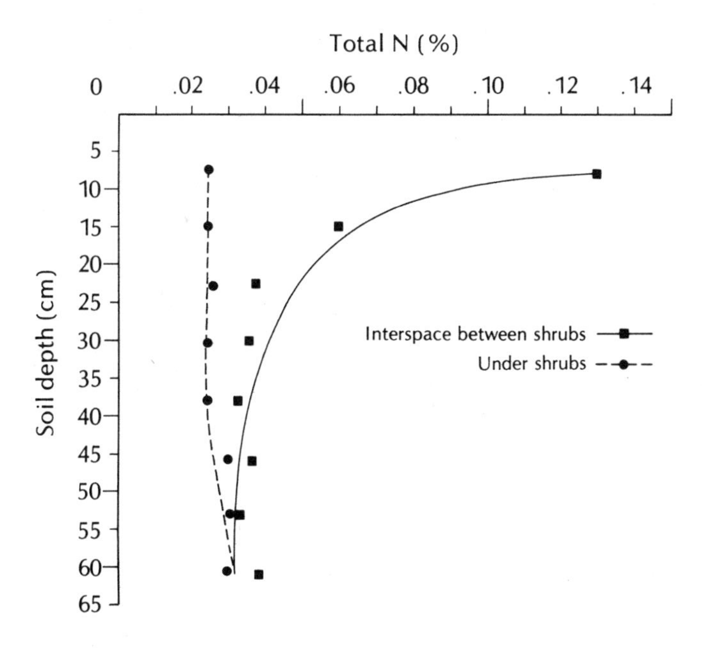

Figure 1-4 *Differences in percent total nitrogen in the soil underneath and in the interspaces between creosotebush* [Larrea tridentata] *shrubs at the Nevada Test Site* [*Nishita and Haug 1973*].

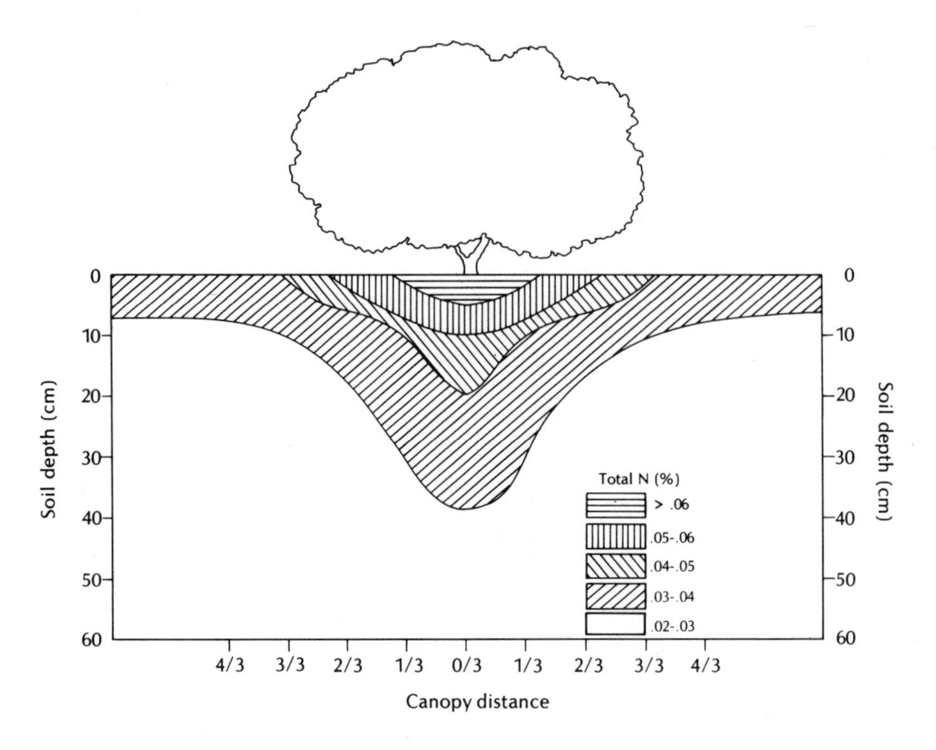

Figure 1-5 *Percent total soil nitrogen concentration beneath an average-size mesquite* [Prosopis juliflora] *shrub, Santa Rita Experimental Range, Arizona, as a function of canopy distance and depth* [*Klemmedson and Barth 1975*].

Atmosphere

The large pool of atmospheric nitrogen overlying deserts is essentially similar to that of other ecosystems (Stevenson 1965). With respect to this pool of nitrogen, deserts differ from other ecosystems primarily in the lower absolute rate of exchange between the atmosphere and the plant-soil system. That is, inputs via precipitation and biological fixation are naturally lower than in more mesic systems, but probably not in direct proportion to rainfall. The importance of physical pathways may be relatively high, however, because of lower inputs through biological means (West, Chapter 11 of this volume). Chemical composition of nitrogen inputs to deserts may vary from that of more mesic ecosystems for a number of reasons (Stevenson 1965), but the significance of this fact is not known. Outflows of nitrogen also may differ in desert systems; direct recycling to the atmosphere via chemodenitrification and volatilization of ammonia are relatively more important than in more mesic ecosystems. Discussion of the specific inputs, outputs and transfers is deferred to following chapters.

INFERENCES FROM DISTRIBUTION TO FUNCTION AND APPLICATIONS

Estimates of relative importance and probably rates of transfer of nutrient elements can be obtained from the structural distribution of nutrients in ecosystems. Presumably those elements in most limited supply will tend to exhibit marked concentration in the places where organisms utilize them. Nutrient elements present in excess of need ordinarily will not display strong patterns of redistribution because of high background values, unless the nutrients are absorbed and released in large quantities (Fireman and Hayward 1952). The marked horizontal distribution of nitrogen observed in most desert situations may be an indirect indication of its short supply. This seems especially true when the various forms of nitrogen and their availabilities are accounted for in ecosystems. Although soils of most ecosystems, even deserts, appear to have large pools of nitrogen, annual turnover of soil nitrogen is very low (Bremner 1965a; Harmsen and Kolenbrander 1965), and the effective pool for rapid mobilization may be but a small fraction of the total soil nitrogen pool (Paul 1970).

Rates of nutrient cycling can be estimated if some further data are at hand showing the distribution of various kinds of nitrogen between various components of the ecosystem. If one assumes successional stability of the biotic community, then the amounts of nutrients taken up by plants and animals should equal the amounts returned in litter, urine, feces and carcasses over the year. Since over 99 percent of the total biomass in most ecosystems is plant material (Rodin and Bazilevich 1968), the annual amount of uptake is approximated by measuring new plant growth. Then by using new growth:old growth ratios, estimates of turnover time can be determined. For instance, Bjerregaard (1971) used this method to estimate a 6.6-year turnover time for nitrogen in plant material for an *Atriplex confertifolia*-dominated ecosystem in Curlew Valley, Utah.

Turnover can also be estimated by dividing the standing crop by annual litter fall, although this is perhaps a less accurate method. The stability assumption is a simplifying premise because early successional stages experience nutrient accumulation in excess of return in the various components of both biomass and soils. Mature ecosystems may have opposite relationships (Vitousek and Reiners 1975). The average annual rate of net accumulation of nutrients can be estimated given age of the system dominants, biomass and nutrient concentration data (Egunjobi 1969; Van Cleve et al. 1971; Klemmedson 1975). The difficulty of estimating annual uptake of nutrients by plants from distribution data depends on the nature of the ecosystem. For systems with annual species or suffrutescent plants the problem is not difficult. For long-lived perennials, however, estimates of annual increment and the concentration of nutrients in these tissues are difficult to make and subject to large errors. Estimates based on annual leaf production and new terminal growth, although not difficult, are of questionable value because of likely seasonal redistribution within the plant (Klemmedson and Barth 1975).

Direct measures of determining nutrient accumulation and turnover yield more definitive data, but at higher cost. Measurement for these functional processes will be discussed in the following chapters. Estimates of the relative importance of these processes can be made by modeling distribution data. We can then set priorities for new research needed to understand the total nitrogen cycle.

Natural and man-induced changes in the structure of desert ecosystems have a distinct effect on the nitrogen cycle. Drought, grazing, fire or tillage, alone or in combination can, at least temporarily, reduce the nitrogen currently in and taken up by producers and subsequently by animals. Considerable changes in the spatial distribution of soil nitrogen may result from movement of soil by water and wind (Beadle and Tchan 1955). Grazing, where intensive, may result in sites almost permanently altered from their original condition (Charley and Cowling 1968). More moderate grazing generally produces less dramatic effects and, as with tillage, largely results in a homogenization of spatial distribution of soil nitrogen (Charley and West 1975). Whether these changes will affect productivity is problematical.

CONCLUSIONS

Accumulation of quantitative data on nitrogen distribution in an ecosystem is a useful first step in understanding the biogeochemical cycle. The marked horizontal and vertical distribution of nitrogen in deserts suggests that it is present in limiting quantities. Most of the biologically available nitrogen is found in plant biomass or litter. Both plants and animals redistribute the element. Animal activity accelerates its rate of cycling. The soil surface close to desert plants is rich in available nitrogen, whereas the interspaces between plants, the deeper soil and the atmosphere are large sinks of less available nitrogen. Examination of the structural distribution of the various forms of nitrogen can lead to estimates of its transfer and movement in ecosystems. Natural and man-made disruptions of this structure can affect productivity of these systems and therefore man's economic gains.

2

NITROGEN FIXATION BY MICROFLORAL-HIGHER PLANT ASSOCIATIONS IN ARID TO SEMIARID ENVIRONMENTS

R. B. FARNSWORTH, E. M. ROMNEY and A. WALLACE

INTRODUCTION

The most popularly recognized input of nitrogen into ecosystems is symbiotic fixation through bacteria on root nodules of legumes. Legumes, however, are not always a major component of desert plant communities (West and Klemmedson, Chapter 1 of this volume). Even when they are, most studies have shown that nodules are often lacking. Other research has shown that nitrogen is definitely fixed in desert soils. Since free-fixing microbes are rare in arid environments, and blue-green algae in lichen soil crusts are not always present, nitrogen fixation must therefore take place in some kind of association with higher plants. We will now review the possibilities.

SYMBIOTIC FIXATION

Legumes

Fixation by *Rhizobium* in root nodules on legumes has been the most thoroughly studied of the biological fixation processes (Vincent 1974). Since there are many legumes in desert regions, it is only reasonable to expect that nodulation of these plants occurs and is important in fixing nitrogen in such systems. Indeed, Beadle (1959) found that a good cover of *Swainsona* (purple pea) may add up to 280 kg of nitrogen per ha/year to desert soils of western New South Wales, Australia. Large flushes of growth by such annuals are rare, however, and we must look to the perennial legumes for more consistent occurrence and therefore a larger possible role in nitrogen fixation.

Several recent studies have not shown nodulation of major North American leguminous shrub genera. No more nitrogen has been found in the proximity of leguminous than nonleguminous shrubs of similar size and form (Wells 1967;

Garcia-Moya and McKell 1970; Klemmedson and Barth 1975). Lack of suitable conditions of temperature and moisture and absence of the proper endophyte frequently prevent nodule formation in legumes of arid regions (Beadle and Tchan 1955). A lack of or a low level of certain nutritional elements in the soil may also be a factor, but this has not yet been demonstrated for arid soils.

Nonlegumes

Recent work disputing the importance of nodule formation for microbial nitrogen fixation (Child 1976) casts some doubts on the importance of nodulated legumes in the nitrogen cycle of all systems. Indeed, for deserts there is abundant evidence that nitrogen fixation takes place in association with the roots of many nonlegumes (Wallace and Romney 1972a; Farnsworth 1975; Farnsworth et al. 1976). Genera occurring in arid to semiarid contexts that have been reported as nodulated are *Artemisia, Opuntia, Purshia, Tribulus, Zygophyllum, Chrysothamnus, Krameria* and *Cercocarpus* (Krebill and Muir 1974).

PARASYMBIOTIC FIXATION

Rhizospheral Associations

Nodulation of plant roots is not the only possibility for effective nitrogen fixation. Wallace and Romney (1972a), Wallace et al. (1974) and Hunter et al. (1975b) surveyed a desert flora at the Nevada Test Site and found positive nitrogen fixation, using the acetylene reduction test, associated with *Artemisia spinescens, A. tridentata, Hymenoclea salsola* and *Tetradymia canescens* of the Compositae; *Coleogyne ramosissima* of the Rosaceae; *Atriplex canescens* and *A. confertifolia* in the Chenopodiaceae; *Krameria parvifolia* of the Krameriaceae; *Larrea tridentata* of the Zygophyllaceae; *Lycium pallidum* and *L. shockleyi* of the Solanaceae; *Menodora spinescens* of the Oleaceae; *Thamnosa montana* of the Rutaceae; *Yucca schidigera* of the Liliaceae; *Bromus rubens* and *Stipa speciosa* of the Poaceae. Snyder and Wullstein (1973b) found similar activity on *Oryzopsis hymenoides.* Since nodules were not obvious, these researchers suspected activity of rhizospheral free fixers. This type of fixation might well involve such microorganisms as *Azotobacter* or *Clostridium* as free fixers (Mahmoud et al. 1964), as well as other organisms such as streptomycetes and *Spirilla* living in the rhizosphere. All are heterotrophic, depending on outside carbon sources to fix nitrogen; thus a loose or parasymbiotic condition on the surface of the root at the soil-root interface, or growing in the outer layers of root tissue as ectotrophic organisms, or endotrophic organisms, could well be involved.

Mycorrhizal Associations

Yet another possibility is that of fixation by mycorrhizae (Fanelli and Albonetti 1972). The term "mycorrhiza" was introduced by Frank in 1885 (Mishustin and Shil-nikova 1971) to describe the complex fungal-root organs on birch and beech.

Since that time, considerable research has been carried on with mycorrhizal associations on forest trees (Harley 1970). Little is known, however, of this relationship on plants of the arid to semiarid regions.

Went and Stark (1968), Khudairi (1969), Khan (1974) and Williams and Aldon (1974, 1976) all have shown that mycorrhizae are abundantly found in association with desert shrubs and forbs in many families around the world. Eleusenova and Selivanov (1975) have shown that there is considerable seasonal variation in the abundance of these fungal associates, with the greatest abundances noted during the wetter seasons. Since nitrogen is often the nutrient most limiting for plant growth in deserts (West, in press c), it might be that mycorrhizae, depending on rhizospheral carbon and vitamin sources, will be found to fix nitrogen under conditions of minimal moisture. For instance, Williams and Aldon (1974) recently found interesting improvements in growth responses of *Atriplex canescens* when it was inoculated with *Endogone mosseae*, a vesicular-arbuscular mycorrhizal-causing organism. This lead requires more extensive and definitive tests; however, in the meantime, these kinds of relationships remain likely.

Phyllospheral Associations

The probable nitrogen fixation in the phyllosphere of arid to semiarid plants is still another mechanism about which little is known. Nodules have been found on the leaves of some tropical dicotyledons (Mishustin and Shil-nikova 1971). Such nodulation has been observed on many other plant leaves (Stevenson 1953; Schwartz 1959; and Silver et al. 1963). The microflora of the phyllosphere are apparently quite effective and important in the nitrogen nutrition of forest trees. The extensive leaf nodulation or galls found on *Artemisia* (Farnsworth 1975) might be a possible locus of such nitrogen fixation in semiarid environments.

CONCLUSION

Although many interesting qualitative leads have developed to implicate a number of possible types of higher plant-microbial and fungal associations in nitrogen fixation within ecosystems of arid to semiarid environments, much quantitative verification of their importance remains to be done. Hopefully, this review will contribute to the development of needed new research.

3

NITROGEN FIXATION BY LICHENS AND FREE-LIVING MICROORGANISMS IN DESERTS

R. RYCHERT, J. SKUJIŅŠ, D. SORENSEN and D. PORCELLA

INTRODUCTION

As the previous two chapters indicate, the conventional modes of nitrogen fixation are not major in arid to semiarid environments. The deserts, in common with some of the harsher environments of the world such as tundras, old fields, poor soils or rock outcrops in more humid areas, have free-living bacteria and blue-green algae in lichens serving as the major sources of biologically fixed nitrogen.

The following treatment summarizes what is known about these organisms' role in nitrogen fixation in arid to semiarid environments around the world. The influences of various environmental factors on the rate of fixation are outlined. Although emphasis is placed on a review of available literature, some new data from Curlew Valley, Utah, are also presented.

NITROGEN FIXATION BY FREE-LIVING ALGAE AND LICHEN SYMBIONTS

While both blue-green and eucaryotic algae are found in the surface crusts of deserts (Lynn and Cameron 1971), it is some of the blue-green algal species that are responsible for biological nitrogen fixation. The great majority of blue-green algae that fix nitrogen are probably heterocystous, although nonheterocystous blue-green algae may fix nitrogen as well (Stewart 1973). *Plectonema boryanum* (nonheterocystous) was shown to fix nitrogen microaerophilically, but not in air (Stewart 1971). Microenvironmental conditions where oxygen tension is reduced might enhance nitrogen fixation by both nonheterocystous and heterocystous blue-green algae. Desert lichen crusts when wet might then provide a situation where mycobiont respiration could reduce oxygen tension so that nitrogen fixation might be enhanced.

While there has been a great deal of work involving nitrogen fixation by blue-green algae and lichens from temperate regions (Henriksson et al. 1972), especially in aquatic systems (Stewart 1973), the significance of nitrogen fixation by blue-green algae-lichen crusts in desert ecosystems has received limited attention.

Desert algal crusts are found on neutral to alkaline soils. Sonoran Desert algal crusts were found to have a net N_2 fixation rate of 0.18 kg/ha per day under continuously wet conditions, and 0.11 kg/ha per day under cycling wet-dry conditions using ^{15}N methodology (Mayland et al. 1966). Atmospheric nitrogen fixation rates increased linearly for 520 days in laboratory studies. Growing algal crusts excreted 1 to 2 percent of the total crust N as extracellular ammonium N. Grass seedlings could take up some of the $^{15}N_2$.

MacGregor and Johnson (1971) found that Sonoran Desert algal crusts from southern Arizona could produce detectable ethylene from acetylene 3 hours after moistening. Premoistened algal crusts could produce 78 $\pm$ 5 nmoles of ethylene/cm^2 per hour, based on the first hour of incubation. It was estimated that following a rainfall, 3 to 4 g N/ha per hour might be fixed. Approximately 4 percent of the surface of an area of desert grassland had crust formations. Most of the nitrogen fixation probably took place during the Sonoran Desert's summer rains.

Mayland and McIntosh (1966) showed that algal crust-fixed $^{15}N_2$ was available to higher plants; 3.4 percent of the crust nitrogen was water soluble, and 1 percent of the crust nitrogen was ammonium N.

Fuller et al. (1960) showed that blue-green algae and lichen crusts from the Sonoran Desert could fix $^{15}N_2$, and that this nitrogen was subsequently available to plants. Sonoran Desert algal crusts were higher in nitrogen and carbon than subsurface layers (Cameron and Fuller 1960). Since *Azotobacter* were absent, nitrogen fixation by blue-green algae was assumed. Species of *Nostoc*, *Scytonema* and *Anabaena* were shown to fix atmospheric nitrogen. There was great variation of algal development in laboratory cultures with varying environmental conditions, and thus it was difficult to identify any particular algae. Blue-green algae from the Negev Desert also have been difficult to culture and identify (Friedmann et al. 1967).

Rogers et al. (1966) found that the lichen *Collema coccophorus* (with a *Nostoc* phycobiont) from the south-central arid zone in Australia could incorporate significant amounts of $^{15}N_2$.

Skujins and West (1974) have shown that $^{15}N_2$ could be fixed in situ by blue-green algae-lichen crusts in the Great Basin Desert. In situ assays in spring showed that the highest period of N_2 fixation occurred during the morning hours. The highest in situ peak value was approximately 10 g ^{15}N fixed/ha per hour and occurred between 0800 and 1000 (Fig. 3-1). It was suggested that morning dew condensation on the surface crust could provide sufficient moisture for crust nitrogen fixation.

Algal crusts in the Sonoran Desert hold soil particles and reduce erosion, increase soil fertility and improve soil structure and water infiltration. *Oscillatoria*, *Nostoc*, *Microcoleus* and *Nodularia* algal species were present in the surface crusts; no *Azotobacter* were found. The crust contained 400 percent more nitrogen than the surface below (Fletcher and Martin 1948).

Microcoleus vaginatus, *Schizothrix californica* and *S. accutissima* have been observed in Nevada desert crusts. *Nostoc commune* and *Scytonema hofmanni* were found to be associated with lichens (Shields and Drouet 1962). The surface 1.2 cm had nitrogen levels twice as high as areas with no lichen crust. Mosses (*Grimmia*) grew on the upper part of the hummock beneath plant canopies, and lichens developed at the lower part where the water drains. It was suggested that mosses appear in the more xeric environment. While the blue-green algae are resistant to desiccation, the Chlorophyceae are not.

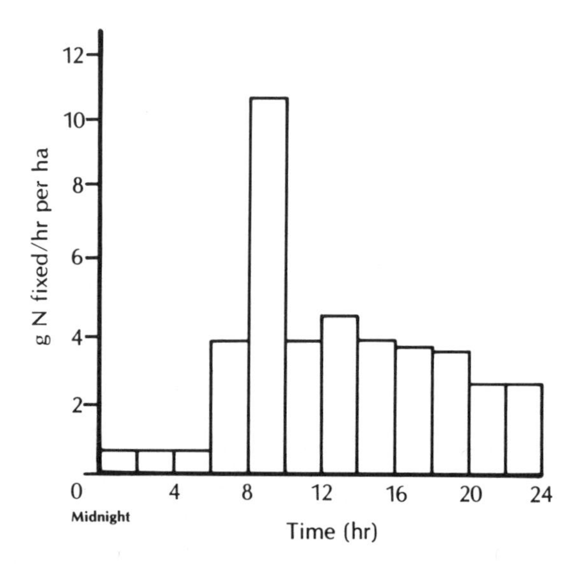

Figure 3-1 *In situ $^{15}N_2$ fixation by blue-green algae-lichen crusts; Great Basin Desert, southern Curlew Valley, Utah, spring 1973 [Skujins and West 1974].*

Cameron (1962) found species of *Nostoc* which had a wide Sonoran Desert distribution. *N. muscorum* along with *Scytonema hofmanni* were common constituents of lichen crusts. *N. microscopicum* and *N. ellipsosporum* were found associated with mosses. The lichens had a high resistance to desiccation and retained moisture for a greater period of time than the surrounding soil. Air-dried algae and lichen crusts were revived after 4 years of desiccation. After 1 year of desiccation there was no change in growth compared to original growth experiments. Apparently, survival of *N. muscorum* does not depend upon sporulation.

Cameron and Blank (1966) have found that desert algae-lichen soil crusts can survive, grow and reproduce in cultures after subjection to extreme cold. This ability would adapt the soil algal microflora to harsh environmental conditions.

Lynn and Cameron (1971) obtained 4.0 x 10^4 nitrogen-fixing bacteria per gram of soil from Curlew Valley, a US/IBP Great Basin Desert site, whereas New Mexico and Arizona desert soils had nitrogen-fixing bacteria roughly 10 times higher than the Curlew Valley site (Lynn and Cameron 1972). Algal cover in various Curlew Valley desert shrub communities varied from 41 to 82 percent depending upon the site (Lynn and Cameron 1973). However, no data are yet available regarding extent of cover by nitrogen-fixing blue-green algae. Table 3-1 shows the blue-green algae nitrogen fixers from a number of desert sites (Lynn and Cameron 1971). A survey of lichens at the IBP Desert Biome southern Curlew Valley intensive study site showed the nitrogen-fixing lichen *Collema tenax* to cover 32.4 percent of the soil surface (Pearson 1972).

Table 3-1 *Nitrogen-Fixing Genera of Blue-Green Algae Observed at US/IBP Desert Biome Sites [Lynn and Cameron 1971]*

Genera	Sites sampled				
	Curlew Valley, Utah	Jornada Uplands, New Mexico	Jornada Playa New Mexico	Silverbell, Arizona	Santa Rita Exp. Range, Arizona
Nostoc	+	+	+	+	+
Scytonema	+		+	+	+
Tolypothrix	+				

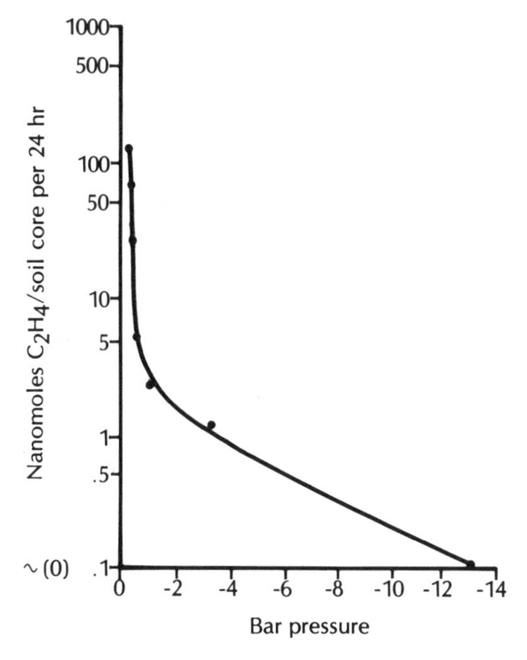

Figure 3-2 *Acetylene reduction vs. water potential for soil cores with intact crust collected from the* Atriplex confertifolia *site [Rychert and Skujins 1974b].*

Rychert and Skujins (1974b) have investigated nitrogen fixation by blue-green algae-lichen crusts in southern Curlew Valley in the Great Basin Desert. Figure 3-2 shows that nitrogen fixation by moistened crusts drops rapidly when the water potential is lower than —⅓ bar. This is in marked contrast to the fact that lichen photosynthesis may occur at relatively low water potentials, < —15 bars (Ahmadjian 1967). Thus, significant nitrogen fixation in desert ecosystems probably occurs only when the blue-green algae-lichen crust is moist; for example, in the Sonoran Desert during the summer rains and in the Great Basin Desert during rainy periods in the fall and spring. The lichen-encrusted areas probably retain their moisture longer than bare areas (Fletcher and Martin 1948) and thus nitrogen fixation may be prolonged by virtue of this property. Sites with cryptogamic crusts on the Colorado Plateau had higher rates of water infiltration than adjacent sites with little or no such crusting (Loope and Gifford 1972). Porcella et al. (1973) also have observed improved filtration of precipitation water at salt desert shrub sites in Curlew Valley that have well-structured lichen crusts.

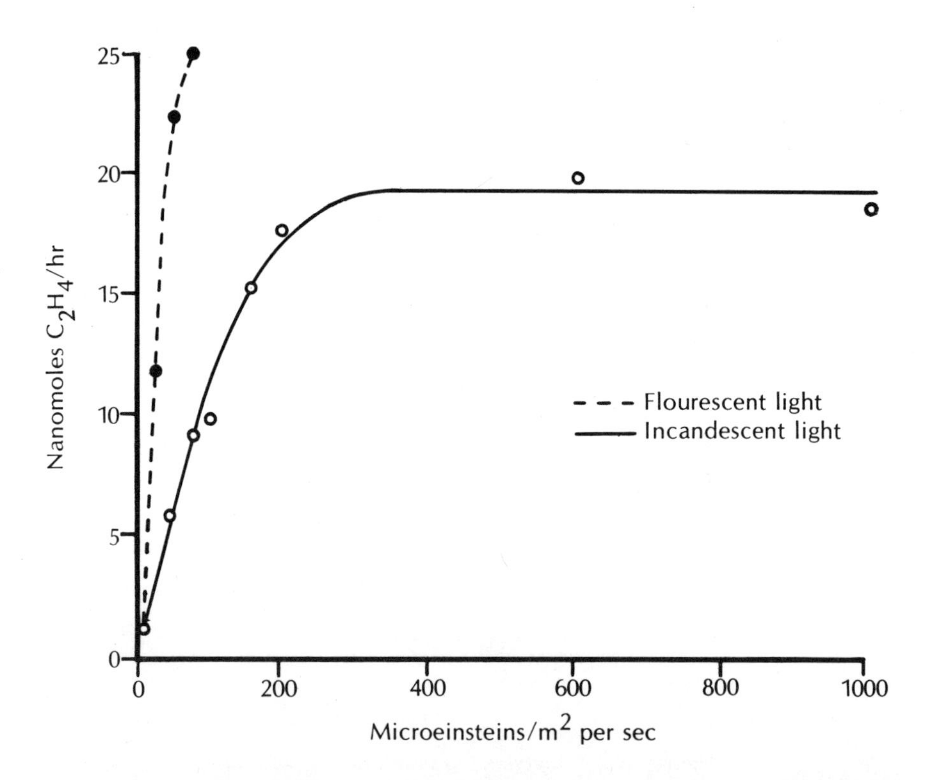

Figure 3-3 *Effect of flourescent and incandescent light intensity upon laboratory assays of acetylene reduction at 23 C by blue-green algae-lichen crusts collected from the* Atriplex confertifolia *site [Rychert and Skujins 1974b].*

Atmospheric nitrogen fixation is optimal at approximately 200 μ einsteins/m_2 per second of incandescent light intensity (Fig. 3-3). While incandescent light is not directly comparable to natural sunlight, the results indicate that nitrogen fixation is optimal at relatively low light intensities. Natural sunlight with a heavy gray cloud cover exhibits light intesities of 50 to 300 μ einsteins/m^2 per second.

Figure 3-4 shows the results of in situ acetylene reduction by blue-green algae-lichen crusts in the Great Basin Desert as a function of temperature and mean light intensity. The in situ assays indicate that up to 13-14 g N/ha per hour may be fixed. The nitrogen-fixation potential of the blue-green algae-lichen crusts is much higher (up to 80 g N/ha per hour), and this potential may or may not be achieved in situ during rainy periods. In the Great Basin Desert, where algae-lichen crusts are common, estimates of annual nitrogen fixation by crusts have been made at 10 to 100 kg N fixed/ha, depending on environmental conditions (Rychert and Skujins 1974b).

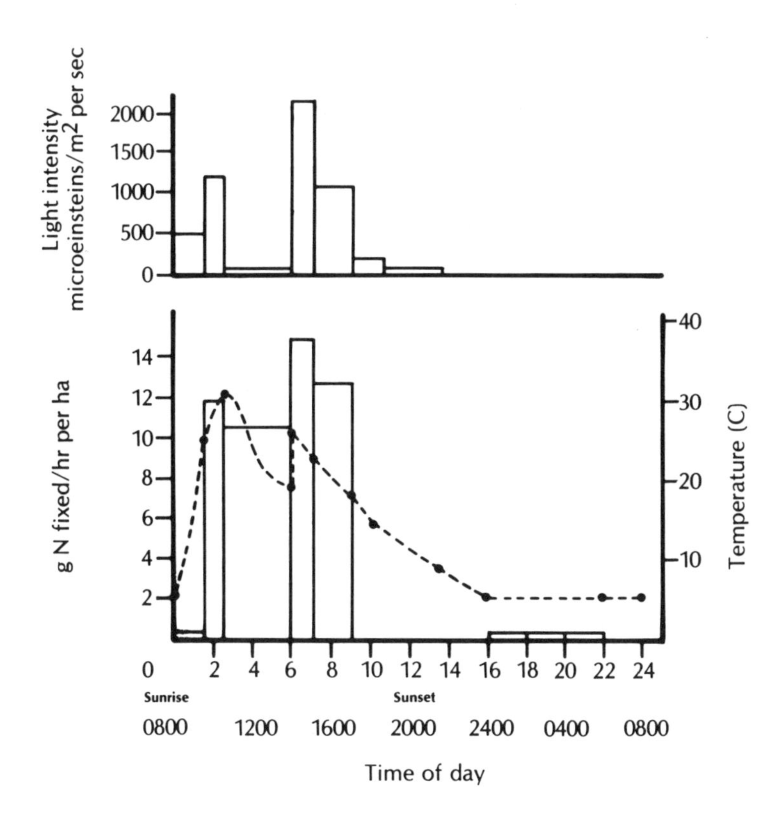

Figure 3-4 *In situ acetylene reduction as a function of mean light intensity and temperature [dashed line] for soil cores with intact blue-green algae-lichen crusts from the* Atriplex confertifolia *site [Rychert and Skujins 1974b].*

Table 3-2 *Nitrogen Fixation at Various Sites in Southern Curlew Valley, Utah [Skujins, original data]*

Site (community dominants)	"Optimum" nitrogen fixation rate (10^{-7} mg N_2/cm^2 per min)
Sarcobatus vermiculatus (no lichens)	3.6
Atriplex confertifolia (lichens present)	3.5
Artemisia tridentata (heavy lichen cover)	9.2
Agropyron desertorum (heavy cover of *Agropyron*, no lichens)	0.26
A. tridentata (lichens present)	6.2
A. desertorum (scant cover of *Agropyron*, with no lichens)	1.9
A. tridentata and *Halogeton glomeratus* (no lichens)	0.76

Table 3-2 shows the results of nitrogen fixation (as measured by acetylene reduction) by blue-green algae and lichen crusts at various sites in south Curlew Valley, Utah, in the Great Basin Desert. *Nostoc* was observed upon microscopic examination of the crusts. Epifluorescent microscopy enabled observation of *Nostoc* colonies in soil crusts in the absence of lichen thalli.

Little knowledge is available concerning lichen crust growth rates, or stability of the crust in terms of disturbance such as grazing. With the slow rates of lichen growth (Ahmadjain 1967), it might be expected that recovery or reformation of lichen crusts would be slow after disturbance. Blue-green algae crusts, such as those in the Sonoran Desert and elsewhere, might then exhibit more resilient properties.

Snyder and Wullstein (1973b) have examined a number of Great Basin Desert cryptogams for nitrogen fixation using the acetylene reduction technique. *Nostoc* was associated with the lichens *Dermatocarpon lachneum* and *Peltigera refescens*, and *Azotobacter* nitrogen fixation was implicated with the moss *Grimmia* (Table 3-3). It was noted that the cryptogam cover was more abundant in fenced enclosures that had been protected from livestock grazing for 20 or more years.

Endolithic blue-green algae (*Gloeocapsa*) are primary producers in the harsh, dry Antarctic "Polar Desert" ecosystem (Friedmann and Ocampo-Paus 1976). This is noteworthy since unicellular, nonheterocystous *Gloeocapsa* may fix nitrogen aerobically and microaerophilically (Fogg et al. 1973). The ability of endolithic blue-green algae to fix nitrogen has not been examined in the Antarctic or desert ecosystems.

Lichens are common on the volcanic rock outcrops in Curlew Valley, Utah. Three of these epilithic lichens, along with some of the rock substrate, were assayed by the acetylene reduction technique as described by Rychert and Skujins (1974b). None exhibited an ability to reduce acetylene.

NITROGEN FIXATION BY FREE-LIVING BACTERIA

The free-living nitrogen-fixing bacteria *Azotobacter* and *Clostridium* are present in relatively high numbers in soils from the Near East, except where the soil is bare or where NaCl salt accumulates (Abd-El-Malek 1971). With cropping or addition of organic matter (maize stalks, wheat straw), along with irrigation, significant increases are seen in the number of nitrogen fixers along with increases in soil nitrogen content. Less nitrogen is gained in calcareous soils than in clay soils. In the absence of organic carbon amendments, heterotrophic nitrogen fixation is negligible. However, microenvironmental conditions, such as those occurring in the rhizosphere and within blue-green algae and lichen crusts, could provide a transient supply of available carbon. In Egyptian desert soils, *Azotobacter* and *Clostridium* exist only at the soil-root interface in soils inhabited by *Moltkia callosa* (Mahmoud et al. 1964). In sandy desert soils, nitrogen fixation within the rhizosphere may be of particular importance (Farnsworth et al., Chapter 2 of this volume).

Table 3-3 *Acetylene Reduction for some Desert Cryptogams* [*Snyder and Wullstein 1973b*]

Sample	Ethylene (nmole/g per ml gas mixture)	
	2 hr	Day 5
	Mean ± SD	Mean ± SD
Parmelia chlorochroa (Awapa Plateau)	0.323 ± 0.082	0.665 ± 0.327
P. chlorochroa (Desert Exp. Sta.)	0.321 ± 0.053	0.754 ± 0.203
Agrestia hispida	0.302 ± 0.075	2.45 ± 1.85
Leoanora peltata	0.383 ± 0.078	0.522 ± 0.107
Caloplaca trachyphylla	0.259 ± 0.016	0.321 ± 0.194
Dermatocarpon lachneum	0.232 ± 0.022	37.6 ± 7.97[a]
Peltigera refescens	3.72 ± 1.99[a]	101.0 ± 14.1[a]
Grimmia sp. (plant plus soil)	3.67 ± 0.583[a]	13.1 ± 8.15[a]
Grimmia sp. (plant)	0.389 ± 0.051	3.11 ± 2.22
Grimmia sp. (soil)	1.07 ± 0.203[a]	0.032 ± 0.119[a]
Tortula ruralis (plant plus soil)	1.50 ± 0.854	1.04 ± 0.901
T. ruralis (plant)	0.367 ± 0.089	0.460 ± 0.441
T. ruralis (soil)	0.478 ± 0.038	0.735 ± 0.380
Moss sp. *X* (plant plus soil)	0.264 ± 0.002	0.454 ± 0.140
Controls (exp. gas mixture)	0.425 ± 0.200	0.240 ± 0.020

[a] $P \leq .01$.

Table 3-4 *Potentiation of Heterotrophic Nitrogen Fixation in Curlew Valley Soils [Skujins, original data]*

Site (community dominant)	Soil sample (cm)	Amendment	nmole C_2H_4/g soil, in 3 days at 30 C
Artemisia tridentata	0-3	Nil	0.0
	0-3	Glucose	1760.7
	0-3, under canopy	Glucose	4.5
	5-20	Nil	0.0
	5-20	Glucose	27.0
	40-50	Nil	0.0
	40-50	Glucose	0.0
Ceratoides lanata	0-3	Nil	0.0
	0-3	Glucose	5740.7
	0-3, under canopy	Glucose	296.1
	5-20	Nil	0.0
	5-20	Glucose	458.8
	40-50	Nil	0.0
	40-50	Glucose	0.0
Atriplex confertifolia	0-3	Nil	0.0
	0-3	Glucose	1696.5
	0-3, under canopy	Glucose	90.0
	5-20	Nil	0.0
	5-20	Glucose	4.8
	40-50	Nil	0.0
	40-50	Glucose	0.0

Using ^{15}N and acetylene reduction techniques, Steyn and Delwiche (1970) estimated that nonsymbiotic nitrogen fixation could account for 2 kg N/ha per year (35-cm depth) with soil from a semiarid site in California dominated by native vegetation (*Avena barbata, Stipa pulchra* and *S. cernua*). It was concluded that nitrogen fixation by free-living microorganisms may contribute significantly to the nitrogen balance in unfertilized range soils.

Pochon et al. (1957) have suggested an absence of nitrogen-fixing bacteria in Saharan Desert soils. However, Hethener (1967) found *Azotobacter* and *Clostridium* in 8 of 15 soils from the central Sahara. The frequency of nitrogen-fixing bacteria was sensitive to changes in moisture.

Azotobacter may be present and fix nitrogen in desert lichen crusts, but definitive evidence is lacking (Snyder and Wullstein 1973b). Snyder and Wullstein (1973a) did find *Azotobacter* associated with Georgia granitic pioneer outcrop systems (lichens) and suggested that nitrogen fixation by *Azotobacter* may contribute to the nitrogen budget of that ecosystem.

Table 3-4 shows that, in the absence of soil amendments (glucose or plant material), heterotrophic nitrogen fixation is negligible in Curlew Valley soils in the Great Basin. There is a potential for heterotrophic nitrogen fixation, however, as seen with soils moistened with 10 percent glucose solutions.

Table 3-5 *Summary of Estimates of Nitrogen Input to Desert Ecosystems from Cryptogamic Crust Fixation*

Ecosystem	Ecosystem component	Estimated annual nitrogen input by fixation	Reference
Sonoran Desert algal crusts	Crust -- cycling wet-dry conditions	7-11 kg/ha per yr	Mayland et al. (1966)
Australian south-central arid zone	Lichen, *Peltigera aphthosa* (*Nostoc*, phycobiont)	1.3 kg/g algal N per yr	Millbank and Kershaw (1969)
Sonoran Desert, southern Arizona	Premoistened algal crusts (estimation based on 12 hr/day)	13-18 kg/ha per yr	MacGregor and Johnson (1971)
Great Basin Desert, southern Curlew Valley	Soil crusts	10-100 kg/ha per yr	Rychert and Skujins (1974b)
South Curlew Valley Validation Site (US/IBP)	Soil crusts in situ measurement	7 kg/ha per yr	Porcella et al. (1973)

Table 3-6 *Inhibition of Nitrogen Fixation by Desert Shrub Extracts [Rychert and Skujins 1974b]*

Desert shrub extract	Ethylene production (nmole C_2H_4/soil core per 24 hr)
Distilled water	24.8
Artemisia tridentata	0
Ceratoides lanata	5.2
Atriplex confertifolia	8.6

COMBINED FIXATION BY FREE-LIVING ORGANISMS AND LICHENS

If free-living bacteria, algae and lichens all occur on the same site it is difficult to separate their activities. In deserts this co-occurrence is probably the rule rather than the exception. Accordingly, any summarization of nitrogen fixation in desert soils must account for the possibility of contributions from several kinds of organisms. Table 3-5 provides a summary of the biological fixation of nitrogen in various desert ecosystems. Blue-green algae and/or lichens with a blue-green algal phycobiont appear to be the major nitrogen fixers in desert ecosystems with clay-containing soils. The potential of blue-green algae to fix N_2 in desert ecosystems is similar to values reported for some mesic biological nitrogen-fixing systems (Hardy et al. 1973).

ALLELOCHEMIC INFLUENCES ON NITROGEN FIXATION

Chemicals given off by plants have recently been shown to adversely affect the rates of nitrogen fixation by free-living microbes in the soil (Kapustka and Rice 1976; Rice 1974). Prior work along these lines focused on mesic systems, especially grasslands. Rychert and Skujins (1974b) have recently shown that these inhibitions are also operative in the cold winter Great Basin Desert (Table 3-6). Both free-living microbes and the blue-green algae in the lichen crust are apparently inhibited, probably by the phenols abundant in plant litter and leachate. Since the deposition of these allelochemics is highly localized, the overall effect is questionable. In the Great Basin Desert the microenvironment under the scattered shrubs and half-shrubs is largely dominated by bryophytes. In the more abundant interspace area, blue-green algae crust dominates with few or no bryophytes. Thus, once again the highly variable and patterned structure of deserts interplays with their dynamics.

CONCLUSION

Blue-green algae crusts and/or blue-green algae-lichen crusts can fix significant amounts of atmospheric nitrogen in desert soils, and are probably responsible for a major input of nitrogen into desert ecosystems. The crusts serve to stabilize the soil surface, to reduce erosion and to increase water retention and infiltration.

Heterotrophic nitrogen fixation in arid soils is probably negligible due to the lack of available carbon.

Desert shrubs appear to possess inhibitors of crust nitrogen fixation, and this inhibition may be important, particularly in the desert shrub canopy microenvironment.

4

DECOMPOSITION OF PLANT LITTER IN TWO WESTERN NORTH AMERICAN DESERTS

P. L. COMANOR and E. E. STAFFELDT

INTRODUCTION

For life to continue, the conversion of organic to inorganic nitrogen is required in the intrasystem organic to inorganic cycle, yet the study of decomposition processes of dead plant material in deserts has been virtually ignored. This is surprising, since one of the noticeable characteristics of the western North American deserts, especially the Sonoran, is the large amount of partly decomposed plant material present, including the very evident 'standing dead' of many species. Dead wood may compose a large percentage of shrub biomass. Obviously part of the availability of nitrogenous compounds for new plant growth in desert ecosystems is a result of their release from nitrogen-containing organic molecules contained in dead plant material. It is therefore essential to study the fate of dead plant material as a prerequisite to the understanding of nitrogen availability from this important source.

The following sections review the literature relevant to this topic and then present some original data from the Great Basin and Chihuahuan deserts.*

PREVIOUS WORK

Above-ground Litter Production

Primary production in western North American deserts has been widely studied (Caldwell 1975); however, the subsequent partitioning of this material into the litter compartment has received little attention. What little is known of the formation, distribution and function of plant litter in desert ecosystems has been recently reviewed by West (in press a).

In one study, Mack (1971) collected 34 kg/ha of leaves and 28 kg/ha of inflorescences of *Artemisia tridentata* annually in mesh litter traps in eastern

*We thank Dr. H. Charles Romesburg of the Desert Biome for the statistical analyses of our data.

Washington. Daubenmire (1975) collected leaves cast from the same species in eastern Washington during another year. Both authors noted a pronounced periodicity in their collections: the bulk of leaf litter falls in June-July, and most of the inflorescences fall in December-January. Mack (1971) found canopy volume was correlated with leaf litter production. This correlation was substantiated for sagebrush in southern Idaho (Murray 1975). In Idaho the above-ground litter production was greater, ranging from 167 to 726 kg/ha, for all sagebrush litter combined.

West and Fareed (1973) found that leaf shedding by three Great Basin Desert shrubs over 4 years was directly influenced by prevailing climatic conditions. *Artemisia tridentata* litter production (per m^3 plant volume) for 1972, a dry year in northern Utah, was: leaves, 21.5 g; stems, 21 g; and reproductive material, 13 g. Larger proportionate amounts of litter:plant volume were found for *Atriplex confertifolia* and *Ceratoides lanata*. Although litter fell throughout the year for these two species, during certain periods litter accumulations were greater.

Rickard and Cline (1970) found the amount of plant litter beneath shrubs (2,346 kg/ha) quite different from the amount of litter between shrubs (157 kg/ha) in an *Artemisia tridentata*-dominated community in southeastern Washington. In a series of studies on desert shrub-dominated communities in Nevada (Blackburn et al. 1969), plant litter was found to cover an average of 35 percent of the ground. The ground covered by litter in *Artemisia tridentata*-dominated communities averaged 48 percent, ranging from 6-97 percent.

DECOMPOSITION OF LITTER

Decomposition of litter is a function of physical and chemical properties of the litter, organisms involved and environmental conditions (Mikola 1958). Environmental conditions affecting decomposition may include aeration, pH, temperature and moisture. Decomposition in the laboratory has been correlated with total nitrogen, water-soluble organic matter and excess bases (Broadfoot and Pierre 1939). The chemical nature of the litter (Waksman and Gerretsen 1931) and age of material are fundamental aspects of degradation; younger plants containing higher quantities of water-soluble carbohydrates are more rapidly decomposed (Waksman and Tenney 1927, 1928). According to Pugh (1974) the nutrients in plant litter are normally sufficient for microbial growth. The portion of the plant decomposing is important; Waksman and Tenney (1927, 1928) demonstrated a greater rate of decomposition for stem vs. root portions of the plant, as evidenced by carbon dioxide evolution. For a review of factors affecting fungal physiology see Pugh (1974).

In the desert many processes are moisture limited, and decomposition of litter at the soil surface-air interface is probably limited by moisture content. Higher plants have evolved many adaptations to the desert environment; microorganisms may also be adapted to some degree. Caution must be used, however, for as Chen and Griffin (1966) point out, desert soil fungi are not necessarily xerophytic populations. In litter with fluctuating moisture levels, the lower levels may be limiting to growth of such fungal populations (Pugh 1974).

Some decomposition work has been done in deserts from an applied, agricultural point of view. In one study (Lyda and Robinson 1969) litter in the form of alfalfa crop residue was incorporated into soil in southern Nevada. Respiration rate and soil organic matter content were closely correlated. In another study Sullivan (1942) incorporated organic material (alfalfa and/or grass) into the soil in southern Arizona. He found, as expected, an initial high rate of decomposition, as measured by carbon dioxide evolution, which decreased with time.

Some important work on decomposition in a semidesert environment was done in Washington by Mack (1971). He trapped *Artemisia tridentata* litter as it fell from the plants. Weighed quantities of leaf litter in nylon bags were placed on the soil surface, under shrub canopies and in adjacent openings. After varying periods of exposure in the field, the bags were recovered and the amount of remaining litter weighed. The final weights of litter remaining under shrubs were approximately 55-58 percent of the initial values 10 months after the study began.

THE PRESENT STUDY

The work reported here, although preliminary in nature, is a first contribution to both in situ monitoring of decomposition through carbon dioxide evolution and analysis of weight loss of "natural" litter in semidesert ecosystems. We were unable to locate any literature dealing with in situ carbon dioxide evolution from litter or soil in nonagricultural hot or cool desert ecosystems. Such work, some of which has been stimulated by IBP, has been undertaken in temperate deciduous and montane coniferous forest ecosystems.

The primary objective of the studies reported here was to determine the rate of decomposition of plant material in two of the western North American deserts: the Great Basin and the Chihuahuan. We followed the conceptual approach outlined by Goodall (1970) in which similar samples are observed at different times (a time sequence) as well as analysis of partial processes associated with, in this case, the decomposition process. This included monitoring the environmental conditions occurring in the field during the study period.

The species selected for study were perennial shrub dominants; *Artemisia tridentata* (big sagebrush), a dominant in the Great Basin Desert, and *Larrea tridentata* (creosotebush), a dominant in the Chihuahuan Desert. The latter is also a dominant shrub species in the Mohave and Sonoran deserts.

The studies were carried out at two locations. The first site (designated as NEV in this chapter) was located northwest of Reno, Nevada, at the foot of the Sierra Nevada Mountains in the extreme western portion of the Great Basin Desert (Fig. 1-1). The second site (designated as JOR in this chapter) was at the IBP Jornada bajada intensive study site (New Mexico) representing the Chihuahuan Desert. NEV is dominated by *Artemisia tridentata* and *Bromus tectorum*; *Tetradymia canescens*, *Prunus andersonii* and numerous grasses are also present in the area. JOR is dominated by *Larrea tridentata*, with *Yucca elata*, *Ephedra trifurca*, and *Opuntia engelmannii* as minor constituents.

Both areas are two-phase communities consisting of isolated or clumped shrubs with much open space. At NEV the shrubs cover 25 percent of the area; at JOR, cover is only 8 percent. As much as possible, the site design followed the guidelines set forth by the IBP (Newbould 1967).

Litter samples were introduced into the field at various times. At JOR, oven-dried (50 C) *Larrea* leaf, stem and root samples were placed in the field in January, May and July 1973. These 3-g samples were used in the carbon dioxide studies during May through August 1973. Weight loss information was obtained from three replications of each type of plant material removed during June, July and August.

At NEV, 12 leaf litter samples (four mesh bags each) were placed in the field each month for 12 consecutive months beginning in March 1972. In successive months, 1 of the 12 samples in the field (from each month) was randomly selected and brought into the lab for analysis. This was done such that each of the prior months' samples was represented in the collection. Thus, in May 1972, 1 sample each from the 12 placed in the field in March and April 1972 was recovered. (Of course only 11 of the 12 samples placed in March were in the field prior to the May 1972 collection date, since 1 sample was collected in April.) In July 1972 one set each from March, April, May and June was recovered, and so on.

In analyses of NEV data which follow, weight losses for samples exposed for 1, 2, 3, . . . , 12 months are compared to environmental conditions during the same period.

Mesh Litter Bags

A widely used method to study decomposition in the field has been the use of mesh bags which enclose the plant materials (Bleak 1970; Bocock et al. 1960; Lemée and Bichaut 1973; Shanks and Olson 1961; Stark 1973; Thomas 1968; and Witkamp 1963); bag material and mesh size vary. Although bagged litter may be more humid than leaves in the undisturbed litter layer (Jensen 1974), the advantages of using this technique outweigh its disadvantages. It allows weighed, identifiable samples to be easily placed in the field and recovered at a later time. Final weights are then obtained and differences between initial and final weights (weight loss) are determined. This is, in effect, the *D* (litter disappearance) term used in calculating litter decomposition over any selected time interval (Medwecka-Kornas 1971). Such weight loss can be correlated with environmental variables which are monitored simultaneously.

Plant materials, such as leaves, are collected after falling (Shanks and Olson 1961; Stark 1973), as they fall (Bocock et al. 1960) or directly from live plants. Freshly fallen leaf samples best represent the actual litter produced, especially in regard to the fungal flora on these aerial parts and the nutrient content remaining in them. However, with plants producing a limited quantity of litter (like the evergreen shrubs used in this study), obtaining a weight of leaves (collected at any one period of time) large enough to allow adequate replicates in the field presented a problem.

For this research, leaves, stems and roots were collected from live plants. This reduced the variability of the experimental material. These materials were air-dried and placed into mesh bags. Use of air-dried weights at NEV minimized distortion of the sample prior to placing it in the field. A correlation coefficient of 0.994 ($n = 36$) between air- vs. oven-dried weight validates the substitution of air-dried samples for oven-dried samples. The use of air-dried samples was made earlier by Bocock and Gilbert (1957). At NEV, the 1-dm^2 bags with their enclosed leaf litter (2 g) were randomly placed under *Artemisia* shrubs within a 12 x 12 m grid. Two grams per dm^2 appeared representative of *Artemisia* leaf litter density beneath shrub canopies.

At JOR, bags were placed both in the open and under *Larrea* shrubs, and were introduced into the field in February, May and July. At this site, a 11.5-cm internal diameter plastic asbestos tube 183 cm long was driven into the ground with the top of the cylinder protruding, and then bags were placed within the tubes. Bags with leaf and stem litter (3 g) were placed on the soil surface within the tubes; those containing roots (3 g) were buried 10 cm in the ground within the tubes. Control tubes were placed in the open in areas where the ground surface was free of litter. All bags at both sites were horizontally positioned (flat side down). These studies used nylon mesh (NEV) and nylon hose (JOR), with openings of about 0.56 mm^2 in both cases.

Carbon Dioxide Monitoring

It is most appropriate to measure decomposition or, for that matter, any biological activity in metabolic terms (Parkinson et al. 1971). A widely used method confines the litter being examined and traps the evolved carbon dioxide in a mild alkali solution (Witkamp 1966).

At JOR, litter bags were placed inside permanently set asbestos tubes that were driven into the soil. Wide-mouthed plastic vials containing 10 ml (when minimum amounts of carbon dioxide evolved) or 30 ml (following precipitation and high carbon dioxide evolution) of 0.6 *M* KOH were placed in the tubes and covered for 24-hr periods. Carbon dioxide was trapped on an alternate-day basis through the summer months. The amount of carbon dioxide evolved was determined by titration with 0.6 *M* HCl following treatment with 10 ml of 0.6 *M* BaCl, to precipitate the absorbed carbon dioxide as carbonate, and 5 drops of thymolphthalein to give a sharp end point. Differences in amounts of carbon dioxide evolved from litter and controls which yielded background carbon dioxide evolution were used to calculate mg CO_2/day.

Meteorological Measurements

Rainfall data were obtained weekly at NEV using a plastic rain gauge. At JOR, rainfall data were obtained from the meteorological grid at the IBP intensive study site. Air temperature and relative humidity data at NEV were taken from a hygrothermograph in a standard weather shelter. Soil moisture and temperature data were obtained on a daily basis at JOR from gypsum blocks; at NEV, balsa hygristers placed in the litter were used to obtain moisture data. The data from both sites were read as ohms resistance and converted to either negative bars of soil moisture tension (JOR) or relative humidity (NEV).

Nitrogen Determinations

Total nitrogen content of the leaf litter was determined by the microKjeldahl method (Jackson 1958; Bremner 1965b) modified for plant material. Oven-dried (45 C) leaves were ground sufficiently to pass a 0.4-mm screen and weighed into 0.5000-g portions. These were wrapped in cigarette paper, dried in the oven for 2 days, then digested using copper (Cal-Pak Powder No. 2—Gunning Method) and a selenium catalyst (Hengar Selenized granules). Distillation was carried out using a boric acid trap; the distillate was titrated with 0.01 *N* sulfuric acid standard using N-point indicator.

Leaching of Plant Litter

Tukey (1970) described leaching to include substances removed from plants by aqueous solutions which included rain, mist, dew and fog. This work extends leaching to also include the removal of substances in aqueous solutions from plant litter, as has been interpreted by King and Heath (1967) and Kowal (1969). In most instances attention has been focused on losses of specific macro- and microelements of organic metabolites from living tissue and changes in decomposition rates of litter due to prior leaching. To augment the interpretation of losses due to precipitation, it was necessary to conduct litter leaching experiments for the Chihuahuan Desert investigation.

Simulated leaching studies were conducted on *Larrea* leaves, stems and roots. The thickness of leaf and stem litter was determined in the field and similar thicknesses were re-established in a Buchner funnel. A simulated rain equivalent to 2.54 cm was passed through the plant materials during four 15-min intervals. These plant materials were then oven-dried at 50 C and weighed to determine any loss of water-soluble substances. Since some litter is exposed to free soil moisture for longer periods of time, the same plant organs were then placed in beakers and exposed to free water for 24 hr. After removal, the litter was again oven-dried and weighed to determine any additional weight loss.

Experimental Results

Weight Loss of Litter Bags

In the material which follows concerning decomposition we will deal specifically with our "litter weight loss" results. Jensen (1974) aptly points out that in discussing decomposition rates, one should distinguish between the rate of disappearance of litter and the rate of complete chemical breakdown of the organic litter components. He indicates the former case to include "disintegration and incorporation into the organic fraction of the soil, whereas in the latter case the mineralization of the organic matter is also taken into consideration." In the studies reported here, the disappearing litter substrate may be incorporated into the soil (as described above) or end up in a gaseous form.

Effects of Duration

Litter bag weights at NEV show an overall decrease in weight with an increase in time in the field (Table 4-1). This applies to all samples, regardless of the time they were placed in the field. An examination of the table reveals that sample weights may seem to increase during certain time intervals (i.e., May, August and September samples between the 6th and 8th months); this is also shown in Figure 4-1.

Table 4-1 *Residual Weights for Bagged* Artemisia *Leaf Litter at the NEV Site from April 1972 to July 1973 [The time intervals for samples in the field were 2, 4, . . . , 12 months. Initial air-dry weight of all samples was 2 g. Samples were placed in the field in successive months, as indicated.]*

Month placed in field	Final weight after indicated time in field (g)					
	2 mo	4 mo	6 mo	8 mo	10 mo	12 mo
Mar 1972	1.89[b]	1.56	1.57	1.36	1.04	0.96
Apr[a]	1.58	1.52	1.39	1.33	1.17	1.14
May	1.76	1.41	1.24	1.26	0.79	1.12
Jun[a]	1.89	1.24	1.36	1.35	1.06	0.94
Jul	1.65	1.19	1.43	1.14	1.01	0.96
Aug	1.67	1.23	1.14	1.18	1.06[c]	
Sep[a]	1.43	1.49	0.95	1.13	0.97[c]	
Oct	1.35	1.26	1.22	1.11[c]		
Nov[a]	1.50	1.29	1.15	1.07[c]		
Dec	1.48	1.39	1.11[c]			
Jan 1973[a]	1.48	1.37	1.21[c]			
Feb	1.73	1.48[c]				

[a]Data are presented in Figure 4-1.

[b]All final weights are means, $n = 3$.

[c]Research site was destroyed by fire at the end of Jul 1973.

This is explained chiefly by the sample design as discussed in a later section. The rates of weight loss are variable for the different samples (Fig. 4-1). All show a steep decrease in weight within the first 3 months in the field. Such rapid weight loss for various kinds of litter has been found by most authors dealing with this subject (Hayes 1965; Melin 1930; Watson 1930; Kucera 1959). When they monitored litter decomposition for longer periods they also found a decreasing rate of decomposition.

Except for the weight increases indicated (especially in the 6-8 month period), the rate of litter weight loss at NEV decreases towards the end of the period in which any sample was in the field. The overall percent weight loss was about 50 percent after 1 year in the field. In another study using litter bags in southern Idaho, big sagebrush leaf litter losses averaged 59 percent over a year (Murray 1975).

In order to more easily compare rates of weight loss, sample air-dry weights were analyzed only in terms of the length of time in the field (e.g., time of placement in the field is ignored). A linear regression of sample weight on time was calculated (Fig. 4-2). The range of weights for all samples in the field for the time (1, 2, . . . , 12 months) is also indicated. The ranges for any 1-month time interval varied from approximately 0.16-0.84 g.

The equation for the regression is

$$Y = 1.72 - 0.067X$$

where

Y = air-dry weight (grams)
X = time samples are in the field (months)

Similarly at the JOR site weight losses were determined from litter bags placed in the field during February, May and July (Table 4-2). Weight loss appeared to be least in the February introductions, intermediate in May and highest in the July placements. For example, the changes in weight for the first 77 days of the February introductions were exceeded in 34 days by the May placement and in 17 days by the July introductions. The rate of loss from roots, stems and leaves varied at different examination times and this was influenced by moisture and microbial populations. Insects and other small invertebrates may speed the process, but we have no data available regarding such activity. Moisture influence was observed in the February placement where the buried roots were associated with soil moisture, while the leaves and stems on the soil surface were dry for substantial periods. Microbial populations were also important in instances when the soil moisture was disappearing. Fungi were observed fruiting at these times, where previously they had not been. When readings were taken on these occasions, the data obtained would usually indicate an addition of carbon over the previous reading. Other influences

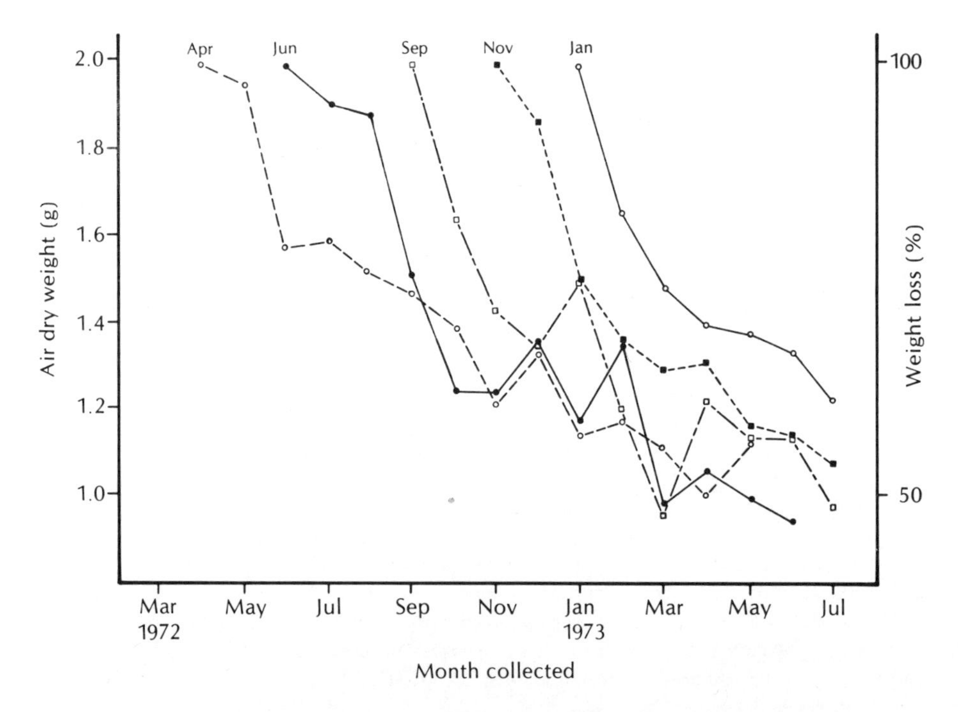

Figure 4-1 *Air-dry weights for bagged* Artemisia *leaf litter samples located under shrubs at NEV during the study period. Twelve samples were placed in the field during each of the 5 months indicated, and one sample from each was collected in subsequent months. Initial air-dry weight for all samples was 2 g. Percent weight loss is also indicated.*

require further and more critical examination of the changes taking place as well as the factors inducing the changes.

Temperature

Weight loss for samples at NEV exposed in the field for 1 to 12 months was correlated with temperature data (daily means) obtained from a hygrothermograph located in a standard weather shelter. Weight loss (grams) was treated as the dependent variable for this analysis. The coefficient of determination and F values were calculated (Table 4-3). For weight loss in the first 6-month period there was a significant correlation with temperature. In the second 6-month period only the weight losses for the 7th month (data not presented) and 8th month were significantly correlated with temperature. The r^2 values are 0.5 for all other months. The r^2 for the 8th month was the highest obtained (0.665). The analysis of data from the 2nd month provides an example of the family of linear regression lines for weight loss for the different time intervals.

The linear regression for weight loss as a function of temperature is

$$Y = 2.076 - 0.026X$$

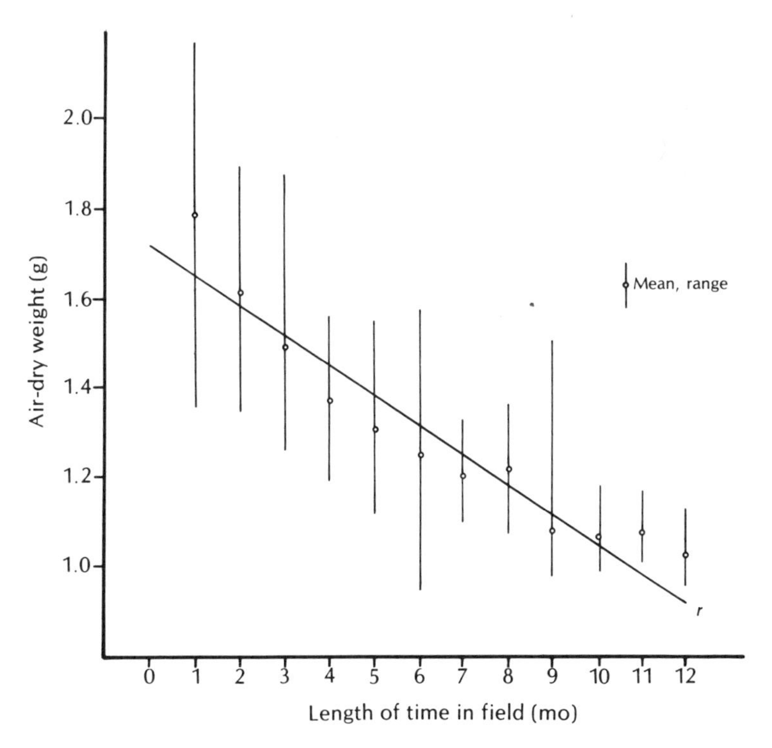

Figure 4-2 *Regression of air-dry weights for bagged* Artemisia *leaf litter samples on time in the field at the NEV site.*

Table 4-2 *Percent Organic Matter Loss from* Larrea *Leaf, Stem and Root Litter Bag Material During Different Time Periods at the JOR Site, 1973*

Month placed in field	Plant part	Month removed, with days exposed in parentheses				
		May (77)	Jun (111)	Jul (141)	Aug	Sep
Feb	Roots	19.8[a]	30.5	28.8[b]	-[c]	-
	Stems	10.8	13.2	24.7	-	-
	Leaves	12.0	20.9	42.8	-	-
			(34)	(64)	(81)	(102)
May	Roots	-	23.8	37.7	29.0[b]	40.2
	Stems	-	17.6	27.4	31.3	41.4[b]
	Leaves	-	39.1	47.4	54.4	37.4[b]
					(17)	(38)
Jul	Roots	-	-	-	21.4	16.8[b]
	Stems	-	-	-	29.6	31.3
	Leaves	-	-	-	43.8	50.3

[a]Each reading is an average of three samples.

[b]Unusual accumulation of fungal mycelia and spores was observed on the decomposing plant material.

[c]No reading was obtained (-).

Table 4-3 *Coefficients of Determination* [r²] *and* F *Values for* Artemisia *Leaf Litter Bag Weights Correlated with Temperature, Relative Humidity and Precipitation at the NEV Site*

Length of time in field (mo)	Temperature		Relative humidity		Precipitation	
	r^2	F	r^2	F	r^2	F
2	0.0638	15.84[a]	0.425	6.64[b]	0.712	24.73[a]
4	0.4607	7.69[b]	0.417	6.43[b]	0.183	2.24
6	0.4489	6.52[b]	0.467	7.00[b]	0.468	7.92[b]
8	0.6645	11.88[b]	0.902	55.23[a]	0.282	2.75
10	0.3427	2.09	0.361	2.26	0.103	0.58
12	0.109	0.24	0.073	0.16	0.153	0.54

[a]Significant at 99% level.

[b]Significant at 95% level.

where

Y = weight loss (grams)
X = temperature (°C)

The changes in decomposition rates were not influenced by fluctuations in temperature during the JOR field studies of 1973. Lack of soil moisture during the spring and fall eliminated the evaluation of temperature as an important variable.

Moisture

Weight loss for NEV samples was correlated with precipitation as well as with relative humidity (daily means) obtained from the same hygrothermograph which provided the temperature data. Sample weight loss was the dependent variable in the first case, sample weight the dependent variable in the second case. The F values for the relative humidity-sample weight correlations were significant ($F_{0.5}$) during the first 6-month period as well as in the 8th month (Table 4-3). The r^2 values for the months exhibiting significant correlation range from 0.42-0.90. The highest r^2 value is in the 8th month, as shown.

The correlation of precipitation with weight loss is less than that with either temperature or relative humidity. Only 2 months show significant correlation (Table 4-3). Coefficient of determination values range from a high of 0.71 to a low of 0.10. The r^2 values for precipitation are generally lower than those for relative humidity and temperature.

The linear regression for weight loss of samples in the field for 2 months on relative humidity as the independent variable is

$$Y = 1.1179 + 0.0096X$$

where

Y = weight loss (grams)
X = relative humidity (percentage)

Carbon Dioxide Evolution

Since litter samples were placed in the field at three different times, CO_2 evolution at JOR could be examined as a function of time as well as under different environmental conditions. The observations suggest that the time a sample was introduced to the soil varied in importance and influenced the decomposition rates (Fig. 4-3). Samples placed in the field in May evolved more CO_2 during June and July than those introduced to the soil in February.

A lag period occurred before CO_2 evolution commenced in some *Larrea* samples. For example, CO_2 evolution in the May samples was reduced until June 14 and then increased in rate into late July. *Larrea* leaf samples introduced to the soil in July evolved CO_2 immediately and continued to do so until August. This CO_2 evolution activity was closely related to soil moisture content (Fig. 4-4). With the exception of one clearly discernible time of CO_2 evolution in late June and early July, the remaining CO_2 evolution corresponded to changes in soil moisture.

Nitrogen

It has been suggested that a shortage of nitrogen may determine microfloral diversity during decomposition (Kononova 1961) and limit rate of decomposition (Melin 1930). Therefore, this may be very important in the desert environment. Total nitrogen decreases as a function of decomposition time (Waksman and Tenney 1927). This is true for total nitrogen in *Artemisia* litter at NEV (Comanor and Prusso 1974). Nitrogen in the March samples decreased from approximately 45 mg/g to 27 mg/g a year later (Fig. 4-5). This decrease in total nitrogen is greater than that for the other samples. The July samples show very little change in nitrogen during the study, actually increasing slightly from the initial 30 mg/g value. Levels of nitrogen inconsistent with the general trend of decreasing total nitrogen content are shown for the March samples during the November-January period, and also for the June samples collected in December and February. This is partly explained by the sample design. The increases in nitrogen content for several samples only partly correspond with certain periods of the year. Similar gains in nitrogen content (as mg N/g dry weight) do not occur for all samples as a direct function of time in the field. The erratic pattern of nitrogen content after several months in the field precludes the development of adequate regression equations relating nitrogen content to decomposition time.

Although H_2SO_4 traps were placed in the field throughout the summer of 1973 at JOR to collect the volatilized ammonia, none was trapped. This does not indicate that NH_3 was not volatilized during 1973, but does show that none was released during the course of this investigation. It is possible that there was sufficient available carbon to maintain and hold any nitrogen released during the decomposition process.

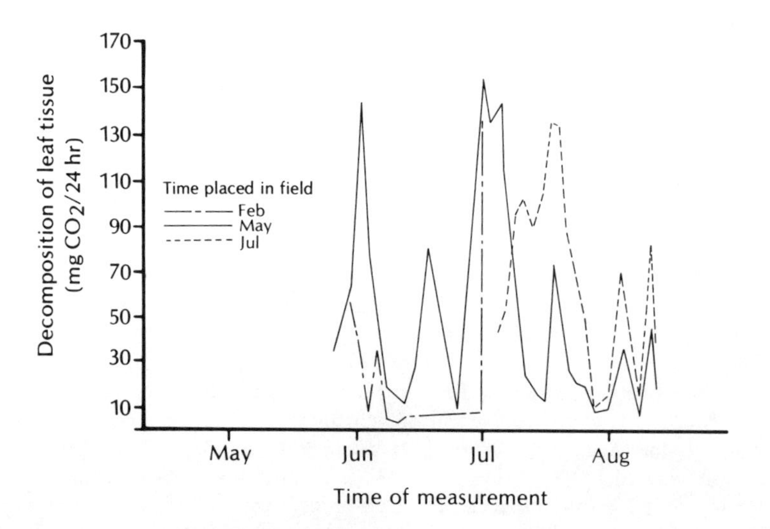

Figure 4-3 *Twenty-four-hour, alternate-day CO_2 evolution measurements from* Larrea *leaves placed in the field in February, May and July at the JOR site.*

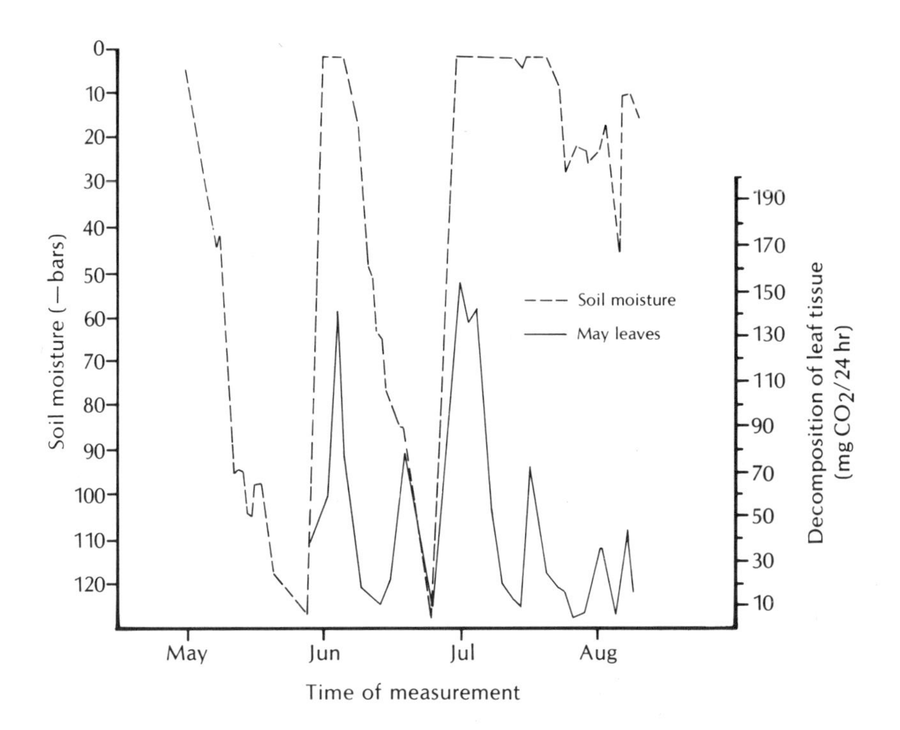

Figure 4-4 *Comparison between soil moisture [daily measurements] and CO_2 evolved [alternate-day measurements] from May-introduced* Larrea *leaves at JOR.*

During investigations conducted at JOR, discrepancies were observed between weight loss and carbon dioxide evolution data in the early testing. Review of the activities and parameters being measured indicated the major change that had occurred in the field was precipitation. Substantial weight losses occurred by the leaching of plant material (Table 4-4). Analyses of these data revealed significant differences between times of plant organ removal from the field, leaching from plant materials and treatments. In fact, when losses due to the simulated rain and standing water (25.5 percent) were added to the loss of carbon dioxide evolved (10.0 percent) from the May-incorporated leaf litter, the total value (35.5 percent) was not drastically different from that recorded for the weight-loss determination (39.1 percent from Table 4-2). Values obtained from leaching plus carbon dioxide evolution were not as precise for stems (leaching at 7.4 percent plus carbon dioxide at 5 percent vs. weight loss at 17.6 percent) and roots (leaching at 8.4 percent plus carbon dioxide at 3 percent vs. weight loss at 23.8 percent).

Weight loss determinations should be supported by carbon dioxide evolution data and leaching loss information because it would be, at present, improper to assume that organic matter loss is synonymous with conversion of carbohydrate to carbon dioxide and water only.

DISCUSSION

Numerous factors differentially influenced the decomposition of litter in the two North American deserts. In the cold winter desert (NEV), where soil moisture was more than adequate during the winter, temperature was an influential variable. In contrast, the Chihuahuan Desert (JOR) was more influenced by soil moisture than temperature because moisture was lacking in the fall, winter and spring when temperature fluctuations might have controlled decomposition rates. During the summer months temperatures were high enough for decomposition, and the activity was related to the presence or absence of moisture in the soil and litter.

Table 4-4 *Measured Changes Due to Simulated Leaching of* Larrea *Leaves, Stems and Roots Prior to Decomposition, May and September 1973, at the JOR Site*

Plant part	Simulated 2.54-cm rainfall			Exposure to free moisture for 24 hr	
	Original Weight	Weight loss (g)	Weight loss (%)	Weight loss (g)	Weight loss (%)
			May 1973		
Leaf	3.00	0.461	15.3	0.773	25.8
	3.00	0.418	13.9	0.739	24.6
	3.00	0.599	20.0	0.783	26.1
Stem	3.00	0.066	2.2	0.196	6.5
	3.00	0.077	2.6	0.239	8.0
	3.00	0.075	2.5	0.234	7.8
Root	3.00	0.064	2.1	0.266	8.9
	3.00	0.078	2.6	0.220	7.3
	3.00	0.160	5.3	0.272	8.1
			Sep 1973		
Leaf	2.999	0.063	2.1	0.511	17.0
	2.999	0.075	2.5	0.511	17.0
	3.000	0.074	2.5	-	-
	2.999	0.076	2.5	0.498	16.0
Stem	2.999	0.058	1.9	0.226	7.5
	3.000	0.049	1.6	0.279	9.3
	3.001	0.053	1.8	0.214	7.1
	3.000	0.043	1.4	0.233	7.8
Root	3.000	0.79	2.6	0.317	10.6
	2.999	0.62	2.1	0.253	8.4
	2.999	0.66	2.2	0.273	9.1
	2.999	0.86	2.9	0.334	11.1

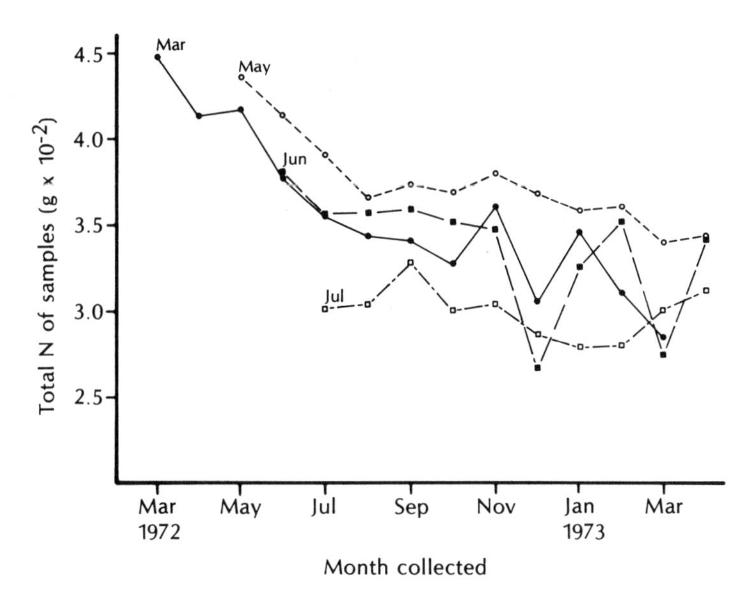

Figure 4-5 *Total nitrogen content for bagged* Artemisia *leaf litter samples at NEV. Twelve samples were placed in the field in the 4 months indicated, and one sample from each was collected in subsequent months. Initial air-dry weight for all samples was 2 g.*

Weight Loss as a Function of Time

Weight loss was analyzed in terms of the length of time samples were in the field. All samples, regardless of the time placed in the field, had weight losses when they were returned. We believe that the anomalous weight gains depicted in Figure 4-1 did, in fact, not occur. Each set of samples recovered from the field was separately located there. Thus environmental characteristics of the sample microhabitats (such as exposure, temperature regime, etc.) could vary. The weight losses for some samples were less (e.g., the June sample recovered in December) than others (e.g., the June sample recovered in November). The overall curves must be considered; the specific weight gains are a result of the required experimental design and are not caused by increases of microbial biomass in excess of weight loss.

Samples were in the field for different lengths of time and during different periods. For example, all samples which were in the field for 1 month were analyzed together. This first month would have been during the cold, wet winter period for some samples. For others, it would have been the cool and wet late fall or early spring. For still others, it would have been the hot, dry summer period. The analysis of weight loss in terms of time in the field does not take into account the different environmental situations. When examining these results it must be borne in mind that the weight losses are averaged over these environmental conditions.

The rate of weight loss decreased with the length of time the sample was in the field (Fig. 4-1). Thus the age of litter seems an important variable affecting decomposition. This was also noted by Witkamp (1966) and is implicit in the findings of many workers who reported that substrates change chemically as decomposition progresses.

The regression line of weight loss on time (Fig. 4-2) shows a decreasing weight of litter with time in the field. The regression presented is linear; this plot accounts for 67 percent of the weight change in the samples. As noted earlier, the weight change of samples is greater in the first few months. This is clearly shown by the means of the litter weights (Fig. 4-3) for samples which were in the field for increasingly longer periods of time. The large range of weights is explained by the range in environmental conditions present for samples during the first 1-3 months in the field; i.e., whether it is cold-dry, hot-dry, cold-wet, etc. (The values above 2 g are attributed to experimental treatment: the techniques for separating sand and other foreign matter were not worked out in the first days of the project.) The mean sample weights are located above the regression line for the last 3 months. It would seem that curves other than linear regression would better fit the data. When a quadratic equation was fitted to the data the r^2 value increased only slightly as a result. Therefore, a simple (first order) model is suggested at this stage of our knowledge.

Percent weight loss for *Artemisia* leaf litter for the first year is comparable to percent loss in forests for various types of litter: 55 percent (Minderman 1968), 60 percent (Witkamp 1966), 44 percent (Thomas 1968), a 15 percent maximum for 4 months (Falconer et al. 1933). The NEV weigt loss data show a considerable range (Fig. 4-2) as do data for tree litter decomposition reported in the literature, whether the trees were angiosperm or coniferous and whether the study was done in the field or in the laboratory.

Weight changes for individual months show some of the variability which can be ascribed to environmental conditions. Samples placed in the field during the cool to cold, moist September-January period (Fig. 4-1) showed a much greater initial weight loss than those placed in the field in the warmer (but drier) June period. The significant relationships of these weight changes with environmental factors are presented in the discussion which follows.

Weight Loss and Environmental Variables

Temperature, when treated in simple linear regression, is a variable which explains as much as 66 percent ($r^2 = 0.66$, Table 4-3) of the observed change in weight. This is not surprising, since soil surface (and therefore litter) temperature is often clearly related to air temperature (Williams and Gray 1974). Air temperature is significantly correlated with weight change in each of the first 6 months; in the last 6-month period it is significantly correlated only with the 8th month. Although the r_2 data are ambiguous (note the low value for month 2, Table 4-3), we interpret the lower correlation with temperature towards the end of the study as indicating the increased importance of other factors in the field. The changed nature of the litter is a variable not handled in the research design at NEV. It is tempting to postulate this as the significant operative factor, since "waves" of microbial activity using different chemical substrates are known to occur (Jensen 1974).

As Figure 4-1 shows, weight loss is rapid in the winter period. At first this appears anomalous; winter temperatures seem limiting (lows of —40 C occurred in NEV in December 1972). However, low winter temperatures are not constantly maintained and many afternoons are warm enough to favor microbial activity. Bleak (1970) found weight loss in his litter samples beneath a snow cover, where surface soil temperatures went just above freezing. Moisture is ususally more than adequate at NEV in the winter since precipitation then is higher than in the summer, and evaporation is reduced. At JOR, which lies much farther south, winter periods are much milder and would not be limiting to biological activity if sufficient moisture were available. Since the winters are dry, decomposition was measured during the warmer months and was found to be closely related to soil moisture.

The periods of rapid weight loss (Fig. 4-1) correspond to periods of high precipitation. The relative humidity in these periods is high also. This latter variable provides a significant correlation with weight change (r^2 values > 0.41 during the first part of the study, Table 4-3). Both temperature and moisture account for less of the weight loss in the later stages of the study.

Precipitation correlates poorly with litter weight changes at NEV; precipitation is significantly correlated with weight loss for the 2nd and 6th months only (Table 4-3). The precipitation data used were the amounts in the 1-month period preceding the removal of samples from the field. Both temperature and relative humidity data show better correlations with weight changes than precipitation, and consequently higher r^2 values (Table 4-3).

The moisture conditions in the litter are obviously critical to microbial life, and litter weight loss has been correlated with litter moisture content (Witkamp 1963). Precipitation is the "input" to the litter system, but later conditions (insolation, wind, temperature) will determine how much of this moisture is available to microorganisms.

These conditions (over a longer period of time than the precipitation "event") should be associated with litter decomposition. It is not surprising, therefore, that temperature and relative humidity are significant variables for circumscribing litter ecological conditions. This holds for these environmental measurements even though they were made in a standard weather shelter, certainly less adequate and a step removed from in situ measurements of litter environmental conditions.

Changes in Nitrogen Content of Leaf Litter

Total nitrogen content in leaf litter is affected by many variables. The time of initial leaf collection at NEV resulted in initially unequal nitrogen levels in the samples (Fig 4-5). The nitrogen content for different samples did not decrease proportionally after equal periods of time in the field. It is difficult to analyze nitrogen content as dependent on time in the field because of the erratic nature of the curves for March and June during the winter period. Similar increases also occurred in the spring for some of these samples. Such variability results in poor correlations of nitrogen loss with environmental variables, especially when one of the samples (July) actually shows a consistent upward trend after a period of loss.

These increases in total nitrogen are difficult to explain. They may be due to errors in laboratory analyses, to unidentified nitrogen inputs or to microhabitat differences, as explained earlier in the discussion concerning apparent weight gains.

Examination of changes in nitrogen content reveals that sometimes they correspond with the "apparent weight gains" and at other times they occur independently. An increase in the relative proportion of nitrogen in desert plant litter may be explained as a result of the decomposition of carbohydrates (sugars, etc.) and other readily decomposed materials, leaving more resistant, nitrogen-containing compounds, such as cellulose. However, this does not explain an increase in the absolute amount of nitrogen (as mg/g) in the litter. It does not appear to be a unique phenomenon, as other investigators have found similar increases in nitrogen during the course of their studies (Lemée and Bichaut 1973).

There seems to be a general downward trend in total nitrogen content with time. It is less rapid at NEV than in a comparable study in Idaho (Murray 1975). The losses there appear consistent with time; the r^2 for leaf litter nitrogen content in the field as a function of time varied from 0.39 to 0.88. At NEV the correlation coefficients for the same relationship varied from about 0.5 to 0.6. Litter nitrogen levels at NEV exhibit more variability than in the Idaho study. Part of this is attributable to the highly variable initial nitrogen levels, which reflect different collection times for leaves used in the study. Initial nitrogen levels at NEV (28-45 mg/g) were much less than those in Idaho (162 mg/g). The final average nitrogen content at the Idaho stations after 1 year averaged 130 mg/g, far in excess of the original levels at NEV. The percent loss of this element also differed. At NEV a wide range of loss occurred (7-40 percent), while the Idaho samples were closer together (about 19 percent). As noted earlier, NEV leaf samples used for the decomposition studies were collected and placed in specific microhabitat locations in the field at different times. The different levels for environmental conditions obtaining, in situ, seem causal to the results obtained. Such variability is undoubtedly the rule in the desert environment. This supports the view that predicting nitrogen losses for the general case is difficult indeed, and may fail without environmental data at the litter-decomposition level.

Another interesting comparison between the NEV and Idaho studies may be made. The calculated litter on the surface of the ground in Idaho was less than at NEV, where litter mounds below the shrubs were substantial. The total nitrogen content of the leaves at Idaho was much greater, however. It is tempting to speculate that the greater litter with its nitrogen pool at NEV compensates for the amount of total nitrogen required to be maintained in the *Artemisia* leaf biomass; i.e., a homeostatic balance exists between litter and leaf nitrogen pools and the former, if available, acts as a nitrogen reservoir in the *Artemisia*-dominated community.

SUMMARY

Decomposition of desert plant leaf litter was studied in two North American deserts. Litter came from two dominant shrubs: *Artemisia tridentata* and *Larrea tridentata* in the Great Basin and Chihuahuan deserts, respectively. Litter was contained in mesh bags, placed and recovered in the field at different times. Environmental variables were monitored during the studies.

Carbon dioxide evolution from the *Larrea* litter at the Chihuahuan site was closely related to soil moisture content; soil moisture stimulated microbiological activity and litter decomposition. Carbon dioxide losses did not correspond totally to weight losses, since leaching accounted for a significant percentage of the weight loss.

In the field, litter weight decreased with time; the final weight change was approximately 50 percent for *Artemisia* after 1 year. A linear regression equation relating weight loss of *Artemisia* litter to time in the field can be developed, as well as regression equations relating weight loss to temperature and moisture. The coefficients of determination for these relationships vary; the highest value was 0.9. Weight loss is significantly correlated more often with temperature and relative humidity than with precipitation. Decomposition during the winter period probably accounts for significant weight loss in the Great Basin Desert.

Apparent weight gains occur on the graphs of litter weight loss. These are believed to be the result of the effects of microhabitat heterogeneity on the decomposition rates of different samples. Nitrogen content varied considerably among *Artemisia* samples in the field. A general downward trend of nitrogen content for many samples is interrupted by increases which are difficult to explain. Several explanations can contribute to our understanding of the pattern of change in nitrogen content of decomposing leaf litter.

5

PROTEOLYSIS AND AMMONIFICATION IN DESERT SOILS

R. T. O'BRIEN

INTRODUCTION

If there is a single environmental characteristic common to all deserts it is moisture impoverishment. Within this general characteristic there are variations from true deserts, such as the Sahara, to semiarid regions, which is the category in which fall the deserts of the southwestern United States.

Rainfall in all deserts is generally concentrated into short time spans. Thus decomposition and other activities are largely concentrated into those periods when moisture and temperatures are adequate for vigorous microbial activity. During dry periods, microbial decomposition activity is at a very low level or quiescent. In spite of the rigorous and somewhat impoverished environments extant in deserts, however, there are remarkable reservoirs of soil microorganisms with a complete array of physiological capabilities to participate in nutrient and energy cycling.

This report deals with results of studies on decomposition of protein and ammonification in Chihuahuan Desert soils (Jornada site of the US/IBP). Unless it was necessary to do otherwise, methods have been omitted from this report and the reader is referred to the original papers for detailed descriptions of procedures.

PROTEOLYSIS IN DESERT SOILS

Nitrogen as amino N of plant and animal protein is the starting point of the process by which nitrogen is returned to the soil. Thus proteolysis is an essential first step in the mineralization process of protein nitrogen. In Table 5-1 the relative densities of proteolytic bacteria in Chihuahuan Desert soils are shown (O'Brien 1972). The data indicate variation in the number of indigenous proteolytic bacteria in soils with respect to soil depth and lateral distribution, although there was no consistent pattern to the variation. The number of proteolytic organisms normally present in soil may not be particularly meaningful since correlation between initial microbial numbers present in soil and microbial activity is generally poor (Schmitt 1972). The values shown in Table 5-1 are comparable to the proteolytic bacteria densities reported for nonarid lands by van Schreven and Harmsen (1968).

Table 5-1 *Density of Proteolytic Organisms in Chihuahuan Desert Soils [O'Brien 1972]*

Sample	CFU[a] at 10 cm	Percent proteolytic bacteria[b] at different depths		
		10 cm	20 cm	30 cm
B-1	10×10^5	77	74	60
B-2	17×10^5	ND[c]	25	33
B-3	8.0×10^5	33	ND	24
B-4	8.0×10^5	50	ND	50
P-11	2.0×10^5	35	38	40
P-15	9.0×10^5	17	29	69
P-24	1.5×10^5	40	60	33
P-42	2.8×10^5		25	62

[a]CFU = Colony Forming Units per gram.

[b]Percentage of CFU.

[c]No data available.

Wet mounts were made from 100 protease-positive colonies and examined by phase microscopy. Spores were observed in 91 of the colony preparations, indicating that the aerobic spore formers of the genus *Bacillus* were the dominant proteolytic bacterial group. Due to heat sensitivity the majority of *Bacillus* in soils were present as spores. Fungal growth was visible in the surface of the protein-amended soil samples after 5-7 days incubation, indicating that soil fungi were also active protein decomposers.

Protein decomposition rates in Chihuahuan Desert soil samples are presented in Table 5-2 (O'Brien 1972). These data clearly show there was little variation in the rates of decomposition in soil samples taken at 0-30 cm depth from different bajada (slope) sites. It should also be noted that the rates at 37 C were approximately three times the rates at 20 C ($P < 0.005$), which suggests that the environment favors organisms with high temperature optima. Indeed, we have observed decomposition of carbon and protein substrates from 10-60 C in Chihuahuan soils, indicating that proteolysis occurs over a broad temperature range.

Adsorption of proteins to soil particles was also tested since adsorption has been suggested as inhibiting protein breakdown. However, as shown in Table 5-3, Jornada soils at 0-30 cm depth exhibited negligible adsorptive capacity for trypsin, casein and gelatin.

Skujins and West (1973) reported somewhat different results from the Great Basin Desert. Some of their data are summarized in Table 5-4, where it can be seen that biological activities, including proteolysis, declined with soil depth in samples from the southern Curlew Valley IBP site in Utah. It is not clear at this point whether the difference in proteolysis results obtained with Great Basin and Chihuahuan soils is due to methodology or to distribution of microorganisms in the soils. It is evident, however, that no obvious correlation can be inferred between bacterial numbers and biological activity since slight variations in plate counts were accompanied by much greater changes in biological activities.

Table 5-2 *Influence of Depth and Temperature on Rates of Proteolysis in Chihuahuan Desert Soils [O'Brien 1972]*

Sample no.	Rate of proteolysis[a] at different depths and temperatures (g casein/100 g soil)					
	0–10 cm depth		10–20 cm depth		20–30 cm depth	
	20 C	37 C	20 C	37 C	20 C	37 C
B-1	87[b]	295	76	169	84	169
B-2	71	328	98	224	87	295
B-3	52	157	46	157	87	328
B-4	93	240	85	218	77	218
B-5	93	218	79	328	60	328
Mean ± SD	79 ± 17.7	248 ± 66.8	77 ± 19.2	219 ± 67.5	79 ± 11.4	268 ± 71.1

[a]Rate in g/wk per m^2.

[b]Mean of three determinations.

Skujins (1973) has proposed using dehydrogenase activity as an index of soil biological activities. His studies show high correlation indices (0.87-0.99) between dehydrogenase activity and other biological activities, including proteolysis. However, the procedure gives only an indication of biological activity potentials in soil rather than a measure of the actual rates of activity.

Haddad (1972) studied protease biosynthesis by *Bacillus* species and by the soil microbiota in Chihuahuan Desert soils. The experiments included a variety of soil amendments in addition to water. Results, which are summarized in Table 5-5, show that pure cultures of *Bacillus* species and the natural microbial populations of

Table 5-3 *Protein Binding by Chihuahuan Desert Soils [O'Brien 1972]*

Experiment[a]	Percent extractable protein		
	Trypsin	Casein	Gelatin
1	91	86	92
2	85	94	95
3	96	89	87
4	89	94	92
5	92	93	93

[a]Each experiment was done with a composite soil sample; 10-g sterile samples were amended with 0.05 g protein in 10 ml water.

[b]Proteins were extracted after 24 hr with 30 ml of water.

Table 5-4 *Some Biological Activities in Curlew Valley Soils in 1972 [Skujins and West 1973]*

Sampling site and depth	Proteolysis (% hydrolysis)	Dehydrogenase (g Formazan)	Aerobic plate count/g
5-1 (0-3 cm)	30	.98	4.8×10^6
5-2 (5-20 cm)	9	.08	3.1×10^6
5-3 (40-50 cm)	9	.04	2.7×10^6
5-4 (70-80 cm)	6	.02	2.0×10^6
6-1 (0-3 cm)	30	1.21	7.2×10^6
6-2 (5-20 cm)	10	.10	10.3×10^6
6-3 (40-50 cm)	11	.08	5.6×10^6
6-4 (70-80 cm)	8	.03	1.8×10^6
7-1 (0-3 cm)	32	1.95	3.8×10^6
7-2 (5-20 cm)	10	.16	3.5×10^6
7-3 (40-50 cm)	10	.06	2.0×10^6
7-4 (70-80 cm)	5	trace	0.7×10^6

Table 5-5 *Effects of Nutrients on Protease Synthesis in Sterile[a] and Nonsterile[b] Soil [Haddad 1972]*

Amendment	Sterile	Incubation time (days)	Enzyme units/2 g soil
H_2O	No	15	0.0138
0.2% casein	No	15	0.0902
0.2% NH_4Cl	No	15	0.0502
0.2% glucose	No	15	0.0208
H_2O	No	30	0.0069
0.2% casein	No	30	0.3200
0.2% NH_4Cl	No	30	0.0000
0.1% glucose	No	30	0.0194
H_2O	Yes	15	0.0694
0.2% casein	Yes	15	0.0486
0.2% NH_4Cl	Yes	15	0.1041
0.2% glucose	Yes	15	0.1111
H_2O	Yes	30	0.0000
0.2% casein	Yes	30	0.0243
0.2% NH_4Cl	Yes	30	0.0277
0.2% glucose	Yes	30	0.0138

[a]One-hundred grams amended with 20 ml of each nutrient, inoculum 5 ml of *Bacillus* sp. insulated from soil.

[b]One-hundred grams amended with 25 ml of each nutrient.

soils synthesized proteases under all conditions studied. These data indicate that protease biosynthesis by proteolytic microorganisms was partially constitutive since the enzymes were produced in the absence of protein amendment. The enzymes were also very stable in sterile soils, as shown by the results summarized in Table 5-6. The persistence of enzymes in soil has been previously reported by Skujins and McLaren (1968).

From the above results on proteolysis it is clear that, with adequate moisture, protein decomposition in soil was fairly rapid. In experiments where protein substrates were introduced into areas with low density of proteolytic soil microorganisms, there was a lag before proteolysis was detectable. However, once the proteolytic microorganisms were established, the rates of proteolysis were comparable to the rates in soils with high indigenous populations of protein-decomposing microorganisms. Thus, the long-term rate of protein decomposition in soil would be controlled primarily by environmental factors such as moisture and temperature rather than by the initial densities or physiological characteristics of the indigenous soil microflora.

Table 5-6 *Protease Stability in Sterile Soil at Ambient Temperature [Haddad 1972]*

Incubation time (days)	Enzyme units/g soil
30	0.16
60	0.16
90	0.16

AMMONIFICATION

Volatilization of amino N as NH_3 was also measured in proteolysis experiments (O'Brien 1972). In these experiments the NH_3 in the gas phase above soil samples was trapped in dilute sulfuric acid and analyzed as ammonium ion. Results shown in Figure 5-1 indicate that amino N was rapidly converted to NH_3 and escaped into the atmosphere above the samples. In a series of experiments like those shown in Figure 5-1, an average of 80 percent of the amino N in the protein amendment was volatilized as NH_3 within 4 weeks. The soil pH ranged between 7.8 and 8.3. It should be noted in Figure 5-1 that a low, but measurable, rate of volatilization of NH_3 from unamended soil samples was also detected; however, the source of the NH_3 was not determined.

The preceding results promoted further study of deamination and ammonification of amino N at both the Jornada site in New Mexico and the southern Curlew Valley site in Utah. Deaminase activities in New Mexico soils are listed in Table 5-7 (O'Brien 1973). These data show that deaminase activities increased as soil moisture increased ($P < 0.01$). It was also found that deaminase activities were highest at 0-5 cm soil depth. The stratification of microbial deaminase activity is in contrast to the results obtained for protease activity. Apparently some of the microorganisms responsible for deamination are different from those involved in proteolysis. No

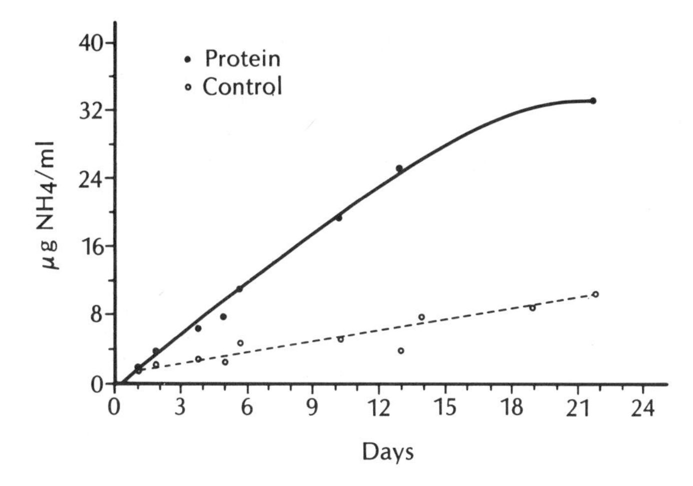

Figure 5-1 *Ammonia formation from Chihuahuan Desert soil. Protein amendment = 2.5 mg [O'Brien 1972].*

attempt was made to determine the density of ammonifiers, so it is not known if they are more numerous than proteolytic microorganisms. At Curlew Valley similar results were obtained by Skujins and West (1973), as shown in Table 5-8. Their data also show that, with casein as the amendment, deamination and ammonia volatilization were maximal when soil moisture was at 75 percent water holding capacity. It is worth noting that, in contrast to the results obtained with casein, ammonia volatilization of plant litter N was lower. Whether this difference was due to recalcitrance of litter to decomposition or to a C:N ratio which favored microbial nitrogen retention is not known at this time.

In field experiments at Jornada the rates of ammonia evolution from soil fluctuated widely. Since the soil pH values were constant, the fluctuations of ammonia evolution were presumably dependent on soil moisture levels and

Table 5-7 *Deaminase Activities in Jornada Soils [O'Brien 1973]*

Date (1972)	Sample site[a] ($mg\ NH_4/24\ hr$)				Mean	SD	Soil moisture (negative bars)
	A	B	C	D			
27 Jun	50	60	60	63	58	± 5.7	92
8 Jul	240	70	220	85	154	± 88.6	20
1 Aug	242	198	198	162	200	± 32.7	0

[a]Sites were on a north-south transect at 100-m intervals.

Table 5-8 *Ammonification Potential at Various Moisture Contents at 0-3 cm in* Atriplex confertifolia-*occupied Curlew Valley Soil*[a] *[Skujins and West 1973]*

Incubated soil	Percent water holding capacity	Ammonium N (mg/g soil) released during incubation for weeks:				Ammonium N trapped (μg)	Organic N at end of experiment (mg/g soil)	Percent total organic N mineralized
		1	2	3	4			
Control[b]	Air-dry	0.056	0.084	0.140	0.084	0	0.945	15.4
Soil + casein[c]		0.154	0.196	0.140	0.168	2	1.302	32.1
Soil + litter[d]		0.098	0.154	0.140	0.168	0	1.8	9.0
Control	25	0.084	0.084	0.112	0.084	2	0.838	24.9
Soil + casein		0.244	0.294	0.420	0.488	29	0.84	56.2
Soil + litter		0.154	0.140	0.140	0.168	2	0.6	19.0
Control	40	0.056	0.042	0.084	0.112	4	0.735	34.1
Soil + casein		0.252	0.420	0.518	0.490	45	0.63	67.1
Soil + litter		0.140	0.168	0.196	0.196	2	1.414	28.5
Control	55	0.056	0.126	0.112	0.084	3	0.54	51.6
Soil + casein		0.196	0.658	0.532	0.532	28	0.84	56.2
Soil + litter		0.140	0.308	0.182	0.252	2	1.54	22.1
Control	75	0.070	0.084	0.084	0.098	5	0.21	81.2
Soil + casein		0.350	0.700	0.578	0.504	28	0.525	72.6
Soil + litter		0.168	0.325	0.210	0.210	10	1.50	24.1

[a]Initial ammonium N in *Atriplex* soil = 0.084 mg/g.

[b]Control = 1.116 mg/g organic N.

[c]Soil + casein = 1.916 mg/g organic N.

[d]Soil + litter = 1.976 mg/g organic N.

evaporation of moisture from soil (Fig. 5-2; O'Brien 1973). Under dry conditions the daily ammonia volatilization rates averaged 7 g/ha. Immediately after precipitation the daily rate was barely measurable, indicating that ammonia was held in the soil in solution. During evaporation of soil moisture, however, the daily efflux rate of ammonia was as high as 230 g/ha. Ammonia volatilization rates were also affected by ground cover (O'Brien 1973; Skujins and West 1973). For example, as shown in Figure 5-3, the rate of ammonia efflux from soil was 10 times higher under mesquite (*Prosopis glandulosa*) canopy as compared to fallow soil. Though no data are available, the ground cover effects on ammonification are presumably due to greater availability of nitrogen substrates in canopy areas.

The ammonia efflux rates calculated for desert soils are somewhat lower than the rates calculated for grazed pastures. Denmead et al. (1974), using acid traps and an ammonia-specific electrode, estimated the average daily ammonia efflux rate was 260 g/ha. A higher rate might be expected in pastures in view of the greater abundance of moisture and ground cover.

CONCLUSIONS

From the data discussed here it is clear that microbial proteolysis and deamination could contribute substantially to ammonification of organic nitrogen in

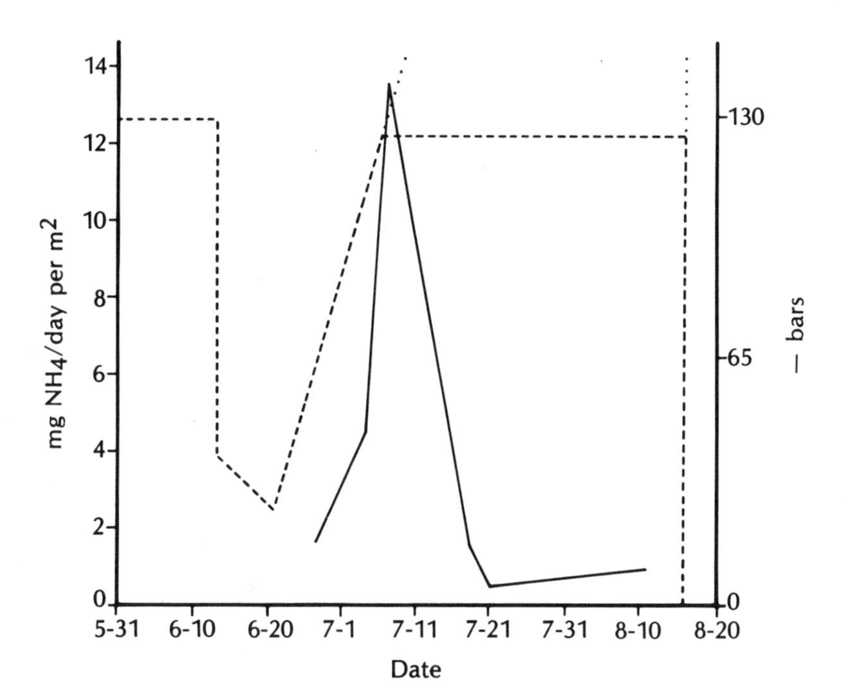

Figure 5-2 *Ammonia evolution [solid line] from Jornada soils [O'Brien 1973]. The dashed line represents soil water stress.*

desert soils. Certainly proteolysis and deamination are important activities in nitrogen cycling, but the flux rates resulting can only be estimated since there are only approximate figures on total organic nitrogen inputs from plants and animals. Whitford et al. (1973) estimated that litter input in the Chihuahuan Desert ranged between 200 and 400 kg/ha per year. Desert plant litter contains 2-3 percent nitrogen (Turner et al. 1973), so that the annual maximum nitrogen input range can be estimated at 6-12 kg/ha.

Ordinarily the dismantling of organic substrates would convert nitrogen to the ammonium ion, which would be bound to anions and subsequently oxidized to nitrate. However, in the relatively alkaline desert soils (Table 5-9) substantial amounts of NH_4^+ not assimilated by the microflora or higher plants are converted to ammonia and escape into the atmosphere. Skujins and Trujillo y Fulgham (Chapter 6 of this volume) have reported nitrification activity in desert soils; however, there is usually a lag of up to 8 days before activity can be detected.

The data in Table 5-7 suggest that litter N is less rapidly ammonified than amino N in protein amendments, although the results are preliminary and need to be verified. The low nitrogen contents of Chihuahuan soils (Table 5-9) clearly show that little nitrogen is retained annually and indicate that microbial proteolysis and ammonification of amino N contributes substantially to nitrogen loss from the soil.

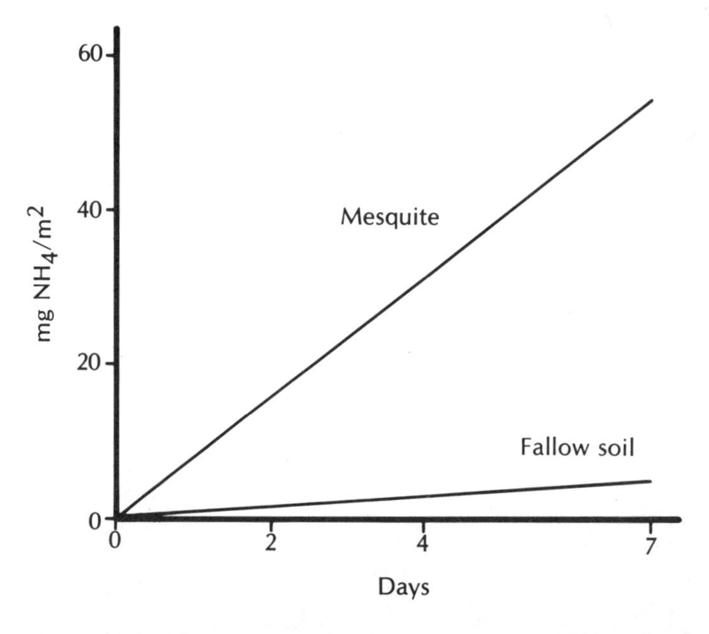

Figure 5-3 *Comparison of rates of ammonia evolution from Jornada soil under mesquite canopy and from bare ground [O'Brien 1973].*

Table 5-9 *Total Nitrogen and Organic Carbon Contents of Chihuahuan Desert Soils [Harding 1975]*

Sample	Depth (cm)	Percent total N	Percent organic carbon
1	0-1	0.014	0.18
	1-5	0.040	0.39
2	0-1	0.022	0.25
	1-5	0.043	0.48
3	0-1	0.020	0.26
	1-5	0.035	0.42
4	0-1	0.018	0.27
	1-5	0.040	0.43
5	0-1	0.027	0.35
	1-5	0.047	0.54

Future studies should be done under field conditions using natural substrate amendments. These studies should, preferably, be done at validation sites where meteorological and other environmental data are available. Hopefully investigations of this type will clarify microbial processes involved in nitrogen efflux from desert soils.

6

NITRIFICATION IN GREAT BASIN DESERT SOILS*

J. SKUJIŅŠ and P. TRUJILLO Y FULGHAM

INTRODUCTION

Nitrification is a biological process by which nitrite and nitrate are formed from ammonium in a two-step oxidation process involving two distinct groups of bacteria. The process is exothermic and the energy produced is utilized in the fixation of atmospheric carbon dioxide.

$$NH_4^+ + 1\tfrac{1}{2}O_2 \xrightarrow{\textit{Nitrosomonas}} NO_2^- + 2H^+ + H_2O + 66\ \text{kcal}$$

$$NO_2^- + \tfrac{1}{2}O_2 \xrightarrow{\textit{Nitrobacter}} NO_3^- + 17\ \text{kcal}$$

The nitrifying bacteria, *Nitrosomonas* spp. and *Nitrobacter* spp., are autotrophic aerobes.

Nitrification has been extensively studied and reviewed by a number of investigators; for example, Aleem (1970), Broadbent et al. (1957), Chase et al. (1967), Dommergues (1960), Justice and Smith (1962), Lees and Quastel (1946) and Quastel and Scholefield (1951). Previous studies have dealt mostly with agricultural, forest and grassland soils. Little has been done to define nitrification processes in soils of arid to semiarid environments.

REVIEW

Since nitrification rates vary with climates and soil types, environmental factors, such as pH, oxygen supply, soil moisture regime, temperature, organic matter content, carbon dioxide content and cation exchange capacity of the soil (Mahendrappa et al. 1966), are relevant. Further, nitrification is obviously dependent on the availability of substrate (i.e., ammonium), which in turn is similarly regulated by environmental factors.

Maximum nitrification rates are achieved by *Nitrobacter* between pH 6.2 to 7.0, whereas the *Nitrosomonas* oxidize ammonium best in soil having pH values above 7.6 (pH of soil paste). Strains from higher pH soils, however, have higher pH optima (Alexander 1961b).

*We are grateful to Dr. Brian Klubek for contributions to this project.

Investigations by Broadbent et al. (1957) have shown that applying elevated amounts of ammoniacal fertilizers may inhibit nitrification completely in poorly buffered soil. The presence of free ammonia selectively inhibits nitrifying bacteria, particularly *Nitrobacter*. Ammonium formed in alkaline soils may result in nitrite accumulating in amounts toxic to plants and further inhibit transformation of ammonium to nitrite. Chapman and Liebig (1952) found that when ammoniacal fertilizers were added to neutral or alkaline soils, substantial quantities of nitrite could accumulate; also, the nitrite persisted longer at low than at elevated temperature.

Martin et al. (1942) added either ammonium sulfate or urea to soil samples having an initial pH of 8.18 or 8.35, respectively. After 107 days of incubation, the pH values had decreased to 7.52 and 7.60, while the nitrate had increased by 337 μg/g and 346 μg/g, respectively. When a soil sample was amended with ammonium hydroxide, however, the pH decreased from 9.51 to 8.36, but no nitrification was found to have occurred. An untreated soil sample (control) accumulated 6 μ g/g of nitrate in 100 days of incubation. It was suggested that high alkalinity may limit or even prevent the oxidation of ammonium to nitrate in calcareous soils.

Using a Gila sandy loam, Martin et al. (1942) indicated that alkaline desert soils seem to have a threshold pH value of 7.7 $\pm$ 0.1. They theorized that above that value the complete oxidation of ammonium to nitrate would not occur, nor would nitrates accumulate in appreciable amounts. The authors (Martin et al. 1942) confirmed the above observation with four experimental systems designed as follows: 1) no treatment; 2) 30 mg of nitrogen as ammonium, plus 10 ml of a saturated $Ca(OH)_2$ solution added to the soil; 3) 30 mg of nitrogen as NH_4OH added; and 4) 30 mg of nitrogen as ammonium, plus 10 ml of 0.5 *N* H_2SO_4. Thus all the pH values were above or below the theorized threshold pH.

After incubation at 30 C, samples with initial pH values above 7.7 had nitrate only after the threshold value had been reached. In those samples with an initial pH below 7.7, nitrate was observed following the lag phase (Martin et al. 1942). The results below showed that nitrate formation in desert soils is a function of the pH value (Martin et al. 1942).

Treatment	*Time [days] for nitrate formation*
$NH_3 + Ca(OH)_2$	25
NH_4OH	19
$NH_3 + 0.5\,N\,H_2SO_4$	8

In all cases, nitrite had built to levels of 70-90 ppm before nitrate began to form. If the pH was raised, nitrate formation decreased, indicating that high alkalinity reduced the activity of the *Nitrosomonas*, although they are better able to function at a higher pH value than are the *Nitrobacter*.

The presence of various salts decreases nitrification, and this factor might be of importance in interpreting results in arid soils. Sindhu and Cornfield (1967) observed that at 50 percent water-holding capacity, 0.5-1 percent chloride suppressed nitrification, whereas 1-2 percent chloride suppressed ammonification. They reported that sulfate is comparably inhibitory at a considerably lower concentration, while phosphate considerably increased nitrifying activity, especially in the presence of calcium. Johnson and Guenzi (1963) demonstrated that osmotic tension reduced both nitrate production and carbon dioxide evolution in a linear manner as the salt concentration of the soil increased, approaching less than 10 percent of maximum activity at soil water potentials less than —30 bars. Sodium chloride was found to be more toxic than sodium sulfate. Microbial populations in calcareous soil had a greater salt tolerance than did populations in noncalcareous soil, as measured by nitrification and carbon dioxide evolution (Johnson and Guenzi 1963).

According to Smith (1964), cation exchange capacity directly influences the process of nitrification. He reported that the nitrification rate was higher in soil with high cation exchange capacity due to the higher adsorbance of ammonium ion on soil particles. Soil with low cation exchange capacity was demonstrated to accumulate nitrite because of the inhibition of *Nitrobacter* by free ammonium ion. Kai and Harada (1969) showed that clay minerals influence the rate of nitrification as a function of their cation exchange capacity. More clay particles equaled faster nitrification.

Soils with low carbon:nitrogen ratios had low nitrate-producing power (Halvorson and Caldwell 1948). Alexander (1961b) stated that the ratio of carbon mineralized to nitrogen mineralized decreased somewhat as the rate of inorganic nitrogen production increased, so that a microflora vigorously forming nitrate tends to release carbon and nitrogen in a ratio of ca. 7:1 while the least active will exhibit ratios near 15:1. The Curlew Valley soils of the Great Basin Desert have carbon:nitrogen ratios of 8:1 to 11:1 (Skujins and West 1974).

Since oxygen is an obligate requirement for nitrifiers, soil structure affects the accumulation of nitrate by influencing aeration. According to Hagin (1955), finely aggregated soils had lower nitrification rates than the coarsely aggregated ones; however, the most favorable aggregate size appeared to be less than 0.5 mm in diameter.

Sims and Collins (1960) found the maximum number of nitrifiers to be 800/g in arid Australian soils. By contrast, those in cultivated soils may reach millions per gram (Alexander 1961b). Drought, high temperature and other variations in environmental conditions had a minor effect on the numbers and distribution of these organisms in Australian soils. According to Alexander (1961b), there is a seasonal variation of nitrifiers in any soil, the largest numbers appearing in early summer during warm, rainy periods. The *Nitrosomonas* population is more stable throughout the year than that of *Nitrobacter*.

Munro (1966) and Neal (1969), among others, showed that substances which inhibited the nitrifying bacteria were present in root extracts of plants present in grasslands. These antibacterial substances may represent an important mechanism for conserving the low amount of available nitrogen present in grassland soils. The pronounced effect of the arid rangeland brush canopy on nitrification has been demonstrated by Rixon (1969, 1971), Tiedemann and Klemmedson (1973a) and Charley and West (1977).

Effective nitrification takes place in a wide temperature range between 10 and 35 C, with a maximum around 25 C. However, apparently there is no real optimum temperature for nitrification. Mahendrappa et al. (1966) have examined the rates of nitrification of different soil samples across the western United States. These authors have reported that the rates for the northern soils were optimal between 20 to 25 C, and for the southern soil between 25 to 35 C. They also showed that microorganisms from the southern soils (Arizona and New Mexico) nitrified added ammonium sulfate faster at higher temperatures than at lower temperatures. The reverse was true for the microorganisms in northern soils (Washington, Montana and Utah). In all cases, however, a lag period of 5 to 7 days was required before any nitrification could be detected. Mahendrappa et al. (1966) also observed that any conditions not favorable to nitrification caused a buildup of nitrite ion, thus indicating that the *Nitrobacter* are more sensitive to environmental conditions (temperature, ammonium ion concentration and moisture) than are the *Nitrosomonas*.

Justice and Smith (1962) detected nitrification in alkaline soils at 2 C. Sabey et al. (1959) showed that as the temperature increased from 0 to 25 C the maximum rate of nitrification increased while the lag periods decreased. In general, these authors noted a 0.1-1.8 week lag period when samples were incubated at 25 C.

Anderson and Boswell (1964) observed that added ammonium delayed or completely suppressed nitrate accumulation at low temperatures. The number of nitrifying bacteria was directly related to soil temperature (Garbosky and Giambiagi 1962). The numbers were highest when the mean temperature of the warmest month was from 14 to 16 C. Extreme variation in temperature resulted in a decreased number of nitrifying bacteria. Schaefer (1964) demonstrated that diurnal temperature variations actually stimulated nitrification.

The most important parameter governing the process of nitrification in arid soils is water availability. The maximum activity of nitrifiers in soils takes place at —0.1 bar water potential; Sabey (1969) showed that at —15 bars the nitrification rate was 13 percent of that at —0.1 bar. Miller and Johnson (1964) found that nitrite ion accumulation in a noncalcareous soil was highest at —0.1 bar potential. Dubey (1968) showed that the nitrification rate increased as the water potential increased from —15 to —2 bars and decreased thereafter. Dommergues (1960) observed that the minimum water content for nitrification is at pF 5.2 (about 90 percent relative humidity). On the other hand, Robinson (1957) indicated that nitrifier activity ceases just below the wilting point.

Undoubtedly there are differences between nitrifying floras of well-irrigated agricultural soils and of arid soils. In addition, some discrepancies may be introduced by the rather difficult methodology, especially before the advent of the thermocouple psychrometer. Low moisture levels seem to inhibit *Nitrobacter* more than *Nitrosomonas*. Arid areas usually have sporadic precipitation, and wetting and drying cycles have a pronounced influence on the mineralization of soil nitrogen (Birch 1958, 1959, 1960; Charley 1972).

Nitrifying activities in arid soils, as in any other soil, are limited by the availability of substrate. It would be of interest to know, however, what the substrate turnover rates would be with a nonlimiting substrate concentration and other environmental parameters optimized. By varying certain known parameters which reflect changes in the ecosystem, we have conducted a comparative study of nitrification potentials in arid soils. The results provided a useful insight into the behavior of nitrification in arid soils.

NITRIFICATION POTENTIAL EXPERIMENT

Samples were taken at seven preselected sampling stations in southern Curlew Valley, Utah. Stations 1 to 4 were at the IBP Desert Biome soil survey pits 21, 15, 22 and 14, respectively. The chemical and physical characteristics of soils of stations 1 to 4 have been described elsewhere (Skujins 1973).

Station 1. Dominant: *Atriplex falcata*, occasional *A. confertifolia* and *Sitanion hystrix*.

Station 2. Dominated by *Artemisia tridentata*, with *Atriplex confertifolia* and *Elymus cinereus*, invaded by *Halogeton glomeratus*.

Station 3. Seeded several years ago and dominated by *Agropyron desertorum* and reinvaded by *Atriplex confertifolia*.

Station 4. Mixed stand of *Atriplex confertifolia, Descurainia pinnata, Lepidium perfoliatum, Atriplex falcata* and *Sitanion hystrix*.

The soils of stations 5, 6 and 7 are from sites dominated by *Artemisia tridentata, Ceratoides lanata* and *Atriplex confertifolia*, respectively. The physical properties of the soils in relation to the prevailing semidesert shrub communities in the experimental area are described by Mitchell et al. (1966). The pH values of soils are between pH 8.5 and 9.2 (1:1 suspension).

The area receives about 18 cm precipitation annually.

At each station samples were taken with an auger from nonvegetated plant interspace areas at 0-5, 5-20, 40-50 and 70-80 cm depths. The samples were placed in sterile Whirl-pak bags, marked for identification and stored in a refrigerator at 3 C until tested. Prior to use, the entire sample was mixed thoroughly with a mortar and pestle, returned to the bag and the desired amount withdrawn for testing. The control used was a cultivated soil, 0-15 cm depth, from an irrigated orchard near Hyde Park, Utah.

Nitrification potential was measured by the perfusion method as follows: 80 g soil was treated with 160 mg VAMA (vinyl acetate-maleic acid copolymer), a soil conditioner to maintain structure; sufficient water was added to make a soil paste; and it then was pressed through a 2-mm diameter sieve. Crumbs were dried at 22 C and then used in a perfusion apparatus as described by Collins and Sims (1956). After drying, 30 g of soil crumbs were placed in a perfusion apparatus containing 250 ml of tap water. The samples were perfused for 24 hours, after which the water in the experimental flask was discarded and replaced with 250 ml of 0.01 *M* ammonium sulfate. The samples were then perfused for 20 days. At the conclusion of the 20-day trials, a fresh 250-ml quantity of 0.01 *M* ammonium sulfate (280 μg ammonium N/ml of perfusate, or 2,333 μg ammonium N/g of soil) was introduced and perfused through the soils for another 20 days. The trials were run at 22 C in the dark and the soil was under conditions of optimal aeration and water saturation, but not water-logging.

The methods described above were also modified as follows: 1) tests in the cold room at 4 C for the same period of time and conditions as described above; 2) tests with addition of 1.25 or 2.50 g chopped fresh leaf and stem material from *Artemisia tridentata, Ceratoides lanata* and *Atriplex confertifolia* to cultivated agricultural crumbed soil (Skujins 1975); 3) gas wash traps containing 0.02 *M* sulfuric acid were attached in line to measure any volatilized ammonia from the perfusate; 4) intact field-collected algae-lichen crust samples (surface 1 cm) were placed in the perfusion apparatus and perfused with distilled water at 200-lumen light intensity.

Aliquots of perfusates were taken every 2 days and analyzed for ammonium, nitrite and nitrate by Nessler's (Allen 1957), by sulfanilic acid (Allen 1957) and by 4-methylumbelliferone methods (Skujins 1964), respectively. At the end of an experiment, trapflasks were analyzed for ammonium by Nessler's method and pH values were determined for each perfusate.

Typical experimental results of single runs are shown in Figure 6-1. The cultivated soil (A) exhibited a normal nitrification pattern with a slight increase in nitrite, followed by an increase in nitrate and no nitrite accumulation. The soil from the *Atriplex* site (B) showed considerable accumulation of nitrite, followed by a considerable lag period for nitrite oxidation to nitrate. Both samples were reperfused (i.e., the perfusate discarded and fresh ammonium sulfate added) after 14 days for the cultivated soil and after 20 days for the *Atriplex* site soils. A soil enrichment of nitrifiers was evident in both cases.

To discern the effects of vegetal cover and soil type, precipitation and other seasonal effects in the natural environment, and of soil depth, the individual results were averaged seasonally as a function of sampling stations, and those of stations were averaged as a function of season and of soil depth.

Soil samples were weighed and extracted with $2N$ KCl + $CaCl_2$ (5 ml/g soil) by shaking for 0.5 hour. The supernatant was used for subsequent nitrite, nitrate and exchangeable ammonium N analysis (Bremner 1965c). For nitrite N analysis the Griess-Ilosvay method with the KCl extract was used. For total nitrite and nitrate N analysis, nitrite and nitrate were reduced with Devarda's alloy and steam distilled

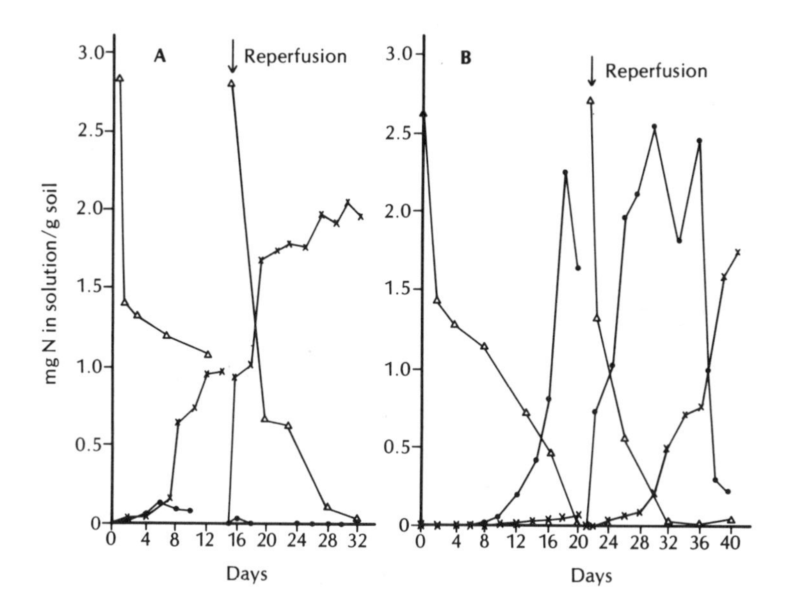

Figure 6-1 *Typical results on nitrification potential from single runs of the perfusion experiment. A = a cultivated soil collected June 1973; B = an* Artemisia tridentata *soil [station 5], 0-5 cm depth, collected April 1973. Triangles represent ammonium, solid circles nitrite and x's nitrate forms of nitrogen.*

with MgO from the KCl extract (Bremner 1965c). Analysis of exchangeable ammonium was performed similarly by steam distillation with MgO from the KCl extract (Bremner 1965c).

EXPERIMENTAL RESULTS

The nitrification potential measurements are based on the results of the first 20-day perfusate tests. A parallel 20-day perfusate test for each sample was used to verify its biological nature: i.e., whether an enrichment had taken place. The measured initial ammonium concentrations in the perfusates were considerably lower than the concentration added. This was due to the considerable ammonium fixation by the clays present in the test soils during the initial stages of the perfusion experiments.

The differences in nitrification potential at the surface layer (0-5 cm) of the *Artemisia tridentata*, *Ceratoides lanata* and *Atriplex confertifolia* stations were averaged over an annual sequence, January to October (Fig. 6-2). Considerable nitrite accumulation during perfusion was evident in all three soils before conversion to nitrate took place. The nitrite production rates appeared the same, but the nitrite

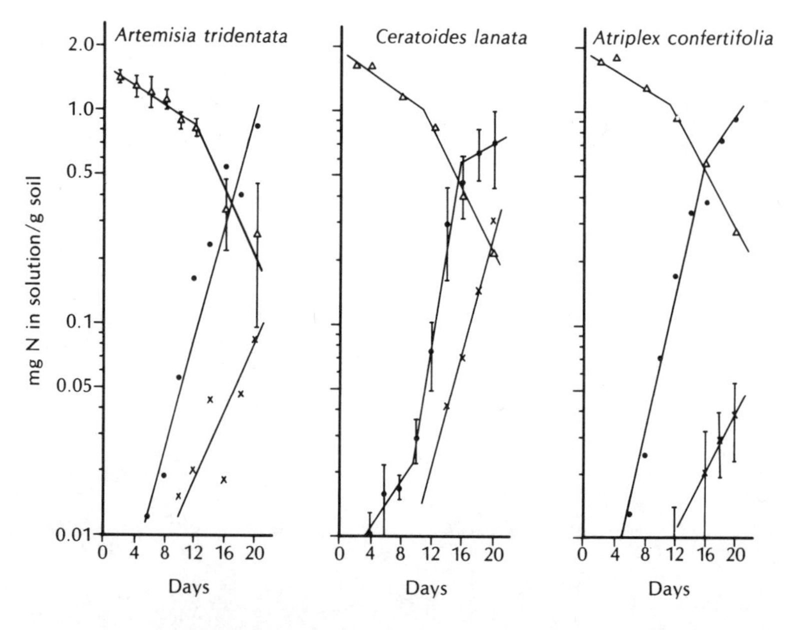

Figure 6-2 *Comparison of nitrification potentials of soils from three vegetation types, 0-5 cm depth [stations 5-7 left-to-right, respectively]. Each point on the graphs is the mean of five samples collected from January to October. Representative standard error bars are shown. [See legend of Fig. 6-1 for identification of symbols.]*

Table 6-1 *Nitrification Potential of a Sample of Curlew Valley Soil from the Surface 5 cm, Sampled September 5, 1973*

Day	Station #1		
	μg NH_4^+-N/g soil	μg NO_2^--N/g soil	μg NO_3^--N/g soil
0*			
2	1436	4	0
4	1282	6	0
8	1110	8	0
12	740	35	6
16	722	226	24
20	199	1056	51

*Initial NH_4^+-N added = 2333 μg/g soil (280 μg/ml perfusate). Initial NO_2^--N = 0 μg/g soil.

oxidation rates decreased with the stations: *Ceratoides* > *Artemisia* > *Atriplex.* Typical standard error ranges are shown in Figure 6-2, but are omitted from further figures to preserve clarity.

Results from the same three stations were averaged for seasons (Fig. 6-3). The January ammonium oxidation rate (soil frozen and partially covered with snow) was the same as that of March/April samples (soil thawed, subjected to intermittent rains and diurnal freeze-thaw cycles). The March/April samples, however, had a pronounced lag period during the first 10 days. The July rate was somewhat slower than the earlier ones (soil dry, soil surface temperature over 50 C). October samples indicated a significant decrease in nitrifying potential. The rate of the oxidation of nitrite to nitrate was minimal in October, somewhat higher in the winter, highest in the spring and slightly decreasing in the summer.

Results of the most active period (April) were averaged according to the depth of sampling (Fig. 6-4). Significant nitrate formation was evident only in the top 5-cm layer. Nitrite was formed in samples to 50-cm depth, but the 70-80 cm samples did not show any nitrification. The 50-cm depth is about the average maximum to which precipitation penetrates.

The results for surface 5-cm soil samples taken in April from stations 5, 6 and 7 were similar to the responses of surface 5-cm September samples from stations 1, 2, 3 and 4. The data set from station 1 is a typical example (Table 6-1). In all four cases there was a considerable accumulation of nitrite after a percolation of 20 days and a minimal appearance of nitrate.

To determine the nitrification potential of these soils at low temperatures, they were percolated with an ammonium sulfate solution at 4 C. The results of these tests showed no apparent oxidation of ammonium and, consequently, no appearance of nitrite or nitrate in any of the *Atriplex confertifolia*, *Artemisia tridentata* and *Ceratoides lanata* station samples.

Two-year averages of exchangeable ammonium in stations 1, 2, 3 and 4 soils showed higher amounts in the wet fall season (Table 6-2) as compared to the spring and summer.

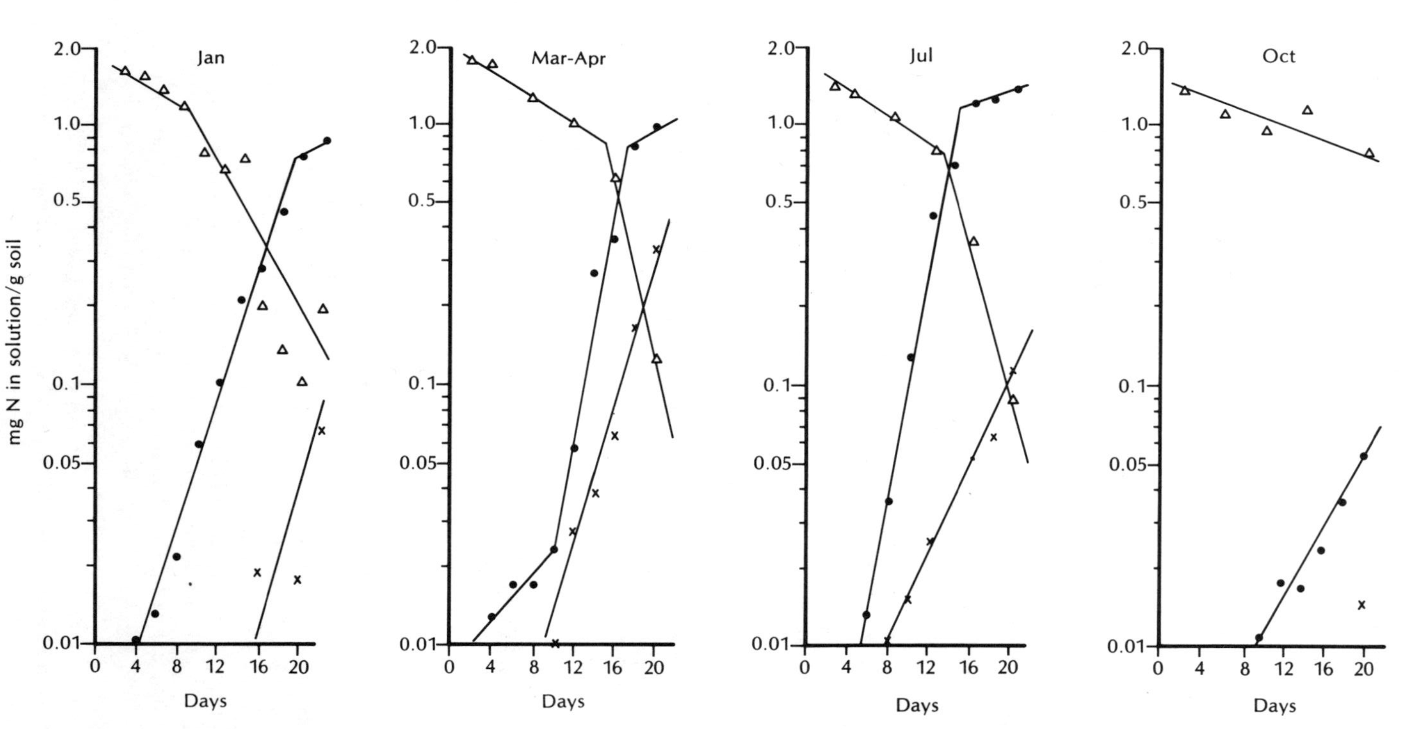

Figure 6-3 *Seasonal changes in nitrification potential as revealed in perfusion runs on soil samples collected in January, March-April, July and October, 0-5 cm depth. Each point on the graphs is the average from the three vegetation types [stations 5-7]. [See legend of Fig. 6-1 for identification of symbols.]*

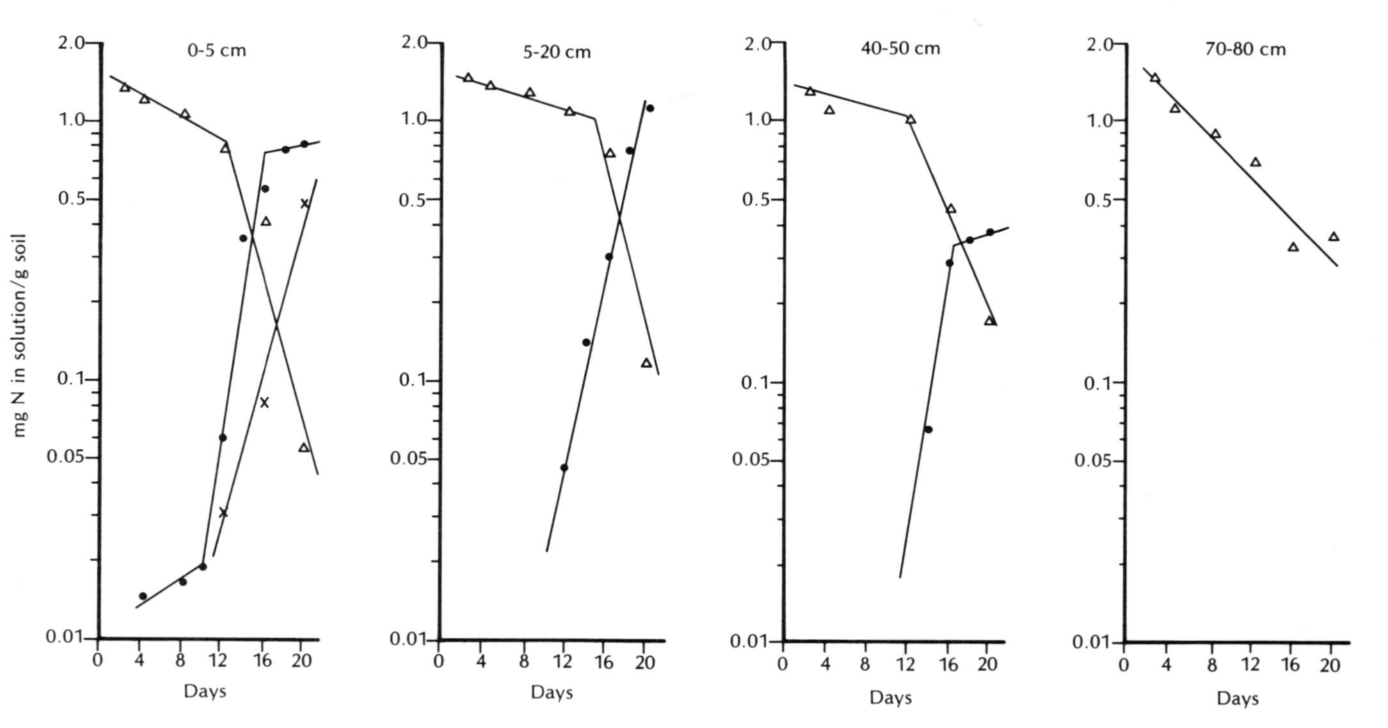

Figure 6-4 *Effect of soil depth on nitrification potential as shown in perfusion runs on April samples averaged across vegetation type. [See legend of Fig. 6-1 for identification of symbols.]*

No significant nitrite accumulation was observed at any time at any of the field sampling stations. The maximum amount detected in the surface 3-cm layer has been 0.3 μg nitrite N/g soil (see Table 6-3).

Nitrate content at all sampling stations usually fluctuated between 0.25 and 2.5 μg nitrate N/g soil. The averages are shown in Table 6-2. There was no significant seasonal variation, but the higher nitrate values were usually reached in the surface soil layer.

DISCUSSION

The first studies of the process of nitrification using a soil perfusion apparatus were performed by Lees and Quastel (1946). They showed that the nitrifying bacteria were sorbed on the surfaces of soil particles and that the rate of nitrification

Table 6-2 *Exchangeable Ammonium and Nitrate Content in Soils [averages of stations 1-4]*

Sampling date	Depth		
	0-5 cm	40-50 cm	70-80 cm
	μg NH_4^+ N/g soil		
1969			
11 Nov	23.0	2.5	2.4
1970			
27 May	7.3	6.5	5.5
21 Aug	3.8	3.8	3.8
1971			
5 Apr	5.5	3.3	4.2
15 May	6.7	5.4	2.7
15 Jun	5.5	2.3	1.5
13 Sep	9.2	3.4	1.6
14 Oct	12.8	4.6	3.0
30 Oct	9.6	0.9	0.9
17 Nov	23.0	16.1	3.0
	μg NO_3^- N/g soil		
1969			
11 Nov	2.50	0.68	0.35
1970			
27 May	0.70	0.65	0.75
21 Aug	0.87	0.58	0.65
1971			
5 Apr	1.05	0.90	0.80
15 May	0.87	0.80	0.37
15 Jun	1.75	0.75	0.39
13 Sep	1.60	0.39	0.39
14 Oct	1.30	0.30	0.30
30 Oct	1.05	1.25	0.39
17 Nov	0.87	0.70	0.42

Table 6-3 *Nitrite Content [μg NO_2^-—N/g soil] in Soils Sampled in 1972*

Station no.	Sampling depth (cm)	Sampling Date		
		15 Sep	14 Oct	14 Nov
5 *Artemisia*	0–3	0.3	0.0	0.1
	5–20	0.1	0.0	0.1
	40–50	0.2	0.0	0.1
	70–80	0.2	0.2	0.1
6 *Ceratoides*	0–3	0.2	0.0	0.3
	5–20	0.2	0.0	0.0
	40–50	0.0	0.4	0.2
	70–80	0.2	0.1	0.1
7 *Atriplex*	0–3	0.2	0.0	0.0
	5–20	0.2	0.0	0.0
	40–50	0.5	0.0	0.2
	70–80	0.1	0.2	0.0

was a function of the degree to which ammonium ions were adsorbed on the base exchange complexes. They also demonstrated that when soil in the perfusion apparatus was well supplied with ammonium, the function representing nitrate accumulation vs. time was sigmoidal. When nitrite was added to the soil, the nitrate accumulation as a function of time was similar in shape, but the rate of conversion of nitrite to nitrate was more rapid than the conversion of ammonium to nitrite. This indicated that the first step in nitrification, the conversion of ammonium to nitrite was rate limiting and seemed to explain why nitrite does not normally accumulate in soils. Under similar experimental conditions, however, Great Basin Desert soils accumulate nitrite in considerable concentrations. Also, recent reports (e.g., Faurie et al. 1975) now indicate that ammonium sorption on clays may hinder nitrification.

Stojanovic and Alexander (1958) showed that adding ammonium in quantities greater than 250 μg ammonium N/mg perfusate depressed the rate of nitrate formation. The range of concentration of ammonium N used in our perfusion experiments was below the depression level. Our results indicated that about half the ammonium was sorbed by the soils in less than an hour, a portion was volatilized from the perfusate, and the ammonium concentration did not affect the rate of ammonium oxidation.

Changes in the nitrate content of a soil may be an indicator of the growth of nitrifying bacteria (Seifert 1966). If environmental factors do not change, the nitrification rate is proportional to the initial quantity of nitrifying bacteria. According to Lees and Quastel (1946), the rate of nitrification in a soil is proportional to the fraction of the total amount of ammonium which is adsorbed or combined in the base-exchange complexes of the soil, and it is apparently independent of the concentration of the ammonium in the soil solution. In studies by Welch and Scott (1960), the presence or addition of potassium ions to illite, bentonite and similar clays blocked the release of ammonium ions from these clays, and hence the rate of nitrification decreased with a decrease in cationic exchange capacity. Recent experimental and theoretical studies on the growth of nitrifiers and substrate utilization rate (Ardakani et al. 1973, 1974) produced results that are applicable to the modeling of nitrification in an arid ecosystem.

Our nitrification potential research using arid soils showed that some of the ammonium added to the soils was fixed on the soil within an hour. A further gradual decrease in ammonium ion concentration was noticeable during the 3-10 day lag period. This loss was determined to be due to ammonia volatilization from the high pH perfusates and to ammonium fixation by the soil clays. There was no perceptible immobilization of nitrogen during perfusion. After 20 days of percolation the perfusates were discarded and replaced with fresh ammonium sulfate solution. Invariably, the nitrification rate then picked up without any lag period. Significant nitrite accumulation occurred throughout all the samples, while conversion to nitrate was often lacking. A classical "textbook" nitrification process was exhibited by a cultivated control soil under the same experimental conditions; i.e., the initial low concentrations of nitrite were followed by considerable increase of nitrate.

The rates of disappearance of ammonium and appearance of nitrite do not appear to differ significantly among our test soils. The much lower and delayed rate of conversion of nitrite to nitrate in soil of the *Atriplex* station, as compared to that of the *Ceratoides* station (*Artemisia* station being intermediate), indicated adverse environmental factors, including the possibility of inhibitors.

The fastest conversion of nitrite to nitrate took place in the spring samples, whereas nitrification was considerably slower in the autumn samples. January and July rates were intermediate. As a corollary, we recorded the highest concentration of ammonium in these soils in the autumn.

Anderson and Boswell (1964) observed that low temperature and increased amounts of added ammonium nitrogen delayed or completely suppressed nitrate accumulation. For a characteristic cool desert soil in northern Utah, the optimum temperature for nitrification was 20-25 C (Mahendrappa et al. 1966). In our studies no significant nitrification was detected at 4 C after either the first or second addition of ammonium sulfate solution, substantiating Anderson and Boswell's findings. At that temperature we observed neither nitrate nor nitrite accumulation, and ammonium slowly decreased.

While our recorded nitrification potential decreased with depth, agreeing with Charley and West (1977), the effect appeared to be more pronounced upon *Nitrobacter* than on *Nitrosomonas*. The inhibitors or other unfavorable factors may affect *Nitrobacter* more. Adding fresh ammonium sulfate solution after the initial 20-day perfusion, however, produced nitrite after a 2-day lag period, but no significant amounts of nitrate during the 3-week experimental period in the samples below 40-cm depth at stations 5 and 7. The available precipitation (including snow- and ice-melt water) normally would not infiltrate below 50 cm.

The seasonal changes in nitrification potential and in the accumulation of nitrite in the perfusate might have been caused by a number of factors. The extended dry season may drastically reduce the numbers of nitrifying organisms, while the high pH of the soils may limit potential conversions of nitrite to nitrate. As a rule, *Nitrosomonas* are able to metabolize in a higher pH medium than *Nitrobacter*.

Large amounts of nitrogen are annually fixed by the blue-green algae-lichen crust of the clay-rich Great Basin Desert soils (Rychert and Skujins 1974b). The fixed nitrogen is ultimately released in these soils as ammonium. Some of it is volatilized

(Klubek et al., Chapter 8 of this volume) but most becomes denitrified (Westerman and Tucker, Chapter 7 of this volume; Skujins and West 1974). Nitrification is an essential process in this pathway:

$$\begin{array}{ccccccc} & & & & N_2O + N_2 & & \\ & & & & \uparrow \text{denitrifiers} & & \\ R{-}NH_2 & \xrightarrow{\text{decomposers}} & NH_4^+ & \xrightarrow{\textit{Nitrosomonas}} & NO_2^- & \underset{\text{denitrifiers}}{\overset{\textit{Nitrobacter}}{\leftrightarrows}} & NO_3^- \end{array}$$

Nitrification is especially effective during active biological periods (i.e., rainy and warm periods), coinciding with the maximum atmospheric nitrogen fixation rate by the soil surface crust organisms (Rychert and Skujins 1974b) as shown by the present results and by previous correlative analysis with enzymatic activities (Skujins 1973).

Nitrification is controlled by the same environmental factors that regulate most biological processes in arid soils; ambient temperatures, available moisture, physical and chemical properties of the soils and prevailing vegetation.

Our studies on the inhibitory effects of desert shrub litter on nitrification (Skujins 1975), among other processes of the nitrogen cycle, have shown that plant material is an important regulatory factor in nitrification and in nitrogen cycling in general (Trujillo y Fulgham et al. 1975). Our observations correlate with the canopy effects described by Rixon (1969, 1971) and Charley and West (1975).

The vegetation (i.e., its litter) controls and inhibits not only nitrification, but also other activities of the nitrogen cycle, for example, nitrogen fixation (Rychert and Skujins 1974a). The inhibitory aspects of the flora on biological activities in the environment have been reviewed in detail by Rice (1974).

Previous nitrification-potential studies indicated that high concentrations of nitrite may easily accumulate in soils. In our work, however, nitrite was easily denitrified in the soil crust. The nitrifiers have a very high affinity for oxygen (Campbell and Lees 1967), whereas some denitrifiers may switch from O_2 to No_2^- as the terminal electron acceptor at oxygen concentrations above 1 percent in the atmosphere. As long as enough carbon is available to denitrifiers as an energy source, both processes may take place simultaneously in the same microenvironment. In our soils, carbon is made available by algae, which assimilate carbon dioxide. As carbon and nitrogen are fixed by the algae-lichen crust they are used by denitrifiers for their heterotrophic activities and again released into the atmosphere as N_2, N_2O and CO_2.

SUMMARY

The results on nitrification potential measurements, the chemical analyses of soils and studies on other aspects of nitrogen cycle in Great Basin Desert soils show that nitrification is a prominent process in nitrogen turnover. Climatic regime has a profound influence on nitrification. Some unique features include pronounced seasonal variation of nitrification potential and its decrease with soil depth, and an absence of nitrification in soil layers not reached by precipitation. A considerable accumulation of nitrite occurs in these soils during in vitro experiments, but no accumulation of nitrite or nitrate in the field is evident. This contrasts strongly with observations of nitrate buildup in Mohave Desert soils (Wallace et al., Chapter 14 of this volume). In the Great Basin Desert, nitrogen is fixed in large quantities by the blue-green algae-lichen crust. The organic nitrogen is rapidly decomposed and oxidized to nitrite. Nitrite in the presence of algal organic matter is rapidly denitrified to N_2 and N_2O. The results indicate that allelopathic substances might inhibit oxidation of nitrite to nitrate.

7

DENITRIFICATION IN DESERT SOILS*

R. L. WESTERMAN and T. C. TUCKER

INTRODUCTION

Nitrogen balance in agricultural soils and in desert ecosystems is receiving increased attention, as evidenced by an increasing number of publications. After accounting for nitrogen removal by agricultural crops or plants grown in desert ecosystems, losses of nitrogen by leaching and runoff, inputs of nitrogen by fertilization or by fixation and additions of nitrogen by irrigation and rainfall, there remains an unaccounted-for deficit of nitrogen. Presumably this deficit results from gaseous loss processes such as ammonia volatilization or biological denitrification.

Numerous papers have been published on denitrification, and an attempt has been made to consolidate some of the pertinent information. Reviews by Delwiche (1956), Broadbent and Clark (1965) and Payne (1973) provide historical as well as other technical information relating to biological and nonbiological losses of nitrogen, but give few references to data gathered in desert contexts.

A brief review of the biochemistry of denitrification and factors that affect gaseous losses is followed by a report of research showing the significance of biological losses of nitrogen in desert ecosystems.

DENITRIFYING BACTERIA

The biological transformation of nitrate to gaseous forms of nitrogen (i.e., nitric oxide, nitrous oxide and elemental nitrogen) is termed denitrification. Numerous bacteria have the potential for denitrification, and an extensive historical review of the subject was prepared by Delwiche (1956). However, fungi and actinomycetes have not been implicated in N_2 production. Any organism which during its growth phase causes a shift in environmental pH from acid to more alkaline or alkaline to acid may cause ammonia volatilization, chemical nitrite decomposition and/or nonenzymatic reaction of ammonium or amino groups with nitrite and, thus, gaseous losses of nitrogen. These reactions have mainly been observed in the laboratory and are not considered biological denitrification. Only a

*This is a contribution from the Department of Soils, Water and Engineering of the Arizona Agricultural Experiment Station, published with the approval of the Director as Journal Article No. 2538.

small number of bacteria present in soil can bring about direct denitrification. A list of genera of chemosynthetic bacteria containing species that denitrify has been compiled in a review by Payne (1973).

Numerous genera of chemosynthetic bacteria have been reported to reduce nitrate to nitrite but only a few have the ability to denitrify. The active species are largely limited to the genera *Pseudomonas, Achromobacter, Bacillus* and *Micrococcus* (Alexander 1961a), although *Thiobacillus denitrificans* and certain *Chromobacterium, Mycoplana, Serratio* or *Vibrio* species will catalyze the reduction. Of the genera present in soils, *Pseudomonas* and *Achromobacter* appear to be the most active in denitrification. Even though there are numerous strains of *Bacillus,* they are rarely important in the overall denitrification process.

Denitrifying bacteria are mostly heterotrophic organisms that grow in both aerobic and anaerobic environments. In the absence of oxygen, nitrate is used as the electron acceptor. Few denitrifiers have been found in soils devoid of both oxygen and nitrate. The existence of a large population of denitrifiers does not indicate denitrification is taking place but does indicate that the potential for denitrification exists. Many factors, such as available organic carbon, pH, temperature, soil moisture, nitrate and redox-potential, determine whether or not denitrification will occur as well as the rate of denitrification. These factors will be discussed in a following section.

BIOCHEMISTRY OF DENITRIFICATION

In an aerobic system, carbohydrates are oxidized in the respiration process to carbon dioxide and water. Using glucose as a carbon source, respiration proceeds as follows:

$$C_6H_{12}O_6 + 6O_2 \rightarrow 6CO_2 + 6H_2O \qquad (7\text{-}1)$$

When oxygen becomes limiting, respiration continues by facultative organisms, some of which use nitrate as an electron acceptor. Anaerobic respiration using nitrate may be expressed as follows:

$$5C_6H_{12}O_6 + 24NO_3^- \rightarrow 6CO_2 + 24HCO_3^- + 18H_2O + 12N_2\uparrow \qquad (7\text{-}2)$$

Pathways for reduction of nitrate in denitrification have been proposed by several investigators and summarized by Delwiche (1956), Bremner and Shaw (1958a), Alexander (1961a) and Payne (1973). Figure 7-1 is a general scheme that outlines the reduction steps of nitrate in soil systems. Nitrate is first reduced to nitrite, then further reduced to a postulated intermediate, hyponitrite (HON =NOH). This intermediate is further reduced to nitric oxide (NO), nitrous oxide (N_2O) and elemental nitrogen, which are sequential gaseous products of denitrification. In case of nitrate reduction for assimilation of nitrogen, reduction steps are similar through the postulated intermediate (hyponitrite). This intermediate is further reduced to hydroxyalamine and ammonia and is utilized in amino acid and protein synthesis.

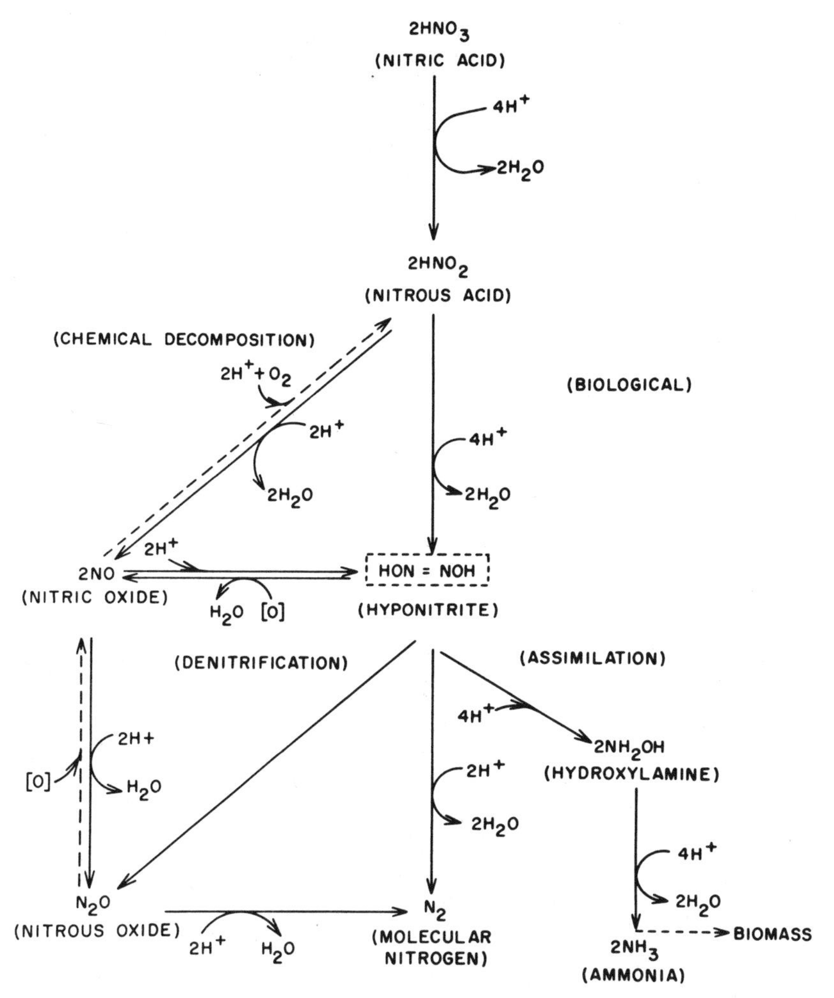

Figure 7-1 *Nitrate assimilation and denitrification in biological systems.*

Cady and Bartholomew (1960) used labeled nitrate in a closed anaerobic glass apparatus that provided continuous internal circulation of the gaseous phase. The closed system was sampled periodically, and mass spectrometer techniques were used to identify the appearance and disappearance of gaseous products. Nitric oxide (^{15}NO) appeared initially but never exceeded 5 percent of the total ^{15}N added. It was postulated that appearance of ^{15}NO was due to the chemical decomposition of $H^{15}NO_3$ and may not necessarily be biological or a precursor to nitrous oxide ($^{15}N_2O$). The proposed reaction is as follows:

$$3H^{15}NO_2 \rightleftharpoons H^{15}NO_3 + H_2O + 2^{15}NO \qquad (7\text{-}3)$$

Later, Cady and Bartholomew (1963) investigated the reactions of nitric oxide in soils and concluded that the hypothesized chemical equilibrium reaction was operating but it did not adequately explain all the reactions involved.

After the appearance of ^{15}NO, $^{15}N_2O$ became evident and increased, and ^{15}NO disappeared. Following the increase of $^{15}N_2O$, $^{15}N_2$ was detected in the gaseous phase. This increased to 83 to 95 percent of the total ^{15}N added, after which $^{15}N_2O$ in the gaseous phase and $^{15}NO_3^-$ in the soil disappeared.

Cooper and Smith (1963) used gas chromatographic and soil analyses techniques to investigate sequential products formed during denitrification of some western soils. They concluded that the sequence

$$NO_3^- \rightarrow NO_2^- \rightarrow N_2O \rightarrow N_2 \qquad (7\text{-}4)$$

which occurred in acid, neutral and alkaline soils was in general agreement with data obtained by Cady and Bartholomew (1960).

There is some disagreement as to the position of nitrous oxide in the biological pathway of denitrification. Using pure cultures of *Pseudomonas stutzerii,* Allen and van Neil (1952) reported nitrous oxide was not the precursor of nitrogen gas.

Wijler and Delwiche (1954) reported nitrous oxide was the major gaseous nitrogen product evolved under most moist soil conditions. However, the N_2O:N_2 ratio was pH dependent. Above pH 7, N_2O could be reduced to N_2; but below pH 7, reduction was strongly inhibited. Below pH 6 the formation of NO was prominent and the rate of denitrification decreased. Under field conditions, gaseous losses of N_2O may occur before reduction to N_2 is completed.

Schwartzbeck et al. (1961) reported evolution of N_2O and N_2 was dependent on the ammonium:nitrate ratio of the nitrogen carrier. Ammonium nitrate favored N_2O evolution; whereas HNO_3 favored N_2 evolution. Broadbent and Clark (1965) suggested that other nonbiological reactions may have been involved.

Although there is some disagreement among investigators, the flow diagram presented in Figure 7-1 generally outlines the sequence of reduction of nitrate and denitrification. The hyponitrite (HON=NOH) intermediate has not been confirmed or ruled out, but it does appear that some intermediate compound containing nitrogen in the +1 oxidation level is operating between nitrite and the gaseous products (Delwiche 1956).

FACTORS INFLUENCING BIOLOGICAL DENITRIFICATION

Many factors, such as the partial pressure of oxygen (pO_2) in the soil atmosphere, available organic carbon, nitrate and moisture content, pH, temperature and redox-potential in the soil and various interactions of these factors, govern the rate of denitrification. The effects of each of these factors are discussed briefly.

pO_2 in Soil Atmosphere

Biological reduction of nitrate occurs when oxygen becomes limiting in the system. Considerable confusion arose with early reports of aerobic denitrification by

Meiklejohn (1940), Korsakova (1941), Marshall et al. (1953), Kefauver and Allison (1957), Broadbent and Clark (1951), Broadbent and Stojanovic (1952) and Allison et al. (1960). Other investigators (Sacks and Barker 1959; Jones 1951) found no evidence of denitrification under aerobic conditions. This confusion was primarily a result of misunderstanding of the terminology used in description of the experiments. Even in soils flushed with air containing oxygen, microzones probably still existed which limited respiration of aerobic microorganisms and denitrification occurred. Thus, two terms — aerobic and anaerobic denitrification — appeared in the literature.

This dilemma stimulated numerous investigations on the influence of partial pressure of oxygen (pO_2) on denitrification. Sacks (1948) demonstrated with resting cells of *Pseudomonas denitrificans* in an atmosphere containing 4 percent oxygen that more than 50 percent reduction in the rate of denitrification occurred. Wijler and Delwiche (1954) reported that an atmosphere of only 5 mm of oxygen (< 1 percent) suppressed denitrification during the early stages of soil incubation to about 12 percent of that obtained under anaerobic conditions. Nommik (1956) reported that denitrification decreased with increased aggregate size, which was speculated to be due to a decrease in micropore volume. Broadbent and Stojanovic (1952) reported that denitrification of added ^{15}N was inversely related to the pO_2, but was appreciable, even under fully aerobic conditions. Allison et al. (1960) reported losses of 10 to 50 percent of added nitrogen in absence and presence of an energy source when the pO_2 was 0.46 percent. Cady and Bartholomew (1961) showed that only under conditions of high levels of carbonaceous material, an available nitrate supply and an oxygen level of < 7 percent by volume would an appreciable amount of nitrate be reduced to gaseous forms of nitrogen.

It is interesting to note that another important nitrogen transformation (nitrification-oxidation of ammonium to nitrate) and denitrification can proceed in soil simultaneously. Loewenstein et al. (1957) reported, in a greenhouse experiment that dealt with different fertilizer and management practices, that nitrate produced in the aerobic soil area moved to oxygen-poor regions and became subject to denitrification. They further stated that the aerobic area may have become anaerobic as a result of rapid oxygen consumption or because of concurrent carbon dioxide evolution by the soil microflora. In either case, nitrification would likely be followed by denitrification.

It appears that a limited oxygen supply with adequate nitrate and an organic energy source will promote denitrification, and true anaerobic conditions are not necessary.

Organic Carbon as an Energy Source

Denitrifying bacteria are facultative microorganisms that require carbon as an energy source. The activity of these microorganisms is directly affected by the pO_2 and the availability of energy as well as other factors. The rate of breakdown of organic matter markedly influences oxygen consumption and denitrification rates. There are numerous reports of the effects of added energy sources on denitrification (Jones 1951; Broadbent and Clark 1951; Broadbent and Stojanovic 1952; Jansson and Clark 1952; Wijler and Delwiche 1954; Hauck and Melsted 1956; Loewenstein et al. 1957; Bremner and Shaw 1958a, b; Carter and Allison 1960; Allison et al.

1960; McGarity 1961; Woldendrop 1962; Cooper and Smith 1963; Meek and MacKenzie 1965; McGarity and Myers 1968; Meek et al. 1969; Broadbent and Tusneem 1971; Stanford et al. 1974; Bowman and Focht 1974). Simple sugars and organic acids are oxidized quickly in soils and are more stimulating than less-readily decomposable plant tissues or residues. Alfalfa tissue generally has been reported to have greater effect on denitrification than small grain straws or grasses. However, straws and grasses are more effective in bringing about nitrate reduction than sawdust and compounds high in lignin. Well-rotted tissue has a minor influence when compared to fresh plant tissue.

The addition of organic carbon not only serves as an energy source for denitrification if other conditions are favorable, but also promotes immobilization of nitrate into organic nitrogen compounds such as amino acids, proteins and nucleic acids by microorganisms. Both immobilization and denitrification can proceed simultaneously, and the rate at which each transformation proceeds is dependent on pO_2, moisture, temperature, pH, etc. Addition of organic carbon may act as a nitrogen-conserving process if immobilization rates are greater than denitrification rates. It has also been observed (McGarity 1961) that the addition of organic carbon may have very little effect on denitrification, provided that the soil has adequate native available organic carbon present. In other cases, however, input of oxidizable carbon as a result of plant root exudation has been shown to be an important energy source for denitrifiers (Woldendrop 1962).

In soils low in organic carbon, losses of nitrate were as high as 10 percent of the added nitrate in absence of an energy source, to 50 to 80 percent in the presence of simple sugars (Allison et al. 1960; Jones 1951).

Bowman and Focht (1974) studied the effect of glucose and nitrate on denitrification in sandy soils. They concluded that denitrification rates closely approximated Michaelis-Menten kinetics in Coachella fine sand, but not in Hanford sandy loam. The K_m value in Coachella fine sand was 500 μg/ml. Michaelis-Menten kinetics were not followed in Hanford sandy loam due to failure to saturate enzyme systems in the denitrifying bacteria with glucose and nitrogen, when each was held constant. Carbon:nitrogen ratios of approximately 2 appeared to provide the greatest denitrification rate.

Stanford et al. (1974) studied denitrification rates in relation to total and extractable soil carbon on 30 soils of different origin that differed widely in organic carbon content and pH. Correlation of first-order denitrification rate constants with total and hot water extractable carbon were significant and r^2 values were 0.69 and 0.82, respectively.

Soil Moisture

Denitrification is affected directly by moisture content in the soil. Bremner and Shaw (1958b) reported that below about 60 percent of the water-holding capacity of soil, little denitrification occurred but increased rapidly up to 450 percent of water-holding capacity with soil amended with nitrate and an energy source. Losses of nitrogen were also shown to increase when the moisture content was adjusted to or greater than field capacity (Mahendrappa and Smith 1967; Meek et al. 1969).

Prior to investigations by Mahendrappa and Smith (1967), the effect of moisture in soils was thought to be primarily one of reducing the volume of air-filled pores which reduced oxygen diffusion rates. By utilizing complete anaerobic systems, they were able to show that soils have specific moisture requirements for maximum denitrification rates. Under fully anaerobic conditions, an increase in moisture content to only 10 percent above field capacity decreased the time required for maximum N_2 production; whereas the trend was reversed with further increases of moisture, and longer periods of time were required for complete denitrification. They speculated that moisture content affects the probability of contact between the organisms and the soluble nitrogen compounds, which function as electron acceptors, as well as controls the movement of gases back into the soil. At moisture levels near field capacity, both soil moisture and electron acceptors are probably quite uniformly distributed throughout the soil. In drier soils, either the nitrogen compounds or the organisms may lack mobility, and toxic effects of products that might occur would be enhanced. In soils with a water content greater than the saturation point, saturated flow is occurring and the organisms may have a different distribution pattern.

Nitrate Concentration

If denitrification is to occur naturally in soils, conditions must have been favorable previously for the oxidation of ammonium to nitrate.

Soils that have accumulated high concentrations of nitrate have the potential for greater gaseous loss of nitrogen than soils low in nitrate. Soils low in nitrate may have had conditions that were not favorable for ammonium oxidation or frequent occurrences of environmental conditions that promoted denitrification, or may have been subjected to leaching by rainfall or irrigation. Due to the sophisticated techniques that must be employed to measure gaseous losses of nitrogen, most of the research in reference to denitrification has been conducted in laboratory incubation experiments. However, there is great need for more research investigating denitrification under field conditions.

In the laboratory experiments, additions of 100 to 1,000 μg nitrate N/g soil have been used and many of the studies have used the heavy isotope, ^{15}N.

Cooper and Smith (1963) reported that decreasing the nitrate level from 300 to 150, 75 and 37.5 ppm N did not influence the overall rate of denitrification. Nitrous oxide in the gaseous atmosphere was reduced with decreased nitrate concentration. The independence of nitrate concentration on denitrification rates (zero-order kinetics) has also been shown by a number of other investigators (Broadbent and Clark 1951; Wijler and Delwiche 1954; Nommik 1956; Bremner and Shaw 1958b; Doner et al. 1974). Contrary to this, some data have been presented that indicate denitrification rates are first-order reactions (Bowman and Focht 1974; Stanford et al. 1974; Ardakani et al. 1975). It appears that nitrate concentration does have an effect on denitrification rates; i.e., at relatively low concentrations first-order kinetics are observed, while zero-order kinetics are observed at relatively high concentrations.

Soil Temperature

Denitrification is markedly affected by temperature and occurs in soils within a temperature range of 2-65 C but not at 70 C (Nommik 1956; Bremner and Shaw

1958b). The transformation proceeds slowly at 2 C but increases rapidly with increased temperature. The rapid release of N_2 in the more elevated temperature ranges suggests enrichment cultures of thermophilic flora (Alexander 1961a). The low temperature effects have considerable economic importance, because the rate of denitrification at 10 C or below indicates the possibility of gaseous losses of nitrogen during the colder part of the year when plants are not assimilating nitrate.

Temperature also affects the relative proportions of N_2O and N_2 evolved. Nitrous oxide is predominant at lower temperatures, with N_2 being dominant at higher temperatures (Broadbent and Clark 1965).

McGarity (1962) used different methods of freezing soils to preserve biological activity and reported that after 28 days of storage at —30 to —78, —5 to 0, 0 to 2, 15 to 19 and 30 to 31 C, there were significant differences in gaseous N_2 evolved in denitrification studies. After storage, all soils were incubated anaerobically at 30 C for 144 hours, and denitrification increased with decreased storage temperature. This could arise from a reduction in numbers of nondenitrifiers resulting in an increased nutrient supply for denitrifiers, or an increase in available organic matter.

Based on the observations of McGarity (1962), denitrification could be expected to change seasonally due to natural freezing and thawing processes and wetting and drying cycles.

Hydrogen Ion Concentration

In pure cultures, the optimal pH for denitrification varies with nitrate concentration, age of culture and the organism concerned. However, denitrifiers are active between pH 5.0 to 9.0 in most systems. The pH has a marked effect on the gaseous nitrogen products evolved as well as on the rate of denitrification. The rate of denitrification is very slow in acid soils and very rapid in alkaline soils (Bremner and Shaw 1958b). The liberation of nitrous oxide is greater in soils with a pH below 6.0 than it is with soil pH values greater than 7.0 and may account for over half of the nitrogenous gases produced. Nitric oxide also appears when the pH of the soil is low, but does not account for a large proportion of the nitrogenous gases evolved. In neutral to slightly acid soils, nitrous oxide appears as the first gaseous product followed by increased concentrations of nitrogen (Cooper and Smith 1963). Molecular nitrogen tends to be the dominant product in soils above pH 6.0. The differences in gaseous composition associated with pH are apparently due to the hydrogen ion sensitivity of the enzyme system involved.

Redox-potential

Redox-potentials have been used to determine if conditions are favorable for denitrification and dependent upon the pO^2 as well as the ions in the system (i.e., $Fe^{+++} + e \rightleftharpoons Fe^{++}$). Denitrification has been reported to occur at a redox-potential of 338 mV (Patrick 1960). Above 350 mV nitrate will accumulate and below 320 mV nitrate becomes unstable (Pearsall and Mortimer 1939). Theoretical electrode potentials of nitrogen and sulfur half-reactions have been reported by Bohn et al. (1969). Thermodynamically, the stability of the intermediate compounds (i.e., nitrite, nitrous oxide) in denitrification decreases with increased pH. This may help account for the observation that larger volumes of nitrous oxide are produced in acid soils compared to alkaline soils.

Meek et al. (1969) reported significant denitrification coincided with a decrease in redox-potential. The correlation was good when the E_h was varied by increasing the soil water content, but an organic matter application at the highest water content increased the loss of nitrogen without further decreasing the E_h. When E_h dropped to 300 mV or below, there was a large loss of nitrogen by denitrification. Whisler et al. (1974) have described the use of redox-electrodes in soil columns intermittently flooded with sewage water. The addition of organic carbon reduced the redox-potential to more negative values (—200 mV) than ordinary sewage water (+200 mV), which indicated that oxidation states of other elements in addition to nitrogen were changing.

KINETICS AND MATHEMATIC MODELS

As mentioned previously, denitrification generally follows first-order kinetics at relatively low concentrations of nitrate and zero-order kinetics at relatively high concentrations. Mathematical models have been proposed to describe nitrogen transformations (Ardakani et al. 1973; Cho 1971; McLaren 1969a, 1969b, 1970, 1971; Doner et al. 1974) and denitrification rates in soil columns undergoing continuous leaching (Ardakani et al. 1974; Starr et al. 1974); miscible displacement in unsaturated soils (Kirda et al. 1974; Misra et al. 1974a); and water unsaturated conditions (Misra et al. 1974b). A zero-order kinetic model for predicting N_2O and N_2 evolution as affected by temperature, pH and aeration was proposed by Focht (1974).

Michaelis-Menten kinetics or models that follow the same general scheme have been used to predict denitrification rates. The following are examples of some of the approaches that have been taken. The dual substrate reactions of carbon and nitrogen that are given by Bray and White (1966) were applied by Bowman and Focht (1974).

$$V = V_{max}\, \mathrm{CN}/(\mathrm{C} + K_c)(\mathrm{N} + K_n) \qquad (7\text{-}5)$$

where V is the denitrification rate, V_{max} is the maximum rate, C is the carbon concentration, N is the nitrate concentration, K_c is the Michaelis constant for carbon and K_n is the Michaelis constant for nitrate. In this case, the rate is not solely dependent upon nitrate concentration if carbon concentration is not high enough to ensure that zero-order kinetics relative to C apply; i.e., $\mathrm{C}/(\mathrm{C} + K_c) \simeq 1$.

Focht (1974) presented a model that shows the effects of aeration, pH and temperature upon the maximum rates of nitrate and nitrous oxide reduction (i.e., V_{max} of the Michaelis-Menten equation). The model is based on the assumption that nitrous oxide is an obligatory precursor in nitrate reduction to N_2 in the following sequence:

$$NO_3^- \xrightarrow{K_1} N_2O \xrightarrow{K_2} N_2$$

Thus the rate of formation for each gas with time (t) is

$$(N_2O) = t(K_1 - K_2) \tag{7-6}$$
$$(N_2) = tK_2 \tag{7-7}$$
$$(N_2O) + (N_2) = tK_1 \tag{7-8}$$

If an intermediate, I, is present and N_2O is not an obligatory metabolite, the model still holds:

$$NO_3^- \xrightarrow{K_1} I \begin{cases} \xrightarrow{K_2} N_2 \\ \rightarrow N_2O \end{cases}$$

Since K_1 equals total gas produced per unit time and K_2 equals rate of N_2 produced per unit time, the production rate of N_2O by this scheme is the same as Equation 7-6.

The effect of temperature on biochemical rate processes is exponential and follows with limits the classical Arrhenius equation

$$K = Ce^{-E_a/RT} \tag{7-9}$$

Where K is the rate constant, C is an integration constant, E_a is the activation energy, R is the universal gas constant, and T is the temperature in degrees Kelvin. For biological systems, T can be expressed directly in centigrade units thus:

$$K = Ce^{\gamma T}$$

Aeration (X_o) and pH (X_p) are both linear functions of K, while the temperature (X_T) is an exponential function. Thus

$$K = f(X_p, X_o, X_T) \tag{7-10}$$

and fX_o, fX_p, fX_T are of the respective forms

$$K = a(A - X_o) \tag{7-11}$$
$$K = b(X_p - B) \tag{7-12}$$
$$K = Ce^{\gamma T} \tag{7-13}$$

when the other two variables are held constant [i.e., $a = f(X_p, X_T)$, $b = f(X_o, X_T)$, $c = f(X_o, X_p)$]. A and B are the X-intercepts for each equation, where $K = 0$, regardless of what the other two functions are. Simultaneous substitution into the above equations and simplifying the product of the respective constants to yield a new constant a gives

$$K = a(A - X_o)(X_p - B)\, e^{\gamma T} \tag{7-14}$$

The numerical values for each constant are:

$$A_1 = 22.2$$
$$B_1 = 3.81$$

$$\gamma_1 = 0.0424$$
$$a_1 = 0.0944$$

for K_1 and

$$A_2 = 12.1$$
$$B_2 = 4.35$$
$$\gamma_2 = 0.0377$$
$$a_2 = 0.226$$

for K_2. The respective a_1 and a_2 for K_1 and K_2 were obtained by substituting the numerical equivalents for the three functions and rate constants. Mathematical restrictions on the models are

$0 \leqslant X_o \leqslant 22.2$ for K_1, $\quad 0 \leqslant X_o \leqslant 12.1$ for K_2
$3.81 \leqslant X_p \leqslant 8.0$ for K_1, $\quad 4.35 \leqslant X_p \leqslant 8.0$ for K_2
$15 \leqslant X_T \leqslant 65$ for K_1 and K_2

Using this model, the variability of N_2O concentrations in the soil atmosphere can be illustrated. The percentage of N_2O in relation to N_2 in the soil atmosphere is not greatly affected by temperature, but aeration percentage and pH have a marked effect.

Ardakani et al. (1975) used the following general rate equations to describe nitrate metabolism:

$$[-d(NO_3^-)/dt] = A(dm/dt) + [(a + \beta)(NO_3^-)]/[K_m + (NO_3^-)] \quad (7\text{-}15)$$

where m is biomass of relevant microorganisms; and A, a and β are parameters for growth, maintenance and wasted metabolism, respectively; and K_m is an affinity constant for nitrate reductase. The left-hand side of the equation is the rate of disappearance of nitrate with time accounted for by growth, maintenance and wasted metabolism for microorganisms. By substituting dX/f (f = flow rate with plug flow) for dt, and assuming that for short periods of time $dm/dt = 0$, the above equation yields upon integration:

$$k_m \ln[(NO_3^-)/(NO_3^-)_o] + (NO_3^-) - (NO_3^-)_o = -K_x \quad (7\text{-}16)$$

which gives the concentration of nitrate at depth x in terms of the Michaelis constant K_m, and $K = (a + \beta)\, m/f$ is the apparent rate constant for maintenance and waste metabolism combined.

Sanitary engineers (Francis and Callahan 1975) have used a modified Michaelis-Menten equation to describe denitrification in packed bed reactors designed for treatment of high nitrate waste water. The equation takes the form:

$$R_s = k_1 S\, X/K_s + S \quad (7\text{-}17)$$

where R_s = mass of substrate removed, mass per unit volume-time; k = maximum specific substrate removal rate per unit weight of microorganisms per unit time; X =

microbial mass concentration, mass per volume; S = substrate concentration surrounding the microorganisms, mass per volume; and K_S = Michaelis-Menten constant, which represents the substrate concentration when the specific substrate utilization rate ($U = R_S/X$, per unit time) equals $\frac{1}{2}K_1$, mass/volume.

The net removal (R_N) in a completely mixed co-recycle reactor, mass per volume-time, may be expressed as

$$R_N = YR_S - bx \qquad (7\text{-}18)$$

where Y = growth yield coefficient, mass of microbial tissue produced per mass of substrate removed; and b = microbial decay coefficient, per unit time. Thus the net specific substrate utilization rate ($U_N = R_N/x$) is

$$U_N = [(Yk_1S)/(K_S + S)] - b \qquad (7\text{-}19)$$

which is equivalent to the reciprocal of the residence time (T_r) of the reactor (Stensel et al. 1973).

$$T_r^{-1} = YRs - b \qquad (7\text{-}20)$$

In this manner, K_1 and K_S can be determined by linear regression from Equation 7-17 transposed to the form:

$$S = k_1 (S/U) - K_S \qquad (7\text{-}21)$$

and Y and b are determined in the same manner using Equation 7-20.

Equations have been derived that describe the transport of nitrogen in liquid and gaseous phases during continuous leaching (Starr et al. 1974). The simultaneous occurrence of nitrification, denitrification and nitrogen movement in porous media has been demonstrated by Cho (1971) using equations having the form:

$$\partial c/\partial t = D(\partial^2 c/\partial x^2) - v(\partial c/\partial x) + \sum_{i=1}^{n} \phi_i \qquad (7\text{-}22)$$

where c is the concentration of the nitrogen compound in the soil solution; D is the apparent diffusion coefficient (cm/day); v is the pore velocity (cm/day); ϕ_i is the rate of production (+ sign) or consumption (—sign) of the nitrogen compound from i sources or sinks; x is the distance within the soil; and t is time. For zero-order reactions, ϕ_i would be invariant. For more dilute solutions and/or where oxygen or carbon supplies are limited, the net rates of oxidation of ammonium c_1, oxidation of nitrite c_2 and reduction of nitrate c_3 would be

$$\phi_1 = k_1c_1 \qquad (7\text{-}23)$$
$$\phi_2 = k_1c_1 - k_2c_2 \qquad (7\text{-}24)$$
$$\phi_3 = k_2c_2 - k_3c_3 \qquad (7\text{-}25)$$

respectively. The rate coefficients above, k, include many environmental factors such as pH, oxygen and carbon supply, temperature, microbial growth kinetics and a host of other parameters.

Equation 7-22, written for the processes of nitrification (Equations 7-26 and 7-27) and denitrification (Equation 7-28) becomes

$$\partial c_1/\partial t = \{[D/(1 + R)](\partial^2 c_1/\partial x^2)]\} - [(v/1 + R)(\partial c_1/\partial x)] - [(k_1/(1 + R)]c_1 \tag{7-26}$$

$$\partial c_2/\partial t = D(\partial^2 c_2/\partial^2 x^2) - v(\partial c_2/\partial x) - k_2 c_2 + k_1 c_1 \tag{7-27}$$

$$\partial c_3/\partial t = D(\partial^2 c_3/\partial x^2) - v(\partial c_3/\partial x) - k_3 c_3 + k_2 c_2 \tag{7-28}$$

where the exchange reaction of ammonium with the soil is included in the $(1 + R)$ term (Gardner 1965; Cho 1971), assuming R to be a constant and equal to the ratio of absorbed to solution ions, and c is the mass of ammonium, nitrite and nitrate per unit volume of soil solution (mg/liter). It is assumed the values of the apparent diffusion coefficients, D, for the three ions are identical. Assuming steady state conditions and that nitrite, c_2, approaches zero as is commonly found in many field soils, Equations 7-26 through 7-28 become

$$D(d^2 c_1/dx^2) - v(dc_1/dx) = k_1 c_1 \tag{7-29}$$

and

$$D(d^2 c_3/dx^2) - v(dc_3/dx) + k_1 c_1 = k_3 c_3 \tag{7-30}$$

If D, v and $c_1(x)$ are measured, the first and second derivatives of c_1 are known and k_1 can be obtained from a plot of the left side of Equation 7-29 vs. c_1. If ϕ_1 is a first-order rate reaction, the slope of such a plot is equal to k_1. In the case of a zero-order reaction, the slope would be identically zero, with the left side of Equation 7-29 equal to the zero-order rate constant. Similarly the value of the rate constant for denitrification, k_3, can be found by plotting the left side of Equation 7-30 vs. c_3 (Starr et al. 1974). For boundary conditions $c_1 = c_1{}^\circ$, $c_3 = 0$, $x = 0$, $t \geqslant 0$ and $c_1 = 0$, $c_3 = 0$, $x > 0$, $t = 0$, the steady state solution of Equation 7-29 (Cho 1971; McLaren 1969a) is

$$c_1(x) = c_1{}^\circ \exp \{x/2D[v - (v^2 + 4Dk_1)^{1/2}]\} \tag{7-31}$$

or when $Dk_1 << v^2$, the approximate solution is

$$c_1(x) = c_1{}^\circ \exp(-k_1 x/v) \tag{7-32}$$

where $c_1{}^\circ$ is the ammonium concentration in the applied solution. The steady state solution of Equation 7-30 (Cho 1971; McLaren 1969a) is

$$c_3(x) = [k_1 c_1{}^\circ(k_1 - k_3)][\exp\{x/2D[v - (v^2 + Dk_3)^{1/2}]\} - \exp\{x/2D[v - (v^2 - 4Dk_1)^{1/2}]\}] \tag{7-33}$$

When Dk_1 is a small area compared with v^2, Equation 7-33 reduces to

$$c_3(x) = [k_1 c_1^\circ (k_1 - k_3)] [\exp(k_3 x/v) - \exp(-k_1 x/v)] \tag{7-34}$$

The mathematics of diffusion of gases in porous media have been summarized by Kirkham and Powers (1972) and show that in the steady state rate a particular gas diffuses in one dimension within the soil as is given by Fick's law:

$$q = -D(\partial c/\partial x) \tag{7-35}$$

where D is the apparent diffusion coefficient assumed constant for a given soil water content. For non-steady conditions, the concentration of the given gas is given by

$$\partial c/\partial t = D(\partial c^2/\partial x^2) + F(x, t) \tag{7-36}$$

where $F(x,t)$ is a source (production) term if it is positive, and a sink (consumption) term if negative. If neither source nor sink terms exist, $F(x,t)$ is zero. For Equation 7-36 to be valid, the gas must not be adsorbed appreciably on the soil particles, and transport must occur by diffusion only.

Under steady-state conditions, c and F vary only with depth, and if D is known, the second derivative of the measured concentration distribution yields values of

$$F(x) = -D(\partial^2 c/\partial x^2) \tag{7-37}$$

A plot of $F(x)$ vs. depth shows the zone of gas production or consumption.

Solutions of equations describing the transport processes and examination of their applicability to nitrogen transformations measured in soils under a series of environmental conditions have been presented by Misra et al. (1974a, b, c).

NONBIOLOGICAL MECHANISMS OF GASEOUS LOSSES OF NITROGEN

Although biological pathways probably account for the major proportion of gaseous loss of soil nitrogen, evidence has been presented that indicates non-biological pathways may be of significance. The nonbiological mechanisms — ammonia volatilization, Van Slyke reaction, decomposition of nitrites and nitrous acid reactions — will be discussed briefly.

Ammonia Volatilization

Of the nonbiological gaseous nitrogen loss processes, ammonia volatilization is probably the most significant. Ammonia volatilization is influenced by soil pH, texture, exchangeable cations, anions present, temperature, moisture content and the species of ammonium salt. It is generally concluded that the higher the pH, the greater the potential for ammonia volatilization (Chao and Kroontje 1964; Du Plessis and Kroontje 1964). Losses of NH_3 were reported by Martin and Chapman (1951) to be greater in soils high in clay content saturated with Na^+ and K^+ than soil saturated with Ca^{++} and Mg^{++}. This was attributed to the higher pH associated with Na^+ and K^+ saturated soils.

Losses of ammonia from soils can occur over a wide range of soil pH values, 4.5 to 7.8 (Du Plessis and Kroontje 1964; Chao and Kroontje 1964). The effects of the ammonium species have been investigated by Terman and Hunt (1964), Martin and Chapman (1951) and Jenny et al. (1945). It was generally concluded that NH_3 volatilization losses are less from NH_4NO_3 than $(NH_4)_2SO_4$, and less from $(NH_4)_2SO_4$ than NH_4OH. The general theory of ammonia volatilization in calcareous soils, effects of temperature and mixture of ammonia compounds, have been discussed by Fenn and Kissel (1973, 1974) and Fenn (1975). They proposed the following mechanism:

$$X(NH_4)_Z Y + WCaCO_3(s) \leftrightharpoons W(NH_4)_2CO_3 + Ca_w Y_x \qquad (7\text{-}38)$$

where Y represents the ammonium anion and W, X and Z are dependent on the valences of the anion and cation. The final reaction product, $(NH_4)_2CO_3$, is unstable and decomposes as follows:

$$\begin{array}{c} (NH_4)_2CO_3 + H_2O \leftrightharpoons 2NH_3\uparrow + H_2O + CO_2\uparrow \\ \upharpoonleft\downharpoonright \\ 2NH_4OH \end{array} \qquad (7\text{-}39)$$

The amount of NH_4OH formed in a given time is dependent on the solubility of $Ca_w Y_x$. If $Ca_w Y_x$ is insoluble, more $(NH_4)_2CO_3$ is formed and consequently more NH_4OH. If no insoluble precipitate is formed, very little $(NH_4)_2CO_3$ will exist. When $(NH_4)_2CO_3$ decomposes, CO_2 is lost from the solution at a faster rate than NH_3, thereby producing additional OH^- ions and an increase in $[OH^-]$. Consequently, more solution NH_4^+ becomes electrically balanced by OH^-, which would favor NH_3 loss as represented by the following reaction:

$$NH_4^+ + OH^- \rightleftharpoons NH_4OH \rightleftharpoons NH_3\uparrow + H_2O \qquad (7\text{-}40)$$

If $Ca_w Y_x$ is soluble, then ammonia loss will be dependent on the resultant pH of the soil. The ammonia-ammonium equilibrium is pH dependent, with lower pH's favoring ammonium forms. They concluded that ammonia volatilization is dependent on the ammonium salt applied, presence or absence of calcium carbonate and temperature. If the ammonium salt reacts with calcium carbonate, greater losses of ammonia can be expected, and a moderate temperature effect is observed. If the ammonium salt does not react with calcium carbonate the temperature effect is greater. Losses of ammonia increase with increased rates of ammonium application from compounds that form insoluble calcium reaction products, but remain constant at a given temperature for compounds that do not form insoluble calcium reaction products.

Van Slyke Reaction

The reaction of nitrous acid with amino acids or ammonia is strictly a chemical one and under suitable conditions yields N_2:

$$RNH_2 + HNO_2 \rightarrow ROH + H_2O + N_2\uparrow \quad (7\text{-}41)$$

This reaction is commonly known as the Van Slyke reaction. Even though this reaction can occur in nature, it is unlikely that it is an important gaseous nitrogen loss process in soils. Allison and Doetsch (1950) and Allison et al. (1952) reported that nitrite reacted with alanine in a pH range of 1.6 to 4.5, with the maximum reaction occurring at pH 3.4. No reaction occurred above pH 5.2. They concluded that conditions in soils are not likely to be favorable for the Van Slyke reaction or nitrite formation by reduction or oxidation. Even if nitrites are formed, they are more likely to be decomposed to nitric oxide or oxidized to nitric acid than they are to react with amino forms of nitrogen.

Nitrite Reaction with Urea or Ammonia

The reaction of nitrous acid with urea or ammonia is very similar to the Van Slyke reaction:

$$HNO_2 + NH_3 \rightleftharpoons N_2\uparrow + 2H_2O \quad (7\text{-}42)$$

Sabbe and Reed (1964) investigated chemical reactions involving urea and nitrite and concluded that this pathway of nitrogen loss in soils is doubtful. Two conditions were necessary for loss of N_2; 1) pH 6.0 or lower and 2) a concentration of each reactant of 0.05 mg N/ml, or greater. The nitrite that accumulates in soil originates from the breakdown of organic matter, via ammonia. Even though concentrations in fertilizer bands may approach the minimum nitrogen concentration necessary for N_2 evolution, conditions are not favorable for nitrite accumulation (i.e., alkaline). Smith and Clark (1960) and Fowler and Kotwal (1924) have drawn similar conclusions.

Decomposition of Nitrous Acid in Soils

Decomposition of nitrous acid occurs at low pH values in soils and yields nitric oxide by the following reaction:

$$3HNO_2 \rightarrow 2NO + HNO_3 + H_2O \quad (7\text{-}43)$$

Nitric oxide may be oxidized further to nitrogen dioxide:

$$2NO + O_2 \rightarrow 2NO_2 \quad (7\text{-}44)$$

and this may react with water to form nitric acid (Broadbent and Clark 1965):

$$3NO_2 + H_2O \rightarrow 2HNO_3 + NO \quad (7\text{-}45)$$
$$2NO_2 + H_2O \rightarrow HNO_3 + HNO_2 \quad (7\text{-}46)$$

The assumption is made in the above reactions that no gaseous loss of nitric oxide occurs. Nitric oxide, however, has been detected in acid soil denitrification studies (Cady and Bartholomew 1960, 1963), and has been attributed to chemical oxidation rather than biological degradation. Nitrous acid is fairly stable at pH values above 5.5 to 6.0, but the breakdown increases with increased acidity. Where nitrous acid is present in acid, sandy soils, Gerretsen and de Hoop (1957) reported that it is highly probable that considerable nitric oxide will escape to the atmosphere before oxidation and hydration to nitric acid. However, in neutral and alkaline soils, loss via this pathway is either very small or not detectable (Wagner and Smith 1958; Tyler and Broadbent 1960; Smith and Clark 1960; Cooper and Smith 1963).

Nitrite Reactions with Organic Constituents

Nitrite is an intermediate in the nitrification process and can accumulate in soils when ammonium or ammonium-type fertilizers are applied to the soil. With the accumulation of nitrite, chemical decomposition and reaction with other constituents become possible under favorable conditions. Clark and Beard (1960) reported that nitrite was unstable with certain organic amendments. Preheating the soil increased stability of nitrous acid, but additions of alfalfa meal and other organic constituents greatly increased the reactivity of nitrous acid. Stevenson and Swaby (1964) and Stevenson et al. (1970) identified N_2, N_2O, methyl nitrite (CH_3ONO) and CO_2 in the gases produced by reaction of nitrite with lignins and humic substances under strongly acidic conditions. Methyl nitrate, however, was not identified in soils at pH values found in productive agricultural soils, but nitric oxide was. The nitrosation process denotes the addition of the nitroso group ($-N{=}O$) to an organic molecule, and is brought about by nitrous acid and other compounds which form species of the type $O{=}N{-}X$. These compounds are labile and react further with the nitrosating agent to form gaseous products. Jansson and Clark (1952) and Meek and MacKenzie (1965) have also observed gaseous loss of nitrogen from soils amended with nitrite and organic materials. The magnitude of loss by this mechanism is small and varies from one soil to another depending on pH, nitrite concentration and soil organic matter content.

GASEOUS LOSSES OF NITROGEN IN DESERT ECOSYSTEMS

The significance of biological gaseous loss of nitrogen from soils in desert ecosystems has been studied intensively at two sites near Tucson in the Sonoran Desert. These sites were part of the US/IBP Desert Biome field program which initially utilized the Santa Rita Experiment Station of the U.S. Forest Service and subsequently established a study area 32 km west and 32 km north of Tucson on the slopes of the Silverbell Mountains. The Santa Rita site is located approximately 50 km south of Tucson, with desert vegetation dominated by *Acacia* shrubs, *Opuntia* cacti, subtrees in the genera *Olneya* (ironweed), *Prosopis* (mesquite) and *Cercidium* (paloverde), and grass species in *Aristida* (threeawn), *Muhlenbergia* (bushmuhly) and *Bouteloua* (grama grass). The soils are designated Anthony (typic torrifluvent) and Sonoita (typic haplargid) profiles. On the Silverbell site the predominant soils are Rillito (typic calciorthids), sometimes in eroded phases, with more arid vegetation characterized by *Larrea* (creosotebush), *Cercidium*, *Opuntia* and *Bouteloua*. Annuals occur at both sites.

Table 7-1 *Nitrogen Forms [in μg N/g soil] in Native Sonoran Desert Soils at Santa Rita*

Depth (cm)	Sonoita sandy loam				Anthony sandy loam			
	NH_4^+	NO_2^- + NO_3^-	Organic N	Total N	NH_4^+	NO_2^- + NO_3^-	Organic N	Total N
0-5	0.7	0.5	264.9	266.1	1.1	0.6	151.3	153.0
15-20	1.2	1.1	256.7	259.0	0.7	1.3	59.6	61.6
30-35	0.7	0.2	169.1	170.0	0.6	0.4	79.3	80.3
60-65	0.6	0.8	152.4	153.8	0.9	1.7	244.7	247.3
90-95	0.6	0.4	65.2	66.2	1.0	1.6	244.9	247.5

Nitrogen forms present in Sonoran Desert soils found at the two sites are reported in Table 7-1 for Santa Rita and Table 7-2 for Silverbell. Inorganic nitrogen (i.e., ammonium, nitrite, nitrate) in soils at the Santa Rita site ranged from 0.9 to 2.7 μg N/g, and organic nitrogen ranged from 60 to 265 μg N/g, depending on depth of the sample. Soils at the Silverbell site were generally slightly higher in inorganic nitrogen and in one case higher in organic nitrogen.

The organic matter content and effect of wetting dry desert soils on the initial biological and/or chemical activity are shown in Table 7-3. Twenty-four hours after wetting, oxygen consumption was highly correlated ($r^2 = 0.94$) with organic matter content in the soils, being greatest in the surface horizons except in the Anthony soil which had sandy material overlying well-defined, buried horizons. During early periods of incubation in three of the four soils, evolution of gaseous products masked oxygen consumption. We speculate that these gaseous products were derived from denitrification or volatile hydrocarbons released upon wetting. The presence of denitrifying organisms in wet and dry surface soils was verified by techniques outlined by Alexander (1965).

Table 7-2 *Occurrence of Nitrogen [in μg N/g soil] in Rillito Loam at Silverbell*

Rillito loam				Rillito loam (eroded)			
NH_4^+	NO_2^- + NO_3^-	Organic N	Total N	NH_4^+	NO_2^- + NO_3^-	Organic N	Total N
2.2	2.1	191.1	195.5	9.1	1.2	355.0	365.3
0.8	0.2	146.5	147.5	1.6	0.3	407.0	408.9
4.6	1.0	171.7	177.3	0.5	0.1	363.0	363.6
3.4	0.4	209.1	212.9	0.6	2.0	262.8	265.4
2.6	0.6	194.9	198.1	--------bedrock--------			

Measurement of Maximum Potential Denitrification

The potential for the Sonoran site soils to denitrify was established using heavy isotope techniques (amendments of 100 μg N, 33 atom percent ^{15}N, as nitrate per gram of soil) with anaerobic incubation. Excess ^{15}N percentages in N_2 and N_2O gaseous phases after 18 hour incubation are shown for Sonoita and Rillito soils in Table 7-4. The presence of excess ^{15}N in the gaseous phases of N_2 and N_2O is positive evidence that the soil will denitrify if conditions are favorable.

The maximum gaseous loss that can occur was not established, but it was shown that the potential for denitrification existed in all depths of the desert soils investigated. Excess ^{15}N in molecular nitrogen did not increase as much as in nitrous oxide but was significantly greater than normal abundance. The addition of organic carbon as an energy source stimulated denitrification, but with addition of labeled nitrate alone, denitrification was detectable and significant.

Table 7-3 *Oxygen Consumption and Gas Evolution [μg/g soil] and Percent Organic Matter Content in Samples of Sonoran Desert Soils after Wetting*

Depth (cm)	Organic matter (%)	Hours			
		1	3	5	24
		Sonoita sandy loam			
0-5	0.46	-5.2	-7.6	-3.6	-41.4[a]
15-20	0.37	-4.3	-7.1	-3.6	-34.0
30-35	0.26	+0.8	+2.0	+0.4	-17.5
60-65	0.23	+0.8	+1.2	+1.2	-12.8
90-95	0.18	+1.2	+4.0	+7.2	-10.8
		Anthony sandy loam			
0-5	0.27	+0.8	+1.2	+2.0	-16.0
15-20	0.16	+0.4	+2.8	+4.8	- 7.2
30-35	0.17	+1.2	-1.2	-4.0	- 9.2
60-65	0.39	0.0	-3.2	-4.8	-19.8
90-95	0.35	0.0	-1.9	-4.0	-26.6
		Rillito loam			
0-5	0.32	-2.0	-6.0	-8.8	-40.2
15-20	0.28	-0.4	-2.4	-2.2	-11.9
30-35	0.34	-0.4	-4.4	-8.0	-20.5
60-65	0.29	+1.2	-1.2	-4.4	-18.3
90-95	0.24	+1.6	-3.7	0.0	-11.2
		Rillito loam (eroded)			
0-5	0.54	-7.7	- 9.0	-6.3	-46.4
15-20	0.53	-6.0	-10.8	-6.0	-32.2
30-35	0.39	-6.1	-10.1	-6.8	-28.5
60-65	0.35	-7.8	-10.3	-5.1	-31.8

Note: - indicates O_2 consumed; + indicates gaseous products released (CO_2 was trapped in KOH).

[a] r^2 = 0.94 for percent organic matter vs. O_2 consumption at 24 hr.

Table 7-4 *Atom Percent Excess ^{15}N in N_2 and N_2O above Freshly Wetted Desert Soils Incubated in an Argon Atmosphere at 36 C for 18 Hours*

Depth (cm)	Treatment	Sonoita		Rillito	
		N_2	N_2O	N_2	N_2O
0-5	$^{15}NO_3^-$	.037	22.441	.027	12.658
	$^{15}NO_3^-$ + glucose	.036	27.007	.084	27.844
15-20	$^{15}NO_3^-$	.009	5.247	.008	1.972
	$^{15}NO_3^-$ + glucose	.469	26.768	.247	30.529
30-35	$^{15}NO_3^-$	.005	1.055	.021	18.945
	$^{15}NO_3^-$ + glucose	2.268	3.789	.237	31.497
60-65	$^{15}NO_3^-$	.003	1.140	.010	6.790
	$^{15}NO_3^-$ + glucose	.738	2.054	.215	29.664
90-95	$^{15}NO_3^-$	.000	---	.018	21.386
	$^{15}NO_3^-$ + glucose	.067	22.468	.150	31.016

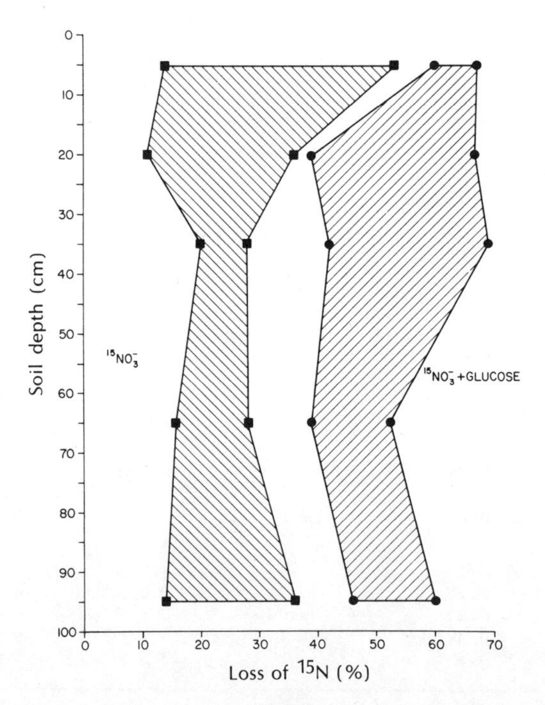

Figure 7-2 *Ranges of denitrification from Sonoran Desert soils amended with nitrate and nitrate plus glucose.*

Composition of nitrogen gases in the atmosphere above soils incubated 18 and 91 hours is reported in Table 7-5. Nitrous oxide comprised a very small percentage of the total nitrogen gas evolved after 18 and 91 hours incubation. Molecular nitrogen increased with time, and nitrate plus glucose had a greater effect on nitrogen loss than nitrate alone. The denitrification potentials of Sonoran Desert soils from the two sites are summarized in Figure 7-2. Gaseous losses of nitrogen per profile averaged from 16.7-31.7 percent and 47.2-62.6 percent in soils amended with labeled nitrate N and labeled nitrate N plus glucose, respectively. Native organic matter content generally decreases in these soils with depth, which was reflected in decreased denitrification.

Table 7-5 *Percent Composition of N_2 and N_2O in an Argon Atmosphere above Incubating Soils*

Depth (cm)	Treatment	Time (hr)	Sonoita		Rillito	
			N_2	N_2O	N_2	N_2O
0-5	NO_3^-	18	4.84	.038	10.47	.034
	NO_3^-	91	11.30	.110	20.60	.067
	NO_3^- + glucose	20	8.00	.138	7.12	.098
	NO_3^- + glucose	91	27.00	.028	17.11	.079
15-20	NO_3^-	18	7.73	.013	2.79	.004
	NO_3^-	91	11.91	.008	15.75	.055
	NO_3^- + glucose	20	5.43	.114	7.10	.123
	NO_3^- + glucose	91	24.69	.047	17.97	.013
30-35	NO_3^-	18	4.58	.007	2.51	.066
	NO_3^-	91	13.92	.007	13.40	.036
	NO_3^- + glucose	20	3.84	.017	4.29	.210
	NO_3^- + glucose	91	13.60	.257	27.40	.035
60-65	NO_3^-	18	5.31	.006	9.66	.016
	NO_3^-	91	18.78	.010	25.17	.024
	NO_3^- + glucose	20	14.30	.061	7.84	.283
	NO_3^- + glucose	91	21.13	.089	31.59	.085
90-95	NO_3^-	18	9.62	.008	1.11	.003
	NO_3^-	91	23.97	.008	10.18	.042
	NO_3^- + glucose	20	8.90	.088	36.70	.166
	NO_3^- + glucose	91	20.96	.377	26.58	.031

Table 7-6 *Effect of Nitrogen, Carbon, Moisture, Temperature and Time on Nitrogen Fractions [μg N/g soil] in the 0-5 cm Depth of Sonoita Sandy Loam [aerobic incubations]*

Time (days)	Amendment		Nitrogen fractions[a]							
			$NO_2^- + NO_3^-$				Organic N			
			20 C		37 C		20 C		37 C	
	C (%)	NO_3^- N (ppm)	8%	15%	8%	15%	8%	15%	8%	15%
5	0	0	1.6	2.1	4.3	2.8	249.9	153.3	126.7	210.7
	0	100	95.4	71.5	98.6	75.9	191.1	186.9	216.2	216.3
	1.5	0	0.0	0.0	0.6	0.3	198.2	164.5	103.6	169.4
	1.5	100	84.1	67.1	66.9	31.7	245.6	186.8	186.8	239.4
10	0	0	3.9	1.5	7.0	4.6	271.6	186.9	147.0	156.8
	0	100	92.0	93.7	99.8	73.4	214.9	244.3	204.3	215.6
	1.5	0	0.6	0.0	0.5	0.0	212.8	196.1	195.2	241.5
	1.5	100	45.6	26.2	10.6	18.7	296.1	243.6	242.8	254.1
15	0	0	3.5	1.9	8.7	5.8	168.7	163.1	106.4	112.7
	0	100	96.0	98.0	101.4	71.2	208.6	179.2	246.3	204.3
	1.5	0	0.0	0.5	0.0	0.0	196.7	185.5	133.0	186.2
	1.5	100	21.1	5.3	4.7	6.3	203.6	221.9	242.8	214.2

[a]8 and 15% moisture by weight approximates field capacity and saturated conditions in the soil, respectively.

Table 7-7 *Effect of Nitrogen, Carbon, Moisture, Temperature and Time on ^{15}N-Labeled Fractions [μg N/g soil] in the 0-5 cm Depth of a Sonoita Sandy Loam [aerobic incubations]*

Time (days)	Amendment		^{15}N-labeled fractions[a]							
			$NO_2^- + NO_3^-$				Organic N			
			20 C		37 C		20 C		37 C	
	C (%)	NO_3^- N (ppm)	8%	15%	8%	15%	8%	15%	8%	15%
5	0	100	89.7	77.3	93.2	73.0	0.1	12.5	0.5	0.6
	1.5	100	79.0	62.9	61.7	27.4	0.9	11.3	9.9	25.8
10	0	100	74.3	87.7	88.4	66.0	0.2	0.1	1.0	0.8
	1.5	100	43.2	24.0	9.5	12.8	42.0	33.5	40.0	41.3
15	0	100	87.0	90.2	89.8	62.1	1.1	0.5	0.9	0.8
	1.5	100	38.2	0.0	4.5	3.0	15.8	38.9	47.5	33.2

[a]8 and 15% moisture by weight approximates field capacity and saturated conditions in the soil, respectively.

Effects of Various Soil Factors on Denitrification

The effects of soil moisture, temperature, nitrate concentration and available organic carbon on denitrification were investigated for a Sonoita sandy loam profile at the Santa Rita site according to the methods described in Tucker and Westerman (1974). Samples were taken to a depth of 30 cm.

The factors investigated did not grossly affect the total ammonium fraction, but did have a significant effect on the nitrite + nitrate and organic nitrogen fractions. Total nitrite + nitrate in all depths of the soils amended with nitrate remained relatively constant with few exceptions within temperature and moisture variables. However, the addition of available organic carbon stimulated transformation of nitrate to either gaseous products or organic nitrogen forms. The disappearance of nitrate was greatest in soils adjusted initially to 15 percent moisture (saturation) amended with nitrate and organic carbon and incubated at 37 C. The effect of additions of organic carbon in soils incubated at 20 C was similar to results observed in the 37 C incubations, except the magnitude of loss of nitrate was lower. Table 7-6 provides an example of the most significant treatment effects for a sample of surface soil.

Only a trace of $K^{15}NO_3$ appeared in the ammonium fraction and it was not directly associated with the variable investigated. Nitrate was reduced by the addition of organic carbon, increased temperature, increased moisture and time of incubation in all soil depths, as illustrated for the surface horizon only in Table 7-7. Only a small portion of the labeled nitrate appeared in the organic nitrogen fraction without the addition of organic carbon. Organic nitrogen in soils amended with organic carbon increased with temperature, moisture and time of incubation.

The effects of soil depth, temperature, moisture, organic carbon and time on denitrification of labeled nitrate expressed on a percentage basis in Sonoita sandy loam are shown in Figure 7-3. Loss of labeled nitrate was highest in the 10-15 cm depth and lowest in the 25-30 cm depth. The magnitude of loss as influenced by soil depth was 25-34 percent of the total amendment. Temperature had a marked effect on denitrification. Losses observed at 37 C (37.7 percent) were 16 percent higher than those observed at 20 C (21.1 percent). The initial moisture content at the beginning of incubation did not have much effect on denitrification losses. Samples adjusted to 15 percent moisture by weight (saturation) resulted in only 2 percent higher losses of labeled nitrate than soils adjusted initially to 8 percent moisture by weight (field capacity). Denitrification losses were 28.4 and 30.4 percent, respectively, for 8 and 15 percent moisture contents. The addition of organic carbon had the greatest effect on denitrification. Samples amended with organic carbon denitrified 40.5 percent of the added labeled nitrate, whereas only 18.2 percent was denitrified in absence of the organic carbon amendment. Loss of ^{15}N in presence of organic carbon represented a 22 percent increase and was shown to have the greatest effect of the factors investigated. Losses of ^{15}N increased stepwise with time of incubation with little additional loss occurring after 10 days.

Similar abbreviated studies have been conducted on Anthony sandy loam from the Santa Rita site and Rillito loam from the Silverbell site to verify that these factors affect denitrification in other soils in the Sonoran Desert in the same general manner.

At the end of 5-day incubation the effects of temperature and soil depth on nitrogen fractions in Anthony and Rillito soils (Table 7-8) were similar to those observed in Sonoita soil. Ammonium N remained at a low level (2 ppm or less), and

Table 7-8 *Effect of Nitrogen and Carbon Amendment, Temperature and Soil Depth on Nitrogen Fractions [μg N/g soil] in Anthony Sandy Loam and Rillito Loam Soils Incubated 5 Days Aerobically*

Soil	Depth (cm)	Nitrogen fractions					
		NH_4^+		$NO_2^- + NO_3^-$		Organic N	
		20 C	37 C	20 C	37 C	20 C	37 C
Anthony	0-5	1.8	1.3	91.9	49.5	258.9	221.2
	25-30	1.2	2.0	81.5	90.0	126.7	146.3
Rillito	0-5	2.0	1.6	76.3	46.9	179.2	250.0
	25-30	1.3	1.3	81.1	62.9	135.8	156.1

Note: Each sample was amended with 96.6 μg labeled N/g soil and 1.5% carbon, and adjusted to moisture saturation (15% water by weight) before incubation.

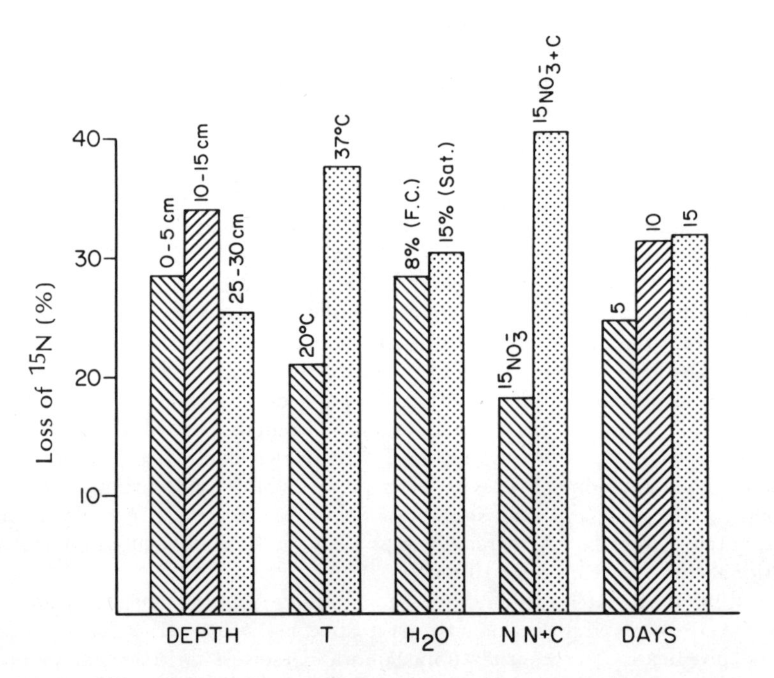

Figure 7-3 *Effects of soil depth, temperature, moisture, organic carbon and time on denitrification in Sonoita sandy loam soil incubated aerobically.*

Table 7-9 *Effect of Nitrogen and Carbon Amendment [see Table 7-8], Temperature and Soil Depth on ¹⁵N-Labeled Fractions [μg N/g soil] and Percent Loss of Nitrogen in Anthony Sandy Loam and Rillito Loam Soils [incubated 5 days aerobically, with initial moisture saturation to 15 percent by weight]*

Soil	Depth (cm)	% loss of N		^{15}N-labeled fractions: $NO_2^- + NO_3^-$		^{15}N-labeled fractions: Organic N	
		20 C	37 C	20 C	37 C	20 C	37 C
Anthony	0-5	0.0	27.0	89.0	45.7	10.0	24.9
	25-30	16.0	5.1	79.0	87.2	2.2	4.5
Rillito	0-5	15.2	23.6	72.9	43.4	9.1	30.4
	25-30	16.6	33.6	78.3	61.1	2.2	3.1

nitrate and organic N varied with temperature and depth of sample. Nitrate N generally decreased with higher temperature, and organic N increased, indicating that both denitrification and immobilization were occurring. In Anthony surface soils, organic N did not increase with temperature. As stated previously, the upper horizon was very sandy and low in organic matter and conditions were more favorable for denitrification than immobilization.

Transformations of labeled nitrate in Anthony and Rillito soils are shown in Table 7-9. Transformation of labeled nitrate into labeled ammonium was not detectable. In general, increased temperature decreased labeled nitrite + nitrate and increased organic ^{15}N. Losses of ^{15}N from these two soils incubated aerobically ranged from 0-34 percent, with the most significant losses occurring from the surface soils incubated at 37 C. This is also similar to results obtained in detailed studies with the Sonoita soil.

Table 7-10 *Nitrogen Transformations and Percent Loss of ¹⁵N from Sonoita Sandy Loam Soils Incubated 5 days Anaerobically at 37 C [with carbon and nitrogen amendments and initial moisture conditions as described in Table 7-8]*

Depth (cm)	^{15}N-labeled fractions (μg N/g of soil): NH_4^+	$NO_2^- + NO_3^-$	Organic N	% loss of ^{15}N
0-5	0.0	0.0	19.1	80.2
10-15	0.0	0.0	23.8	75.4
25-30	0.0	0.0	19.5	79.7
$\bar{x}$			20.8	78.4

Effect of Depth

Transformations of labeled nitrate and maximum losses of ^{15}N from Sonoita surface soils incubated anaerobically for 5 days at 37 C are reported in Table 7-10. Under conditions that were favorable for maximum denitrification, there were no detectable amounts of ^{15}N remaining in the nitrite + nitrate fraction or as ammonium. In the 5-day period approximately 20 percent of the added labeled nitrate was immobilized into the organic nitrogen fraction, and the remaining 80 percent was denitrified. The amount of ^{15}N immobilized under anaerobic conditions was slightly lower than was observed under aerobic conditions (20 vs. 25 percent), but gaseous losses of nitrogen were increased markedly. Gaseous losses observed in similarly amended soils under anaerobic incubations were 25 percent greater than observed under aerobic conditions. Thus, under moisture conditions representative of field capacity and saturation, conditions are favorable for about 55 and 80 percent of the maximum denitrification potential providing other factors such as a carbon source and temperature are favorable.

Nonbiological contributions to gaseous losses of ^{15}N have not been quantified. Some loss may have occurred in these studies by nonbiological mechanisms such as volatilization but were thought to be small relative to the total gaseous losses of nitrogen.

Skujins and West (1974) have investigated ammonia volatilization and denitrification in the cool desert in the Great Basin. After 8 days, minimum and maximum volatilization losses were only 2 and 6 percent, respectively, from soils that were incubated at 22 C and —3 bars moisture tension. They concluded that the major loss of nitrogen from desert soils was by denitrification, and a tentative comparison (Fig. 7-4) showed denitrification was greater from a Sonoran Desert soil (Silverbell) than from a Great Basin Desert soil (Curlew) and a Mohave Desert soil (Rock Valley).

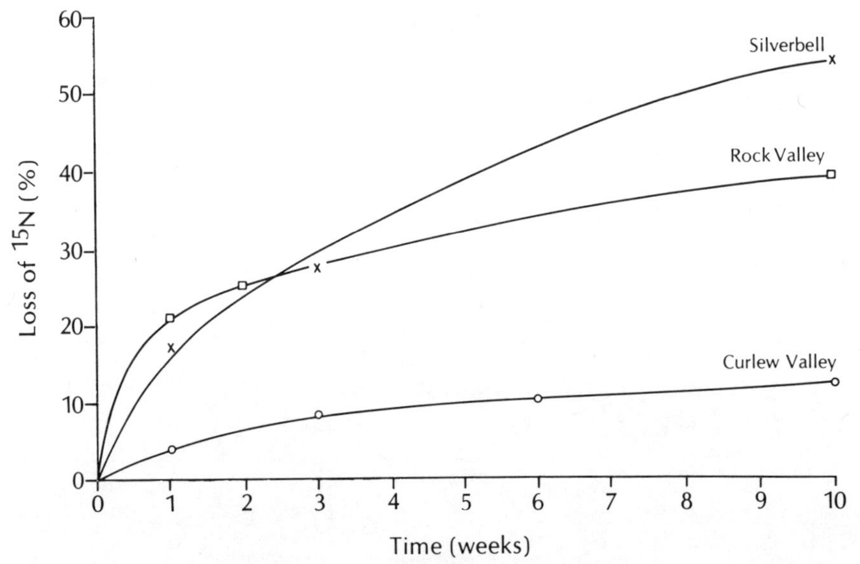

Figure 7-4 *Denitrification loss of ^{15}N-enriched soils selected from Curlew Valley, Rock Valley and Silverbell sites incubated at 22 C and —1 bar moisture for 10 weeks.*

Denitrification in Native Desert Soils

Previous laboratory incubation studies have shown the potential for denitrification in Sonoran Desert soils is high. The magnitude of loss of applied nitrogen was dependent upon available organic carbon, temperature and moisture. After investigating the factors that affect denitrification and establishing the maximum potential denitrification under known conditions, gaseous losses of applied ^{15}N to field sites under natural environmental conditions were investigated.

Two field experiments were established before summer rains on 20 and 21 June 1973 at two locations within the Santa Rita Validation Site on Anthony and Sonoita sandy loam soils. Aluminum cylinders (4.8 cm ID) were driven into the soils to a depth of 25 cm and various treatments introduced on a randomized block design (Tucker and Westerman 1974). The treatments comprised different amendments of labeled nitrate, labeled ammonium, ground wheat straw and ground barley grain, singly and in combinations mixed into the upper 1 cm of soil inside the cylinder. These treatments were allowed to incubate under natural conditions for 3, 6, 9 and 12 months. At the end of specified intervals, the soils in the cylinders were collected, subdivided into 0-5 and 5-10 cm depths and analyzed for soil nitrogen forms (Tucker and Westerman 1974).

During the first month of the experiment, a total of 3.25 cm of rainfall was recorded in three events. The next 270 days were extremely dry. Only trace amounts of rainfall were recorded after 300 days.

Microbial Respiration

Microbial respiration data from field experiments were collected 13, 21 and 37 days after initiation of the experiments and are reported in Table 7-11. Microbial respiration varied with soil temperature, moisture content, soil type and treatment. At the end of 13 days, microbial respiration was higher in all treatments than observed in the control. Addition of nitrogen as $(^{15}NH_4)_2SO_4$ or $K^{15}NO_3$ did not have a stimulatory effect on microbial respiration as evidenced by CO_2 evolution. Microbial respiration was increased by wheat straw and further enhanced by increasing nitrogen amendments, but ground barley grain stimulated microbial respiration more than any other amendment investigated. Similar trends were observed at the end of 21 and 37 days. Increased microbial respiration with additions of wheat straw and barley grain indicates that immobilization of inorganic nitrogen occurred. This point will be discussed in detail in the next section.

There was no detectable labeled ammonium collected in acid traps at these time periods. Therefore, it is assumed that gaseous loss by ammonium volatilization in these soils was very small if it occurred and did not contribute significantly to the total gaseous loss of nitrogen.

Nitrogen Transformations

The two soils behaved very similarly in response to the treatments. Data from the Anthony soil are used to illustrate the distribution of labeled nitrogen forms remaining in the upper 10 cm of soil after 3 and 12 months incubation under natural

Table 7-11 *Carbon Dioxide Evolution [mg CO_2/m^2 per hour] from Field Experiments*

Conditions and treatments	13 days		21 days		37 days	
	Anthony	Sonoita	Anthony	Sonoita	Anthony	Sonoita
Ambient air temperature (°C) at 0800	32	31	27	24	37[a]	37[a]
Soil surface temperature (°C) at 0800	35	30	30	25	49[a]	48[a]
Soil subsurface temperature (°C), -2 cm, at 0800	39	34	33	29	45[a]	59[a]
Soil moisture (%)	0.37	0.46	10.3	8.6	0.91	1.07
	mg CO_2/m^2 per hour					
Control	81.4	104.0	166.2	206.8	105.2	65.6
6 mg N ($^{15}NH_4SO_4$)	87.2	109.7	156.8	157.2	97.0	83.7
6 mg N ($K^{15}NO_3$)	83.3	107.8	152.7	152.3	102.7	81.8
1 g wheat straw (W.S.)	131.1	123.0	254.3	304.6	102.7	89.2
1 g W.S. + 3 mg N ($^{15}NH_4SO_4$)	154.0	170.3	410.5	460.7	117.8	87.4
1 g W.S. + 3 mg N ($K^{15}NO_3$)	154.0	140.1	362.6	360.7	108.4	78.1
1 g W.S. + 12 mg N ($^{15}NH_4SO_4$)	148.3	155.2	677.8	639.9	112.1	80.0
1 g W.S. + 12 mg N ($K^{15}NO_3$)	138.8	151.4	633.2	557.0	110.3	98.5
1 g barley grain	270.5	335.5	1,116.4	1,011.5	131.0	105.9

[a]Temperature and moisture measured at 1500.

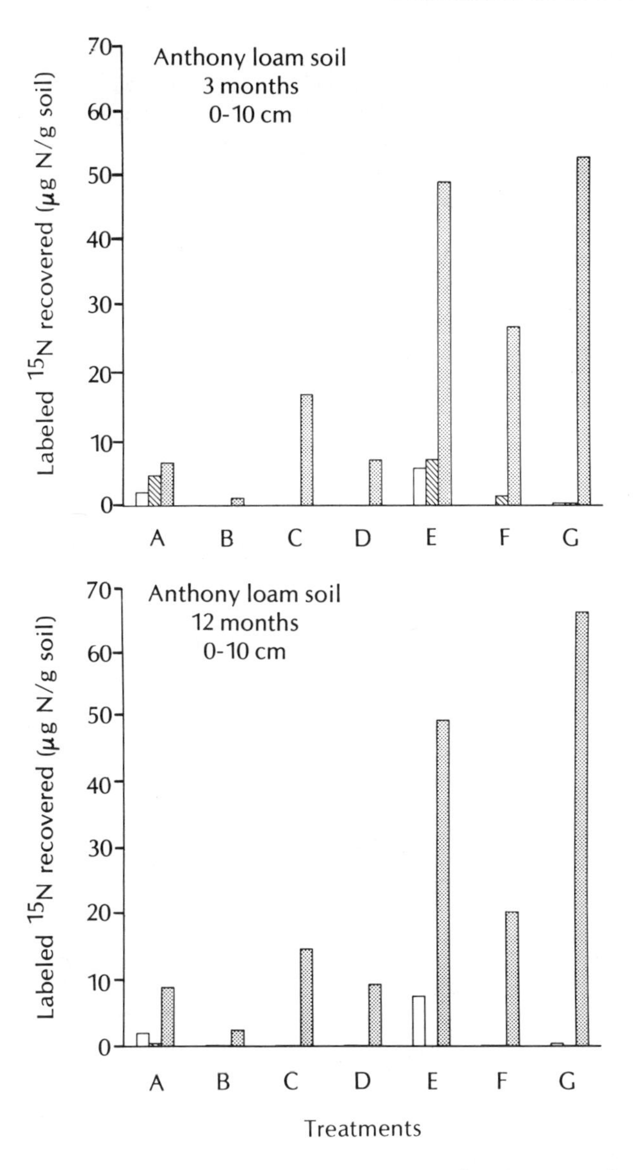

Figure 7-5 *Nitrogen-15 forms remaining in the 0-10 cm depth of Anthony sandy loam after 3 and 12 months under natural environmental conditions. Open columns represent ammonium N, hatched columns nitrate plus nitrite N, and stippled columns display organic N values. Nitrogen was added via the following treatments:* A, *6 mg ammonium N;* B, *6 mg nitrate N;* C, *3 mg ammonium N plus 1 g wheat straw;* D, *3 mg ammonium N plus 1 g wheat straw;* E, *12 mg ammonium N plus 1 g wheat straw;* F, *12 mg nitrate N plus 1 g wheat straw;* G, *10 mg organic N as barley grain.*

environmental conditions (Fig. 7-5). At the end of 3 months very little of the added inorganic nitrogen remained as such, regardless of source. However, there was less inorganic ^{15}N detected when the nitrogen source was nitrate.

This indicated that added nitrogen in the form of nitrate is more susceptible to denitrification and/or immobilization than nitrogen added in the form of ammonium. Significant immobilization of added inorganic ^{15}N did occur in all treatments except added nitrate alone. Organic nitrogen in soils at the end of 3 months increased with increased nitrogen applications and with wheat straw amendments. The organic nitrogen forms present in the soils comprised from 50-95 percent of the total nitrogen remaining in the inorganic nitrogen amendments and 79-100 percent in inorganic nitrogen plus wheat straw amendments.

Mineralization of ^{15}N-labeled barley grain occurred but apparently at a very slow rate. Inorganic nitrogen accumulated at the end of 3 months comprised only 1.1-1.7 percent of the total ^{15}N remaining in the soil. This indicated that the inorganic nitrogen mineralized was either lost from the system as quickly as mineralized and/or immobilized. Analyses of the subdivided soil sections (0-5 and 5-10 cm depths) showed that 95-100 percent of ^{15}N remaining in soil did not move below the 5 cm depth.

During the last 9 months of the study only trace amounts of rain were recorded, and this occurred during the last month of the study. Therefore, little change in nitrogen form or gaseous losses was observed at the end of 6 and 9 months. There were only slight changes in nitrogen forms and gaseous losses observed at the end of 12 months compared to 3 months (Fig. 7-5). These additional denitrification losses ranged from 5-10 percent. Most of the nitrogen remaining in the soil was in organic nitrogen form and only traces of inorganic nitrogen were detected when $K^{15}NO_3$ was used as a nitrogen source. There were no significant accumulations of inorganic nitrogen as a result of mineralization of organic ^{15}N from barley grain.

Denitrification losses calculated by the difference method were naturally also very similar between the Anthony and Sonoita soils. Losses for the Anthony sandy loam are illustrated in Figure 7-6 for 3 and 12 months. Percent loss of added labeled nitrogen at the end of 3 months ranged from 25-97 percent of the amount applied in the two soils. The largest loss (95-97 percent) was observed when nitrogen was added as ^{15}N-labeled potassium nitrate.

Significant losses of nitrogen from ^{15}N-labeled ammonium sulfate did occur but the magnitude of loss (55-60 percent) was much lower than observed with ^{15}N-labeled potassium nitrate. Lower denitrification losses were observed with the addition of wheat straw. This was attributed to the enhancement of immobilization of added ^{15}N. Approximately 10 percent of the mineralized nitrogen from ^{15}N-labeled barley grain was denitrified. Subsequent observations at 6, 9 and 12 months indicated lower losses than were calculated for the 3-month period. Perhaps this can best be explained by heterogeneity of the sample for analysis, or other biotic factors such as insects (ants, termites) physically transporting the material away from the site.

Denitrification losses at the end of 12 months were 73 and 94 percent of added ^{15}N-labeled ammonium sulfate and potassium nitrate. The addition of wheat straw reduced gaseous losses. However, gaseous losses increased with nitrogen applications, and a greater loss was observed when the ^{15}N-labeled barley grain amendment was 19 percent. Since there was no accumulation of inorganic ^{15}N from the mineralization of the organic ^{15}N from barley grain, it is assumed that denitrification occurred soon after or simultaneously with nitrification.

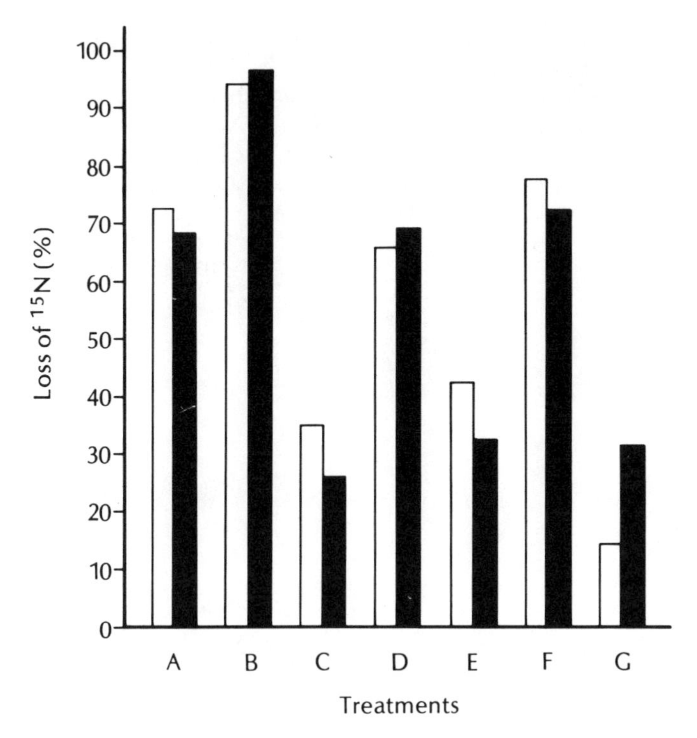

Figure 7-6 *Denitrification losses of ^{15}N, applied with various forms of N amendment [A-G], from Anthony sandy loam after 3 months [solid black column] and 12 months [open columns] under natural environmental conditions. See Fig. 7-5 for identification of treatments A-G.*

SUMMARY AND CONCLUSIONS

The denitrification potential for Sonoran Desert soil profiles is high. The greatest denitrification potential is in the surface horizon and/or horizon with the highest organic matter content. Denitrification increases with increased temperature, organic carbon and moisture content in the soil. However, complete saturation of the soils is not necessary. Apparently there are enough microsites in the soil that are saturated at lower moisture contents (i.e., field capacity or —0.3 bar) to allow significant denitrification to occur. This would become extremely important after rainfall events in desert ecosystems.

Significant denitrification losses have been observed in field studies in native desert ecosystems. The magnitude of loss is influenced mainly by the type and quantity of the organic carbon available, form of inorganic nitrogen present, moisture content in the soil and degree of competition for nitrogen by other organisms. Nitrate N is lost more readily than ammonium N, and losses of both increased with rate of application. Denitrification of both sources tends to be depressed by wheat straw amendments. The observed depression of denitrification was primarily due to enhanced immobilization of the nitrogen sources.

Mineralization of organic nitrogen from ground barley grain occurred, but there was no significant accumulation of inorganic nitrogen. Therefore it was assumed to be denitrified after mineralization, when conditions became favorable.

Denitrification in desert ecosystems plays a very significant role in the nitrogen cycle. Transformations of nitrogen in soils, uptake of nitrogen by plants and potential leaching losses as well as other gaseous loss processes in desert ecosystems cannot be thoroughly interpreted without considering denitrification.

8

AMMONIA VOLATILIZATION FROM GREAT BASIN DESERT SOILS

B. KLUBEK, P. J. EBERHARDT and J. SKUJIŅŠ

INTRODUCTION

Nitrogen escaping from soils in the form of ammonia is termed ammonia volatilization. This process has been reported as a significant source of nitrogen loss from soils with high pH values (Acquaye and Cunningham 1965; Blasco and Cornfield 1966; MacRae and Ancajas 1970). Most of these losses occur following fertilizer application, especially urea, as reported by a number of authors (Baligar and Patil 1968; Cornforth and Chesney 1971; Dhar and Banarjee 1966; Terman and Hunt 1964; Hamissa and Shawarbi 1962; Larsen and Gunary 1962; Loftis and Scarsbrook 1969; MacRae and Ancajas 1970; Misra and Singh 1970; Overrein 1969; Puh 1964; Shankaracharya and Mehta 1969; Vojinovic and Sestic 1967; Volk 1970).

Western American deserts have soils of high pH and periods of wetting and drying that may be expected to increase volatile ammonia losses. Nevertheless, literature reports of ammonia losses from desert soils are essentially nil. Dutt and Marion (1974), however, have reported on ammonia volatilization from a desert soil treated with a high rate of fertilizer nitrogen.

Although ammonia volatilization is thought of as a chemical process, it may depend on microbiological processes in the soil. The basic chemistry of these possible reactions is reviewed in the preceding chapter (Westerman and Tucker, Chapter 7 of this volume). Under natural conditions, ammonia must be supplied by ammonification from decomposing organic materials containing nitrogen. Conversely, when urea-containing fertilizers are applied, the hydrolysis of urea by soil urease produces ammonium and, eventually, ammonia. When the ammonium ion is present, the rate of nitrification may reduce the ammonia level and thereby reduce volatile losses.

Surface application of urea can greatly increase losses of ammonia. When 440 kg/ha of urea N were surface applied, 58 and 34 percent losses were recorded, respectively, at 15 and 75 percent of the moisture-holding capacity (Shankaracharya and Mehta 1969). When these same amounts were mixed into the soil, the losses were nearly zero.

Losses of surface applied urea usually increased when the pH was increased by liming (Shankaracharya and Mehta 1969; Terman and Hunt 1964). Watkins et al.

(1972) found that greater air movement and rising temperatures also increased volatile losses. These losses ranged from 6 to 30 percent on mineral soils and from 27 to 46 percent on forest soils. The loss from forest floors may have been via denitrification, however, since loss was determined by subtraction.

Ammonium salts have been extensively used in ammonia volatilization research. In general, the losses from ammonium salts have not been as high as from urea. For example, 5.5 percent of applied ammonium sulfate was lost when incubated at 45 C for 10 days (Puh 1964). The ammonia loss and water evaporation were linearly related. Nemeryuk and co-workers (1965) also found that evaporating water was accompanied by ammonia. Conversely, Chao and Kroontje (1964) determined that the rates of volatilization and evaporative losses follow different functions. Furthermore, they indicate that ammonia loss is proportional to the original soil pH. However, Blasco and Cornfield (1966) state that ammonia volatilization is not correlated with pH.

The carbonate content and cation exchange capacity of soils are accepted as influencing ammonia volatilization. Coarse-textured soils with their corresponding low cation exchange capacity have the highest reported volatile losses (Chao and Kroontje 1964). By contrast, soils with montmorillonite clay have low volatile losses due to high cation exchange capacity (Puh 1964).

Larsen and Gunary (1962) found their highest volatile losses came when $(NH_4)_2SO_4$ was applied to a calcareous soil. The same researchers stated that after 8 days, ammonia volatilization losses from soil treated with $CaCO_3$ were much greater than those following $(NH_4)_2SO_4$ application to a calcareous soil, or from soils treated with $(NH_4)_2HPO_4$ or NH_4NO_3. Soil first treated with $CaCO_3$ had a greater loss after NH_4NO_3 application than from the sulfates and phosphates. Soil first treated with $MgCO_3$, however, had its greatest loss following applications of sulfates and nitrates.

The anion influence in volatilization has been studied by Fenn and Kissel (1973). Again, the greatest rate of volatilization resulted from surface applied fertilizers. Up to 68 percent nitrogen losses were noted during 100 hours of incubation. The following reaction sequence was proposed: when the ammonia fertilizer dissolves in a calcareous soil, ammonium carbonate forms with a soluble calcium salt. The decomposition of this ammonium carbonate releases carbon dioxide, which escapes at a greater rate than the ammonia. This causes formation of ammonium hydroxide which increases the soil pH and concurrently creates conditions favoring a greater ammonia loss.

EXPERIMENTAL PROCEDURES

Materials and Methods

Figure 8-1 shows the basic scheme for collecting volatilized ammonia for analysis. Soil (100 g) was placed in a 500-ml Erlenmeyer flask wrapped in aluminum foil (to block photosynthetic nitrogen fixation) and subjected to the experimental treatment. The volatilized ammonia was captured in the gas washing bottle in 250 ml of 1 percent H_2SO_4. The captured NH_3 was analyzed by alkaline distillation with 40 percent NaOH and collected in boric acid containing Tashiro's indicator. After titration the samples were redistilled by alkaline distillation and captured in 0.05 *N* H_2SO_4 and evaporated to dryness for later analysis for ^{15}N content. This redistillation

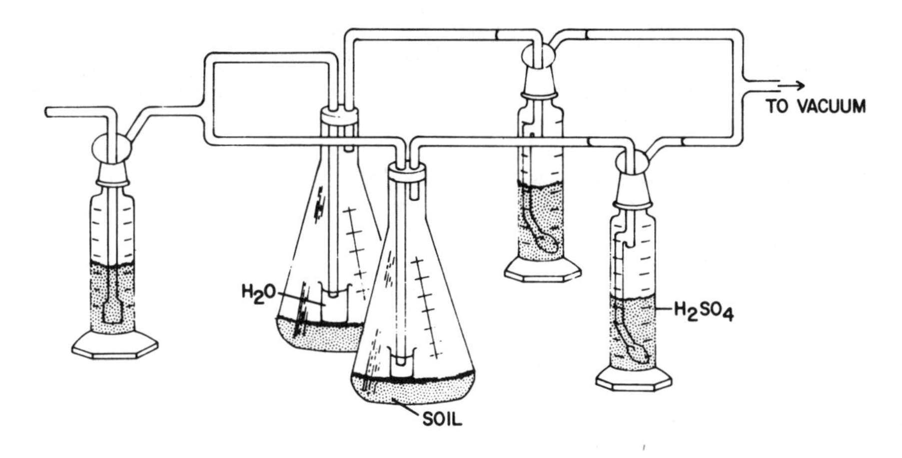

Figure 8-1 *Ammonia volatilization trap.*

process was used for all ^{15}N samples to prevent microbial contamination (Bremner 1965b, d) during an unavoidably lengthy storage.

Ammonia Volatilization Losses from Exogenously Supplied ^{15}N-Labeled Ammonium Sulfate

Samples of surface soil (0-3 cm) were tested for ammonia losses using the volatilization apparatus described above. The soils were collected in April 1974 from between plant canopies on sites dominated by *Artemisia*, *Ceratoides* and *Atriplex* in Curlew Valley, Utah (Mitchell et al. 1966). The procedure, described in detail by Skujins (1975), involved the addition of ^{15}N-labeled nitrogen to the soil samples, which were under differing soil water regimes (—1 bar, —15 bar and air-dried). In a parallel experiment, 1 g of fresh ground stem and leaf material from the dominant shrub was added to the appropriate soil and volatilization examined at —1 and —15 bar soil water potential. Volatilized ammonia was assessed for up to 5 weeks and then the soils were measured for water content, water potential, pH and for the presence of various forms of nitrogen: organic N, fixed ammonium, exchangeable ammonium, nitrite and nitrate. The analytical methods are described in Skujins (1975).

Ammonia Volatilization Losses from Decomposing Algae-Lichen Crust

It was recently reported (Fenn and Kissel 1973) that losses from exogenously supplied ammonium may vary depending on the companion anion in solution, with the sulfate allowing for greater ammonia loss than other anions such as nitrate.

Based on that information, ammonia loss had to be determined by supplying the ammonium endogenously, as it is provided in the field by algal crusts. To accomplish this, labeled nitrogen gas was introduced into the atmosphere above an algal crust of 63.6-cm^2 surface area. The dry crust (weighing 48.2 g) was moistened to —0.3 bar moisture tension and allowed to fix $^{15}N_2$ from the 95 percent ^{15}N-enriched atmosphere for 2 days. The crust was illuminated with 1,000 lumens at 24 C. The fixation of atmospheric nitrogen by the algae-lichen crust under such environmental conditions has been demonstrated by Rychert and Skujins (1974b).

After 2 days the crust, now enriched with ^{15}N, was placed in the volatilization chamber and the ammonia loss was monitored over a 10-week period. The soil in the chamber was allowed to dry out as the experiment progressed, but was intermittently moistened to —0.3 bar moisture each week. At the end of 10 weeks the soil was analyzed for exchangeable ammonium, fixed ammonium, nitrite plus nitrate and organic nitrogen.

Ammonia Volatilization Losses from *Artemisia* Undergoing Decomposition

Dried leaves and stems of *Artemisia tridentata* previously grown in ^{15}N-enriched solution cultures were added to the soil in the volatilization chamber. The leaves and stems were 20-mesh ground and analyzed for total nitrogen and labeled nitrogen. One percent of plant material by weight (i.e., 0.5 g *Artemisia*/50 g soil) was added to the soil. This amounted to an addition of 120 μg ^{15}N to the soil. The plant material was mixed with between-the-canopy soil samples from the *Artemisia tridentata* site. Chamber illumination and soil water conditions followed the procedure employed for the algal crust. The soil was analyzed at 10 weeks for the nitrogen fractions.

Rate of Release of Ammonium from Algae-Lichen Crust Subject to Subsequent Ammonia Volatilization

Since ammonium ions from nitrogen gas fixed by algal crust would have to be liberated into the soil in order for volatilization to occur by this pathway, an experiment was designed to determine how fast the fixed labeled nitrogen gas appeared as exchangeable labeled ammonium in the soil. Three areas, each encompassing 1 m^2, were allowed to fix $^{15}N_2$ under simulated field conditions as described above. A sample from each area was removed after 1-, 4- and 12-hour periods. The samples were mixed and 100 g were extracted with 150 ml 2 *N* KCl. The entire sample was steam distilled with MgO and titrated with standard .01 *N* $KH(IO_3)_2$. Analysis for $^{15}N_2$ was also made on the same sample. Presence of ^{15}N indicated exchangeable $^{15}NH_4^+$ in the soil from $^{15}N_2$ fixation by the algal crust. Such labeled ammonium may be available for ammonia volatilization.

RESULTS

Labeled Ammonium Sulfate Amendments to Soil

Table 8-1 shows the analysis of the different forms of nitrogen and the pH when the experiments were begun and at the fifth week when they were terminated. Upon inspection of the data, three distinct patterns become apparent. The soils amended

Table 8-1 *Soil Analysis of Organic N, Total NH_4^+, NO_2^- and NO_3^-, for Soils Treated with $[^{15}NH_4]_2SO_4$ [All values expressed as μg N/g soil]*

N fraction	Soil water status								
	-1 bar			-15 bars			Air dry		
	Artemisia	*Ceratoides*	*Atriplex*	*Artemisia*	*Ceratoides*	*Atriplex*	*Artemisia*	*Ceratoides*	*Atriplex*
Initial (time 0)									
Exchangeable NH_4^+	8.87	5.13	8.63	8.87	5.13	3.97	2.8	0.47	3.97
Fixed NH_4^+	77.76	61.20	57.0	77.76	61.2	63.46	73.97	84.7	63.46
Organic N	1084.2	898.2	927.9	1084.2	898.2	824.8	709.1	712.8	824.8
Total N	1170.8	965.6	993.5	1170.8	965.6	892.2	785.8	798.0	892.2
NO_2^- and NO_3^-	7.93	1.6	0.0	7.93	1.6	1.63	1.4	0.93	1.63
(pH)	8.52	8.58	8.91	8.52	8.58	8.80	8.71	8.53	8.72
Final (time 5 weeks)									
Exchangeable NH_4^+	0.7	0.0	5.95	1.75	2.45	2.8	2.8	2.1	4.55
Fixed NH_4^+	96.25	82.95	64.55	102.9	89.75	61.9	57.7	67.5	55.1
Organic N	932.0	848.0	755.5	1067.1	927.0	751.5	862.1	979.0	996.6
Total N	1029.0	931.0	826.0	1171.8	1019.2	816.2	922.6	1046.5	1056.3
NO_2^- and NO_3^-	129.1	86.8	38.1	79.1	46.5	58.1	71.05	80.1	90.6
NO_2^-	10.15	2.1	3.5	0.0	0.0	1.4	0.0	0.0	8.05
NO_3^-	119.0	84.7	34.6	79.1	47.9	56.7	71.2	80.1	90.6
(pH)	8.28	8.38	8.39	8.20	8.42	8.32	8.52	8.27	8.51

with 912 μg ^{15}N and allowed to air-dry had similar patterns of nitrogen cycling. Fixed ammonium had decreased, while organic nitrogen (16-57 percent) had increased substantially. In addition, a 39- to 93-fold increase of nitrite and nitrate (primarily nitrate) was noted. Exchangeable ammonium showed slight increases in both experimental systems, except for the *Artemisia* soil treated with ^{15}N-labeled ammonium sulfate, which showed no change.

The soils treated with ^{15}N-labeled ammonium sulfate and maintained at either —1 bar or —15 bars had similar results, except for the *Ceratoides* soil at —15 bars. All of the soil types (*Artemisia*, *Ceratoides* and *Atriplex*) at —1 bar water tension, and the *Artemisia* and *Atriplex* soils at —15 bars, showed increases in fixed ammonium and in nitrite and nitrate (10- to 54-fold, again predominantly as nitrate) while decreases in exchangeable ammonium and organic nitrogen were noted.

Ceratoides soil maintained at —15 bars water tension was the only exception. As in the other soils, exchangeable ammonium had decreased, while fixed ammonium and nitrite and nitrate (29-fold) had increased. The difference was in the organic nitrogen fraction, where a small increase was observed (Table 8-1).

Figures 8-2 to 8-6 give the data on ammonia volatilization from the previously mentioned experimental systems. For the ^{15}N amended, air-dried soils, ammonia volatilization was plotted with the changes in pH, soil water potential and percent moisture (Figs. 8-2 to 8-4). As can be seen from these figures, ammonia volatilization from the three different soils is as follows: *Artemisia* soil < *Ceratoides* soil < *Atriplex* soil. In all cases, the percent ammonia volatilized did not exceed 1.33 percent, this value being from the ^{15}N-labeled ammonium sulfate amended, air-dried soil of *Atriplex*. Last, the soils which were maintained at —1 or —15 bars water tension (Figs. 8-5 and 8-6) fluctuated in volatilized ammonia, indicating the effect of nitrification on this process at a constant water potential. However, when the soils were air-dried and amended with ^{15}N, a generally smooth curve was observed (Figs. 8-2 to 8-4), indicating a limited or removed nitrification influence. The nitrification process appeared to affect the soil pH as well. After rising at first, the pH decreased and fell below the initial value by 0.2-0.5, presumably due to the concurrent appearance of hydrogen and nitrite ions.

^{15}N Analysis of the $(^{15}NH_4)_2SO_4$ Amended Soils

Table 8-2 lists the amount of ^{15}N-labeled volatilized ammonia from $(^{15}NH_4)_2SO_4$ amended soils maintained at either —1 bar or —15 bars water tension, or air-dried, during a 5-week incubation period. The amount of volatilized ammonia ranged from 9.83 to 12.64 μg ^{15}N for soils at —1 bar; 6.33 to 9.66 μg ^{15}N for soils at —15 bars; and 0.23 to 21.44 μg ^{15}N for air-dried soils. In general, the total loss of nitrogen through ammonia volatilization amounted to approximately 1 percent.

Tables 8-3 to 8-5 provide data concerning the ^{15}N-labeled soil nitrogen fractions of the ammonium sulfate amended soils. For the soils maintained at —1 bar moisture tension, 74.74 to 78.31 percent of the applied $(^{15}NH_4)_2SO_4$ was lost; at —15 bars, 61.51 to 80.28 percent; and air-dried soils, 81.87 to 84.86 percent. In all three experimental systems, the nitrate N had the greatest atom percent ^{15}N enrichment (0.522-1.946) and the highest level of ^{15}N (63.9-190.7 μg ^{15}N). The *Ceratoides* air-dried soil, however, did have a lower level of nitrate N (52.20 μg ^{15}N) as compared to its organic nitrogen content (72.25 μg ^{15}N).

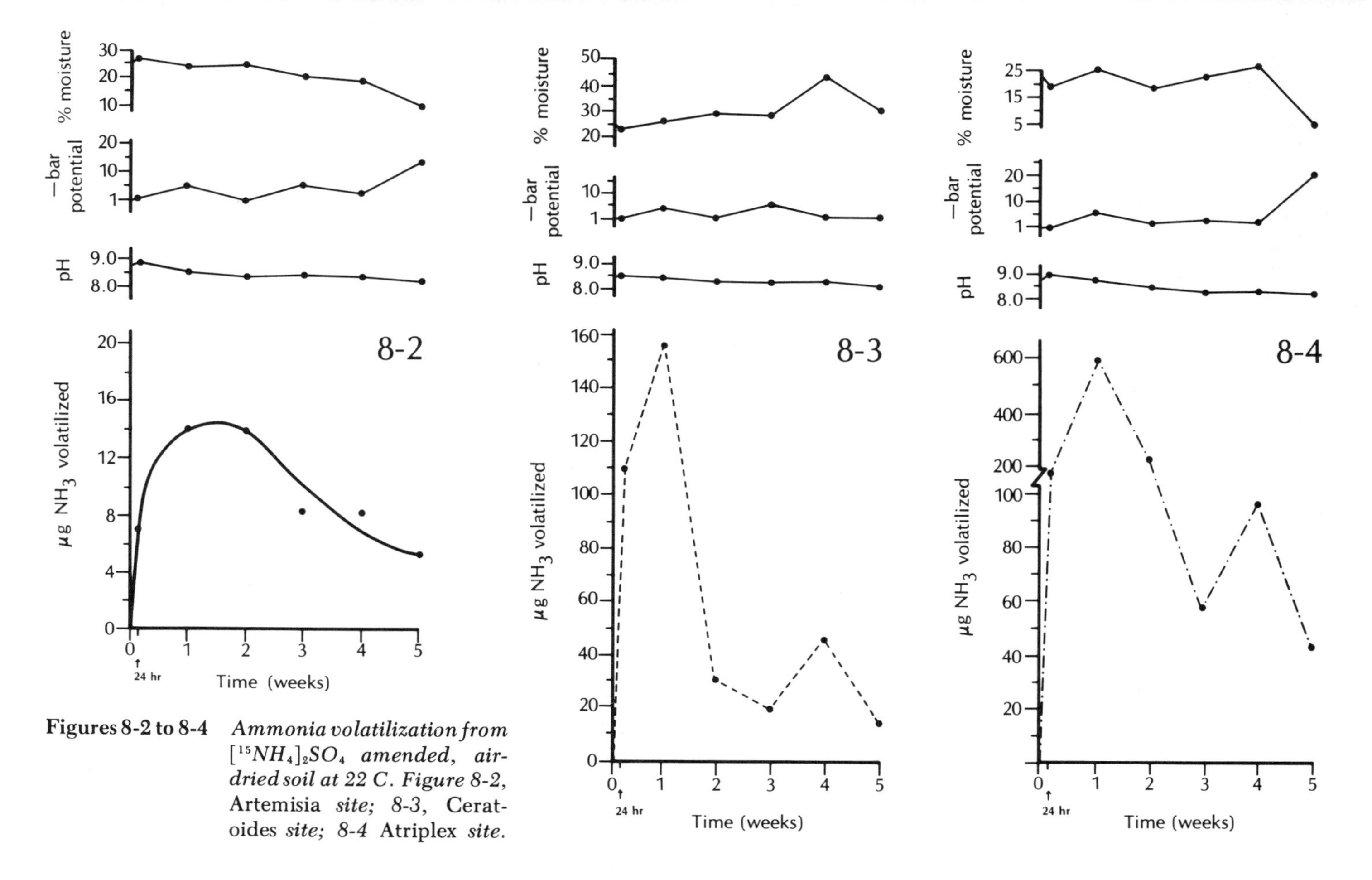

Figures 8-2 to 8-4 *Ammonia volatilization from* $[^{15}NH_4]_2SO_4$ *amended, air-dried soil at 22 C. Figure 8-2,* Artemisia *site; 8-3,* Ceratoides *site; 8-4* Atriplex *site.*

Table 8-2 *Total ^{15}N Lost Over a 5-Week Incubation Period as Volatilized NH_3 [$\mu g\ ^{15}N$] from [$^{15}NH_4$]$_2SO_4$ Amended Soils from the Three Plant Communities*

Soil water status	*Artemisia*	*Ceratoides*	*Atriplex*
-1 bar	12.64	9.85	9.83
-15 bars	6.33	8.44	9.66
Air dry	0.23	8.38	21.44

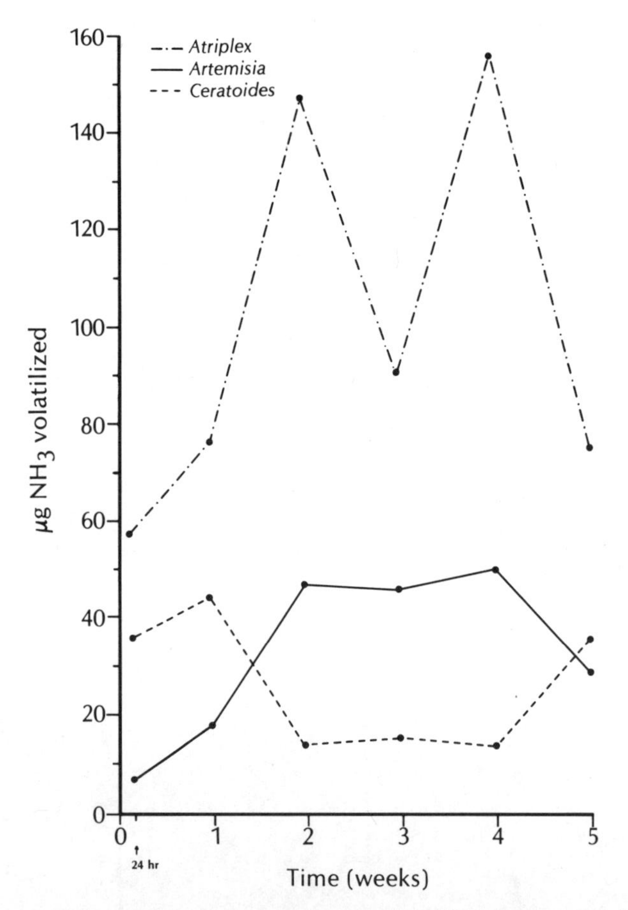

Figure 8-5 *Ammonia volatilization of [$^{15}NH_4$]$_2SO_4$ amended soils; soil water potential = —1 bar, temperature = 22 C. [All 3 curves start from zero at time zero.]*

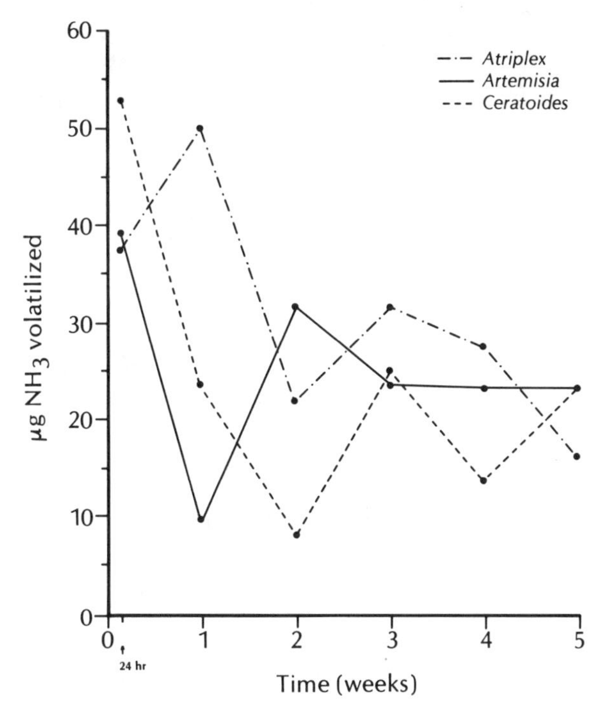

Figure 8-6 *Ammonia volatilization of [$^{15}NH_4$]$_2SO_4$ amended soils; soil water potential = —15 bars, temperature = 22 C. [All 3 curves start from zero at time zero.]*

Organic N was second to nitrate N in the amount of ^{15}N (41.17-118.56 μg ^{15}N) but was very low in the amount of ^{15}N enrichment (0.043-0.127 atom percent excess).

Clay-fixed ammonium N was the only other important source of ^{15}N, with a range of 21.57-37.52 μg ^{15}N, or an atom percent excess of 0.378-0.437. However, it should also be noted that the *Ceratoides* soil at —1 bar moisture tension had a nitrite N level of 32.9 μg ^{15}N (.360 atom percent excess).

^{15}N Plant Material Amendments to Soil

Table 8-6 lists the data concerning the fractional analysis of nitrogen in the plant material amended soils. Soils amended with ^{15}N plus plant material did not display similar patterns under identical water tension regimes as they did with the ^{15}N ammonium sulfate amendments. In general, the *Ceratoides* and *Atriplex* soils were similar in that exchangeable ammonium decreased, while organic nitrogen and nitrite and nitrate inci eased (10- to 17-fold, again primarily in the form of nitrate). The main differences between these two soils were 1) that the *Ceratoides* soil decreased in fixed ammonium while the *Atriplex* soil increased and 2) that the

Ceratoides soil underwent a 25.7 percent net loss of nitrogen via the denitrification process, whereas the *Atriplex* soil had a zero net denitrification rate, with a gain in nitrogen (Table 8-7).

The *Artemisia* soil was identical to the *Atriplex* soil in reactions except that both nitrite (57 percent) and nitrate (43 percent) increased, indicating an allelopathic inhibition of nitrification by the fresh plant material. The *Artemisia* soil also had a 2.56 percent net loss of nitrogen via denitrification (Table 8-7).

The ^{15}N plant material amendments to soil held at —15 bars water tension produced a different set of results (Table 8-6). Unfortunately, the *Ceratoides* soil generated no data after the 5-week period, because the sulfuric acid had backed up during sampling and acidified some of the soil. *Artemisia* and *Atriplex* soils were identical in having increases in fixed ammonium, organic nitrogen (18.5 and 15.3

Table 8-3 *Soil ^{15}N Fractions from $[^{15}NH_4]_2SO_4$ Amended Soils Maintained at —1 Bar*

Site	N fraction	µg ^{15}N/ 100 g soil	Atom % excess ^{15}N	% loss ^{15}N
Artemisia	Exchangeable $^{15}NH_4^+$	0.05	.000	
	Fixed $^{15}NH_4^+$	36.46	.378	
	Organic ^{15}N	70.09	.075	
	$^{15}NO_2^-$	11.50	.023	
	$^{15}NO_3^-$	99.60	.836	
	$^{15}NH_3$	12.64	---	
	Total	230.34	---	74.74
Ceratoides	Exchangeable $^{15}NH_4^+$	0.00	.000	
	Fixed $^{15}NH_4^+$	34.50	.415	
	Organic ^{15}N	56.62	.066	
	$^{15}NO_2^-$	32.90	.360	
	$^{15}NO_3^-$	63.90	.754	
	$^{15}NH_3$	9.85	---	
	Total	197.77	---	78.31
Atriplex	Exchangeable $^{15}NH_4^+$	0.01	.000	
	Fixed $^{15}NH_4^+$	28.25	.437	
	Organic ^{15}N	41.17	.054	
	$^{15}NO_2^-$	0.00	.000	
	$^{15}NO_3^-$	122.60	1.447	
	$^{15}NH_3$	9.83	---	
	Total	201.86	---	77.86

percent, respectively) and in nitrite and nitrate (2.5- to 31-fold). They differed, however, in details of the nitrite and nitrate data. In the *Artemisia* soil, only a 2.5-fold increase was noted, with most of the oxidized nitrogen in the form of nitrite (60 percent), while in the *Atriplex* soil, nitrate was the predominant form (85 percent). Exchangeable ammonium decreased in the *Artemisia* soil, while it increased in the *Atriplex* soil.

Table 8-8 provides data on ammonia volatilization from the soils amended with ^{15}N plant material. As in the ^{15}N ammonium sulfate treated soils, ammonia volatilization proceeded as: *Artemisia* soil < *Ceratoides* soil < *Atriplex* soil. In addition, nitrification did not appear to influence volatilization. Smooth trends were obtained in all three cases, with the peaks of volatilization reached in the first week, followed by a decline. This decline, however, was due to the nitrification process. Last, in the *Artemisia* and *Atriplex* soils, the pH by the fifth week had increased by 0.1 when held at —1 bar moisture tension, but decreased (.06 and 0.3, respectively)

Table 8-4. *Soil* ^{15}N *Fractions from* $[^{15}NH_4]^2SO_4$ *Amended Soils Maintained at —15 Bars*

Site	N fraction	μg ^{15}N/ 100 g soil	Atom % excess ^{15}N	% loss ^{15}N
Artemisia	Exchangeable $^{15}NH_4^+$	0.11	.006	
	Fixed $^{15}NH_4^+$	37.52	.418	
	Organic ^{15}N	63.38	.059	
	$^{15}NO_2^-$	0.00	.000	
	$^{15}NO_3^-$	72.50	.915	
	$^{15}NH_3$	6.33	---	
	Total	179.84	---	80.28
Ceratoides	Exchangeable $^{15}NH_4^+$	0.01	.000	
	Fixed $^{15}NH_4^+$	34.03	.379	
	Organic ^{15}N	118.56	.127	
	$^{15}NO_2^-$	0.00	.000	
	$^{15}NO_3^-$	190.70	1.946	
	$^{15}NH_3$	8.44	---	
	Total	351.04	---	61.51
Atriplex	Exchangeable $^{15}NH_4^+$	0.01	.000	
	Fixed $^{15}NH_4^+$	24.39	.394	
	Organic ^{15}N	58.24	.077	
	$^{15}NO_2^-$	0.00	.000	
	$^{15}NO_3^-$	94.60	.886	
	$^{15}NH_3$	9.66	---	
	Total	186.90	---	79.50

Table 8-5 *Soil ^{15}N Fractions from $[^{15}NH_4]_2SO_4$ Amended, Air-Dried Soils: 5 Weeks*

Site	N fraction	µg ^{15}N/ 100 g soil	Atom % excess ^{15}N	% loss ^{15}N
Artemisia	Exchangeable $^{15}NH_4^+$	0.46	.001	
	Fixed $^{15}NH_4^+$	24.40	.425	
	Organic ^{15}N	45.34	.052	
	$^{15}NO_2^-$	0.00	.000	
	$^{15}NO_3^-$	67.60	.676	
	$^{15}NH_3$	0.23	---	
	Total	138.03	---	84.86
Ceratoides	Exchangeable $^{15}NH_4^+$	0.20	.001	
	Fixed $^{15}NH_4^+$	28.74	.425	
	Organic ^{15}N	72.25	.073	
	$^{15}NO_2^-$	0.00	.000	
	$^{15}NO_3^-$	52.20	.522	
	$^{15}NH_3$	8.38	---	
	Total	161.77	---	82.26
Atriplex	Exchangeable $^{15}NH_4^+$	6.50	.014	
	Fixed $^{15}NH_4^+$	21.57	.391	
	Organic ^{15}N	42.95	.043	
	$^{15}NO_2^-$	8.11	.081	
	$^{15}NO_3^-$	64.79	.647	
	$^{15}NH_3$	21.44	---	
	Total	165.36	---	81.87

at —15 bars. On the other hand, the *Ceratoides* soil experienced a 0.2 decrease in pH at —1 bar. This *Ceratoides* decrease correlated well with having the highest amount of nitrite and nitrate (83.77 μ g/g). Table 8-7 gives the total nitrogen balance sheet for all of the experimental systems.

^{15}N Analysis of the ^{15}N Plant Material Amended Soils

The extent of ammonia loss from ^{15}N ammonium sulfate plus plant material amended soils maintained at either —1 bar or —15 bars water tension ranged from 0.039 to 2.12 percent (Table 8-8). In both experimental conditions the *Atriplex* soils had the greatest loss, while the *Artemisia* soils had the least.

Tables 8-9 and 8-10 list the soil ^{15}N fractions for these soils. In both experimental systems, the loss of applied ^{15}N-labeled ammonium sulfate was high

(80 ± 2 percent for —1 bar soils and 83 ± 1.5 percent for —15 bars). In addition, the organic N fraction had the highest level of ^{15}N (39.63-132.50 μg ^{15}N), but did have a lower atom percent excess when compared to clay-fixed ammonium N.

The clay-fixed ammonium N varied from 13.50 to 40.70 μg ^{15}N, with the lower values found in the soils maintained at —1 bar, and the higher values in —15 bar soils. Similarly, the atom percent excess of this soil fraction was lower in the —1 bar and higher in the —15 bar soils.

In general, nitrite N predominated over nitrate N in both experimental systems. At the —1 bar moisture tension, the *Artemisia* soil had more labeled nitrite N than labeled nitrate N, but the *Ceratoides* and *Atriplex* soils had greater levels of labeled nitrate N. However, the atom percent excess labeled nitrite N was greater than labeled nitrate N in all three cases.

Table 8-6 *Soil Analysis of Organic N, Total* NH_4^+, NO_2^- *and* NO_3^- *for* $[^{15}NH_4]_2SO_4$ *+ Plant Material Amended Soils [All values expressed as μg N/g soil]*

N fraction	Soil water potential					
	-1 bar			-15 bars		
	Art.	*Cer.*	*Atr.*	*Art.*	*Cer.*	*Atr.*
Initial (time 0)						
Exchangeable NH_4^+	8.8	3.7	17.7	2.8	0.4	3.9
Fixed NH_4^+	96.8	83.8	62.4	73.9	84.7	63.4
Organic N	1169.7	1183.6	647.8	709.1	712.8	824.8
Total N	1275.4	1271.2	728.0	785.8	798.0	892.2
NO_2^- and NO_3^-	3.8	4.9	7.3	1.4	0.9	1.6
Plant material[a]	11.9	13.0	13.0	11.9	13.0	13.0
Total N/100 g soil[a]	142.8	143.6	89.5	93.6	95.5	105.4
(pH)	8.05	8.21	8.23	8.58	8.60	8.79
Final (5 weeks)						
Exchangeable NH_4^+	3.0	3.5	3.5	2.1	ND[b]	9.1
Fixed NH_4^+	125.7	87.9	78.5	86.95	ND	72.6
Organic N	1190.0	891.3	789.73	1016.95	ND	1129.3
Total N	1318.8	982.8	871.73	1106.0	ND	1211.0
NO_2^- and NO_3^-	71.6	83.7	75.83	3.5	ND	49.7
NO_2^-	40.8	2.1	0.0	2.1	ND	7.35
NO_3^-	30.8	82.6	77.35	1.4	ND	42.35
(pH)	8.13	8.01	8.31	8.52	ND	8.51

[a]Expressed in mg.
[b]ND = no data.

Table 8-7 *N Balance Sheet [All values expressed in mg N]*

	^{15}N ammonium sulphate amendments								
	At -1 bar			At -15 bars			Air dry		
	Artemisia	*Ceratoides*	*Atriplex*	*Artemisia*	*Ceratoides*	*Atriplex*	*Artemisia*	*Ceratoides*	*Atriplex*
Initial total N	121.8	99.7	102.3	121.8	99.7	92.3	81.7	82.8	92.3
Final total N	116.0	101.9	87.0	125.2	106.7	87.6	99.4	113.0	115.9
mg of N denitrified	5.8	0	15.3	0	0	4.7	0	0	0
% denitrification	4.8	0	14.9	0	0	5.1	0	0	0
mg N_2 fixed	0	2.2	0	3.3	7.0	0	17.6	30.1	23.5
% N_2 fixed	0	2.2	0	2.2	7.0	0	21.6	36.4	25.4

Table 8-7 (Continued)

	^{15}N ammonium sulphate + plant material amendments					
	At -1 bar			At -15 bars		
	Artemisia	*Ceratoides*	*Atriplex*	*Artemisia*	*Ceratoides*	*Atriplex*
Initial total N	142.8	143.6	89.5	93.6	95.5	105.4
Final total N	139.2	106.8	95.6	111.0	ND	121.6
mg of N denitrified	3.6	36.7	0	0	ND	0
% denitrification	2.5	25.7	0	0	ND	0
mg N_2 fixed	0	0	5.	17.3	ND	16.1
% N_2 fixed	0	0	5.	18.5	ND	15.3

Initial total N = Total N + NO_2^- and NO_3^- per 100 g soil + 3.0 mg $(^{15}NH_4)_2SO_4$ for plant material amendments, with 1 g plant material per 100 g soil.

Finial total N = Total N + NO_2^- and NO_3^- + volatilized NH_3 per 100 g soil.

N_2 fixed = Heterotrophic nitrogen fixation.

ND = No data.

Table 8-8 *Ammonia Volatilization from $[^{15}NH_4]_2SO_4$ + Plant Material Amended Soils [All values expressed as μg NH_3 volatilized from 100 g of soil]*

Time	Water potential (bars)	*Artemisia*		*Ceratoides*		*Atriplex*	
		NH_3	$^{15}NH_3$	NH_3	$^{15}NH_3$	NH_3	$^{15}NH_3$
24 hr	-1	45.5	3.49	67.9	2.94	60.9	5.16
1 wk	-1	50.4	2.2	93.1	3.18	369.6	8.76
2 wk	-1	26.6	0.36	25.2	0.72	342.3	4.18
3 wk	-1	14.0	0.07	11.2	0.16	112.7	1.18
4 wk	-1	14.7	0.13	4.9	0.03	16.8	0.14
5 wk	-1	23.8	0.12	16.8	0.23	12.6	0.16
Total NH_3		175.0	6.37	219.1	7.26	914.9	19.58
% volatilized		0.12	0.7	0.15	0.8	1.04	2.12
24 hr	-15	5.6	0.23	71.4	0.25	295.4	7.57
1 wk	-15	8.4	0.00	56.0	0.07	148.4	1.97
2 wk	-15	7.0	0.00	16.8	0.02	25.2	0.03
3 wk	-15	11.2	0.01	8.4	0.00	16.8	0.01
4 wk	-15	4.2	0.00	8.4	0.00	30.8	0.04
5 wk	-15	9.8	0.08	9.8	0.00	19.6	0.01
Total NH_3		46.2	0.34	170.8	0.34	536.2	9.63
% volatilized		0.04	0.03	0.17	0.04	0.50	1.05

The soils maintained at —15 bars water tension had labeled nitrite N in excess of or equal to their levels of labeled nitrate N. However, the *Artemisia* soil had a higher atom percent excess labeled nitrite N, while the *Atriplex* soil had a greater atom percent excess labeled nitrate N.

Ammonia Volatilization from Fixed ^{15}N by Algae-Lichen Crusts and from Decomposing ^{15}N-Labeled Plant Material

Little ammonia was lost from the nitrogen fixed by algae-lichen crust organisms during the 10-week incubation time. Table 8-11 shows the data for ammonia released from algae-lichen crust during incubation. The largest ammonia loss occurred during the 3-6 week period. The amount of ammonia released over 10 weeks was less than 1 percent of the total $^{15}N_2$ fixed by the algae-lichen crust. Analysis of the soil for ^{15}N after the crust was incubated in the light for 2 days at the end of 10 weeks showed that little of the ^{15}N fixed by the crust remained, thus suggesting little denitrification loss.

The volatilization losses from *Artemisia* material undergoing decomposition are shown in Table 8-12. The volatilization rate decreased at a fairly constant rate over the 10-weeek period. After 10 weeks, most of the ^{15}N remained in the still undecomposed organic fraction of soil nitrogen. Fixed ammonium accounted for the major share of ^{15}N released from the litter and still remaining in the soil.

Percentage-wise, more ammonia was volatilized from the decomposing plant material than from the algae-lichen crust (Fig. 8-7). Slightly over 7 percent of the nitrogen from decomposing litter was lost as ammonia over 10 weeks, while less than 0.4 percent of the nitrogen gas fixed by algae-lichen crust was lost as ammonia. The entire balance sheet for both ammonium sources is shown in Figure 8-8.

Within 1 hour, labeled ammonium fixed from $^{15}N_2$ in the atmosphere by the algal crust appeared in the exchangeable fraction of soil nitrogen (Table 8-13). This amounted to nearly 19 percent of the N_2 fixed by algal crust (Table 8-14). As fixation continued over several hours, this dropped to about 7 percent of the fixed N_2 being released into the soil as ammonium ion. In the field this could be available for plant uptake.

Table 8-9 *Soil ^{15}N Fractions from $[^{15}NH_4]_2SO_4$ Plant Material Amended Soils Maintained at —1 Bar*

Site	N fraction	μg ^{15}N/100 g soil	Atom % excess $^{15}N^a$	% loss ^{15}N
Artemisia	Exchangeable $^{15}NH_4^+$	0.42	.001	
	Fixed $^{15}NH_4^+$	17.68	.140	
	Organic ^{15}N	122.08	.106	
	$^{15}NO_2^{-b}$	17.52	.429	
	$^{15}NO_3^-$	13.20	.004	
	$^{15}NH_3$	6.39	---	
	Total	177.29	---	80.57
Ceratoides	Exchangeable $^{15}NH_4^+$	0.59	.001	
	Fixed $^{15}NH_4^+$	14.91	.169	
	Organic ^{15}N	132.50	.150	
	$^{15}NO_2^{-b}$	7.65	.557	
	$^{15}NO_3^-$	40.72	.493	
	$^{15}NH_3$	7.26	---	
	Total	203.63	---	77.68
Atriplex	Exchangeable $^{15}NH_4^+$	0.60	.001	
	Fixed $^{15}NH_4^+$	13.50	.172	
	Organic ^{15}N	99.48	.130	
	$^{15}NO_2^{-b}$	4.53	.429	
	$^{15}NO_3^-$	28.03	.362	
	$^{15}NH_3$	19.58	---	
	Total	165.73	---	81.83

[a]Average of three determinations.

[b]NO_2^- was determined by subtracting the value of NO_2^- from NO_2^- and NO_3^-. Therefore, the atom % excess given is that of the NO_2^- and NO_3^- determination.

Table 8-10 *Soil ^{15}N Fractions from $[^{15}NH_4]_2SO_4$ + Plant Material Amended Soils Maintained at —15 Bars*

Site	N fraction	μg ^{15}N/ 100 g soil	Atom % excess ^{15}N	% loss ^{15}N
Artemisia	Exchangeable $^{15}NH_4^+$	0.30	.001	
	Fixed $^{15}NH_4^+$	40.70	.468	
	Organic ^{15}N	51.86	.051	
	$^{15}NO_2^-$	0.43	.001	
	$^{15}NO_3^-$	0.17	.001	
	$^{15}NH_3$	0.35	---	
	Total	93.81	---	89.71
Atriplex	Exchangeable $^{15}NH_4^+$	1.57	.001	
	Fixed $^{15}NH_4^+$	33.19	.457	
	Organic ^{15}N	39.63	.035	
	$^{15}NO_2^-$	3.10	.001	
	$^{15}NO_3^-$	3.10	.007	
	$^{15}NH_3$	9.64	---	
	Total	90.23	---	90.1

Table 8-11 *Ammonia Volatilization Losses from Algae-lichen Crust Enriched with 933 μg ^{15}N, Initially 1.064 Atom % Excess ^{15}N*

Time (weeks)	mg N	Atom % excess ^{15}N	μg $^{15}NH_3$-N
0-1	0.160	0.108	0.170
1-3	.218	0.095	.207
3-6	.077	3.744	2.88
6-10	.048	0.216	.125
			μg ^{15}N
After 10 weeks			
Exchangeable $^{15}NH_4^+$		0.039	0.079
NO_2^- + NO_3^-		0.014	.129
Fixed NH_4^+		0.007	.260
Organic N		0.003	1.810

Total recovery after 10 weeks < 1%.

Table 8-12 *Ammonia Volatilization Losses from* Artemisia *Litter Decomposing in Soil for 10 Weeks, Initial Litter Values of 1.46 Atom % Excess ^{15}N and 120 μg ^{15}N*

Time (weeks)	mg N	Atom % excess ^{15}N	μg $^{15}NH_3$-N
0-1	0.277	0.728	2.02
1-3	.235	.883	2.08
3-6	.333	.812	2.70
6-10	.127	1.587	2.02
			μg ^{15}N
After 10 weeks			
Exchangeable NH_4^+		.139	0.10
$NO_2^- + NO_3^-$		.187	.40
Fixed NH_4^+		.128	5.39
Organic N		.094	58.1

Total recovery after 10 weeks, 60.8%.

Table 8-13 *Release of NH_4^+ Ion by Algae-Lichen Crust Actively Fixing $^{15}N_2$ Under Simulated Field Conditions [Determined as KCl-extractable NH_4^+]*

Length of $^{15}N_2$ fixation (hr)	Total ^{15}N fixed, (μg/100 g soil)	^{15}N-NH_4^+ (μg/100 g)	NH_4^+/total N (percent)
1	4.07	0.78	19.2
4	5.23	.38	7.3
12	10.70	.70	6.5

Table 8-14 *Proportion of Fixed $^{15}N_2$ Found as Exchangeable NH_4^+*

Time (hr)	Total N (mg/100 g)	Atom % excess ^{15}N	^{15}N-NH_4^+ (μg/100 g)	N released (g/ha)
1	0.38	.20	.78	6.9
4	.52	.07	.38	3.9
12	.32	.21	.70	6.7

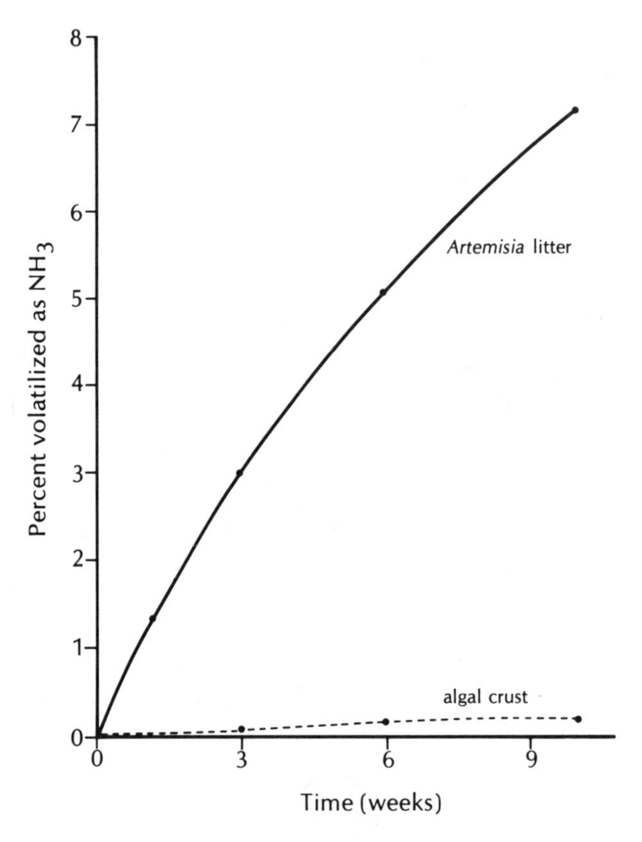

Figure 8-7. *Ammonia volatilization from decomposing* Artemisia tridentata *leaf material and algae-lichen crust during a 10-week wetting and drying cycle.*

DISCUSSION

In all cases, the loss of ^{15}N was approximately 75-80 percent, regardless of experimental conditions. This indicates that as inorganic nitrogen is applied to these Curlew soils a priming effect occurs, probably resulting in a rapid increase in the activity of *Nitrosomonas* and *Nitrobacter*. Furthermore, the greatest amount of ammonia volatilization (in general) occurred during the first week, followed by a rapid decrease. During the first week of incubation the *Nitrosomonas* and *Nitrobacter* populations may be undergoing a lag phase and, therefore, their capability for oxidizing the ammonium substrate would be low. This view is supported by data from Skujins and West (1974) that showed the nitrification potentials of these same soils to have a 10-12 day lag period prior to the oxidation of ammonium to nitrate N. In addition, the pH (in general) increased during the first week followed by a decrease during the remaining 4-week incubation period, suggesting the formation of nitrous or nitric acid, or both, due to the appearance of nitrite and nitrate anions.

The level of nitrate N also has other significance with respect to allelopathic inhibition. In this report, plant amended soils had a lower atom percent excess labeled nitrate N (0.001-0.493) than labeled nitrite N (0.001-0.557). Although the values for atom percent excess of the *Ceratoides* and *Atriplex* treated soils were high (0.493 and 0.362), they were less than the percent enrichment of the non-plant material amended soils (0.522-1.946). Furthermore, no substantial increase in ^{15}N-labeled organic nitrogen has been shown. These data, therefore, strongly suggest that the ground plant material exerted an allelopathic inhibition on the *Nitrobacter* population. The lower levels of labeled nitrate N cannot be explained on the basis of nitrate reduction since the organic nitrogen fraction fails to incorporate large amounts of ^{15}N. Therefore, as the *Nitrosomonas* population is primed and oxidizes ammonium N to nitrite N, the *Nitrobacter* population is inhibited, thus allowing the nitrite to be reduced to nitrogen or nitrous oxide by the denitrifying population; hence, the observed 80 percent loss of ^{15}N. It should also be noted, however, that the higher levels of nitrate N in the *Ceratoides* and *Atriplex* treated soils indicated temporary inhibition of *Nitrobacter*. As the plant material decomposed, the inhibitory material was apparently removed, allowing the *Nitrobacter* population to oxidize nitrite to nitrate nitrogen.

The ^{15}N-labeled ammonium sulfate amended soils maintained at —1 bar, —15 bars or air-dried had substantial losses of the applied ^{15}N (61.51-84.86 percent). This suggests that denitrification will occur even in the absence of an allelopathic inhibitor. Whether the applied ^{15}N is oxidized to nitrite N and subsequently denitrified, or completely oxidized to nitrate N and denitrified, is not really of importance. The occurrence of the denitrification is the crucial point.

Table 8-7 summarized the nitrogen balance sheet for the ^{15}N experiments just described. The most significant result is definition of a threshold value of organic nitrogen for heterotrophic nitrogen fixation. When the initial total nitrogen of an experimental system was below 900 μg N/g soil, nitrogen fixation occurred with an increase in organic nitrogen (Tables 8-1 and 8-6). When the initial total nitrogen ranged from 90-110 μg N/g soil, nitrogen fixation or net denitrification might or might not occur, with low levels of nitrogen fixed or lost. When the total nitrogen values exceeded 110 μg N/g soil, net denitrification occurred, and nitrogen fixation was turned off. This suggests a dependence of heterotrophic fixation and denitrification on the carbon:nitrogen ratio in the surface soil. If the carbon:nitrogen ratio was high, nitrogen was limiting and fixation began. As the level of organic nitrogen increased, the carbon:nitrogen ratio decreased until nitrogen was no longer limiting, but in excess. Assuming a loss of carbon due to microbial respiration, the nitrogen would then be denitrified and nitrogen fixation would continue to be turned off until the level of nitrogen again becomes limiting. Thus, a higher carbon:nitrogen ratio would subsequently occur.

It is important to note that we are considering heterotrophic fixation. Thus, a high carbon:nitrogen ratio is important because the carbon represents the energy source for nitrogen fixation and for denitrification. The source of this carbon would be the gelatinous sheaths of the blue-green algae *Lyngbia* and *Microcoleus* (Sorensen

1975), since these two genera have been observed frequently. Hence, a rhizosphere effect is created by the input of carbon by the sheathed blue-green algae, allowing heterotrophic nitrogen fixation and denitrification to occur.

In the overall picture, carbon (from the blue-green algae) is the energy source driving the reactions of heterotrophic nitrogen fixation and denitrification. Allelopathy makes its contribution to the loss of nitrogen as the *Nitrobacter* population is inhibited, thus allowing the nitrite N to be directly denitrified to either nitrogen gas or nitrous oxide. Last, if nitrogen is exogenously supplied, e.g., as ammonium sulfate, the nitrifying and denitrifying populations are primed and substantial losses of the applied nitrogen occur.

Although large amounts of nitrogen gas were fixed by algal crust, very little was lost as ammonia (Table 8-11). Most of the ammonia was lost during the 3-6 week period. This indicates that the bulk of nitrogen gas fixed had been ammonified during this period, although some ammonia was released immediately. Of the large amount of $^{15}N_2$ fixed, little remained at the end of 10 weeks. Since it was not lost via ammonia volatilization, it can only be concluded that the loss was due to denitrification (Skujins and West 1974).

The ammonia loss from *Artemisia tridentata* plant material showed no period of increased ammonia loss, in contrast to the algae-lichen crust; the loss gradually tapered off with time. Much larger ammonia losses were associated with decomposing plant material than with the algae-lichen crust (Table 8-12, Fig. 8-7). Much less nitrogen was lost by denitrification from the decomposing plant material.

A comparison between fixed $^{15}N_2$ and ^{15}N-enriched plant material is shown in Figure 8-8. Denitrification accounted for nearly all the ^{15}N loss from the algae-lichen crust. Denitrification associated with decomposition of plant material, while still significant, accounted for slightly less than 40 percent of the loss. It is interesting that clay fixation of ammonium was nearly as important as the volatilization loss. Very little exchangeable ammonium remained after 10 weeks.

Volatilization losses from algae-lichen crust were remarkably low, especially since ammonium ion is released into the soil environment by the algae-lichen crust (Table 8-13). After only 1 hour, 19 percent of the $^{15}N_2$ fixed was found as exchangeable ammonium. After 4 hours this had been reduced to about 7 percent, after which it remained fairly constant up to 12 hours. Upon wetting, rapid N_2 fixation produced an excess of ammonium ion, which regulated the N_2 fixation process. After several hours of N_2 fixation the rate became more constant and the ammonium remained constant.

SUMMARY

The major loss of nitrogen from Great Basin Desert soils occurs through denitrification. Volatilization of ammonia is of lesser importance, but is significant in the case of decomposing higher plant residue in the soil. Fixation of ammonium on clays is an important nitrogen-conserving process in the case of decomposing plant material and serves as an eventual ammonia source.

Volatilization losses from exogenously supplied ammonium sulfate may be higher than endogenously produced ammonium nitrogen. While this would not be important under natural conditions, it may be important when nitrogen fertilizer is applied.

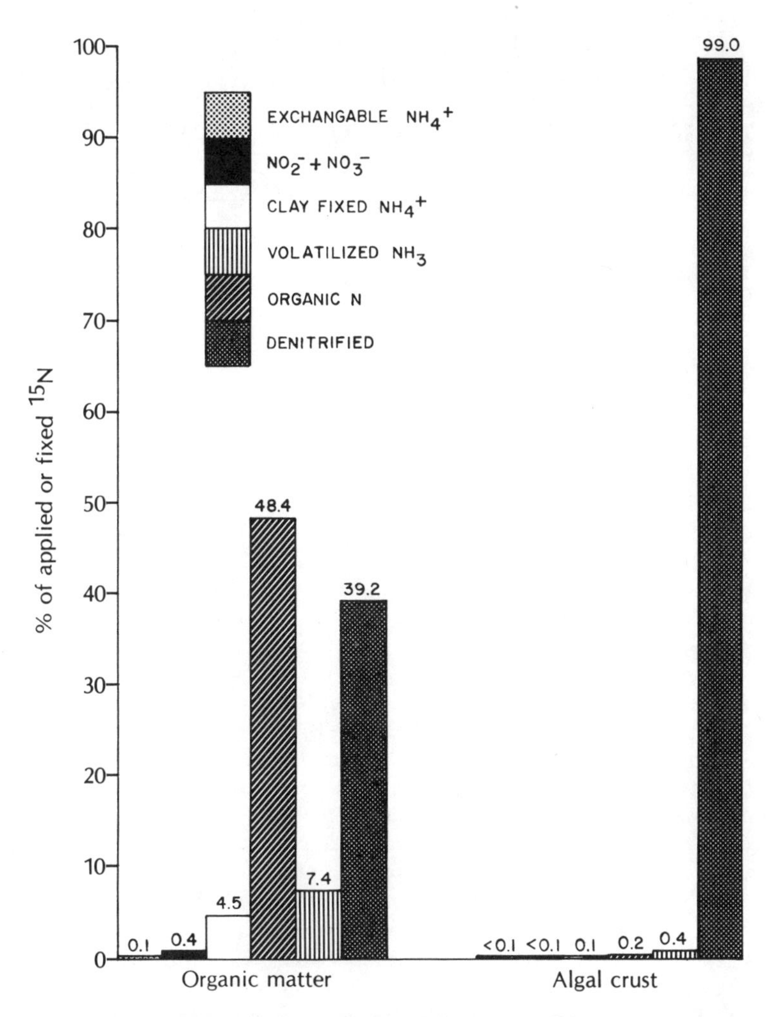

Figure 8-8 *Nitrogen balance sheet for decomposed* Artemisia tridentata *and algae-lichen crust after 10-week incubation in a lighted chamber.*

Ammonium ion is released directly into the soil environment by the algal crust. If not used by plants it is denitrified fairly rapidly with minimum volatilization of ammonia.

The canopy of desert shrubs is an important parameter in controlling the nitrogen supply available to desert plants. As the plant residue in the canopy area decomposes, much nitrogen is mineralized and may be utilized by the plants. Volatilization is less under a plant canopy and amounts of clay-fixed ammonium are high.

9

UPTAKE OF MINERAL FORMS OF NITROGEN BY DESERT PLANTS*

A. WALLACE, E. M. ROMNEY, G. E. KLEINKOPF and S. M. SOUFI

INTRODUCTION

Many studies of nitrogen uptake by plants have been made, but very few have been done on desert species. A critical question concerning these desert plants is whether they differ significantly from other plants in their nitrogen uptake characteristics.

Most plants take up nitrate and ammonium N with almost equal facility when these N sources are equally available (Kirkby 1970). There are differences due to pH of nutrient medium (Weissman 1950; Wallace and Mueller 1957) and carbohydrate supply (Michael et al. 1970). The mineral nutrition of plants varies considerably for nitrate vs. ammonium as N sources (DeKock 1970).

Because soil nitrate N concentrations are generally low under desert conditions, it might be anticipated that desert plants make greater use of exchangeable ammonium N since its concentration is greater than that of nitrate N (Nishita and Haug 1973), and that the affinity of Mechanism 1 (Epstein 1972) for nitrogen is very high for roots of desert plants. The Michaelis-Menten K_m would need to be 10 μM or lower, for great efficiency of uptake. In fact, values have been found in that range by several workers for plants in general. Becking (1956) observed that with intact corn plants the relationship between the concentration of ammonium supplied and its rate of absorption was a hyperbolic function with a maximum uptake of between 50 and 100 μM. The K_m was 13 μM. Van den Honert and Hooymans (1955), also working with corn plants, found maximum uptake of ammonium and/or nitrate at between 120 and 240 μM with a K_m of 23 μM. Eppley and Thomas (1969) and Eppley et al. (1969), in studies of marine phytoplankton grown in seawater where nitrate and ammonium ion concentrations are very low, found K_m values of the order of 1 μM or even less. If desert plants possessed equivalent affinity for NO_3^- and NH_4^+, their nitrogen uptake behavior could be better explained. All the characteristics of nitrogen uptake in the studies quoted are indicative of active uptake. Even so, there is need for further work of this type with desert plants.

---

*The study reported here was supported in part by Contract AT(04-1) Gen 12 between the U.S. Energy Research and Development Administration and the University of California and the US/IBP Desert Biome and the University of California. S. M. Soufi's participation was supported in part by the Lebanese National Council for Scientific Research, Beirut, Lebanon.

In the following sections the subject of mineral nitrogen uptake by desert plants is considered from several points of view. The tentative conclusions are examined in each section where appropriate.

EXPERIMENTAL DATA

Nitrogen Fractions in Desert Soils

Even though quantities of nitrogen are low in desert soils, nitrogen fractions change continually with seasons and with plant and animal activity. Information concerning the quantities of available forms of nitrogen and their flux through soil at any given time is vital to any consideration of nitrogen uptake in a desert ecosystem. The amounts and forms of soil nitrogen do much to regulate its uptake by plants. The study of Nishita and Haug (1973) is a good example of a one-time evaluation of nitrogen forms in a desert soil.

Nitrogen uptake by desert plants is, in part, regulated by the concentrations of nitrate and ammonium N. Nitrate N is probably freely available to plants, but amounts in desert soils are generally low at any one time (Bjerregaard 1971; Nishita and Haug 1973; Rychert and Skujins 1973; Skujins and West 1973; Charley and West 1975). Mineral nitrogen concentrations in desert soils, as in other soils under native conditions, are greatly dependent upon season (Rychert and Skujins 1973). From near zero to 4 ppm nitrate N in desert soils are common (Table 9-1). The two sample profiles in Table 9-1, calculated from Nishita and Haug (1973), illustrate conditions in which depth of rainfall penetration and plant position dominate N distribution in soil. These soil profiles also represent different fertility status. Both were in relatively fertile areas under shrubs, and the fixed ammonium N concentrations throughout both profiles are very uniform. On the other hand, nitrate and organic N concentrations reflect root distribution and average annual penetration of rainfall. Both roots and organic N drop off at the point where rainfall in most years ceases to penetrate.

The process of mass flow in soils (Barber 1971) is very important in moving nitrates through soils to root surfaces where they are taken up by plants. A partial indication of the importance of nitrate in the economy of higher plants in desert systems can, perhaps, be estimated from the nitrate concentration of the soil solution and the amount of water transpired by a given plant. Perennial plants in the northern Mohave Desert have about 3 percent nitrogen in leaves on an average dry-weight basis (Wallace et al. 1974; Romney et al. 1974). If the transpiration rate for 3 months of the year is 4 ml water/g dry weight per hour for 4-6 hours/day, then mass flow would bring sufficient nitrate N to the root surfaces (assuming an average of 2 ppm nitrate N in dry soil and 10 percent soil moisture) to supply the leaves with over 100 percent of their nitrogen needs. The data in Table 9-1 show quantities of nitrate N much larger than 2 ppm. If as much as 50 percent of the nitrogen need for perennial plants was supplied by conservation and retranslocation of the previous year's nitrogen already within the plant (Wallace et al. 1974), the value would be about twice the plant's need.

If 2 ppm nitrogen supplies twice as much nitrogen as needed, available nitrate N concentrations must be at least 1 ppm on the average if nitrate is the predominant source for uptake by perennial plants which can exercise internal conservation of nitrogen. Moderate concentrations of nitrate in the rhizosphere and continued production of nitrate, either directly by nitrification from ammonium or through atmospheric fixation, would have to take place to maintain the nitrate supply as it is being taken up. The nitrate pool may be continuously recharged from ammonium oxidation by nitrifying bacteria as plants remove what is formed. The ammonium N pool itself could simultaneously be recharged from other nitrogen sources. Organic nitrogen yields ammonium N, which in turn yields nitrate N by soil microbiological processes. A small quantity of ammonium N would also be continually available to plants by the process of mass flow, according to data in Table 9-1.

The argument for mass flow as a major factor in nitrate N movement to roots requires higher soil concentrations to meet the needs of annual plants than is the case with perennials. If their major growth period were 1 month and they were photosynthetically active 5 hours/day, the needed soil concentrations would be 4 ppm, continuously maintained, resulting in leaves with 2 percent nitrogen. This is usually achieved, but nitrogen concentrations generally are lower in the foliage of annual plant species than in perennial plants.

Table 9-1 *One-Time Sampling of Nitrogen Fractions in Two Soils from the Nevada Test Site [calculated from data of Nishita and Haug 1973]*

Depth increment (cm)	Fixed ammonium N (ppm)	Soluble ammonium N (ppm)	Nitrate N (ppm)	Organic N (ppm)	Total N (ppm)
In Atriplex-Sphaeralcea-Oryzopsis *vegetation*					
0 - 7.6	53	2	29	490	574
7.6-15.2	56	1	11	250	318
15.2-22.9	59	1	8	170	238
22.9-30.5	54	2	8	130	194
30.5-38.1	61	4	9	100	174
38.1-45.7	61	2	6	110	179
45.7-53.3	58	1	4	110	173
53.3-61.0	55	1	4	110	170
61.0-68.6	59	1	3	40	97
68.6-76.2	57	1	11	12	81
76.2-83.8	65	1	19	20	104
83.8-91.4	55	2	26	28	111
Total	682	19	138	1,570	2,409
In Larrea-Ambrosia *vegetation*					
0 - 7.6	67	1	12	270	350
7.6-15.2	78	1	8	80	168
15.2-22.9	79	1	2	90	172
22.9-30.5	77	0	2	80	159
30.5-38.1	76	2	2	40	120
38.1-45.7	73	1	1	50	125
45.7-53.3	65	1	0	80	146
53.3-61.0	63	0	0	80	143
61.0-68.6	62	1	0	30	98
68.6-76.2	64	1	0	50	115
76.2-83.8	57	1	0	60	118
83.8-91.4	56	1	0	30	87
Total	818	11	27	940	1,796

Table 9-2 *Response of Three Shrub Species to Nitrogen Fertilization of Soil from Mercury Valley under Glasshouse Pot Culture Conditions, Dry-Weight Basis [Romney et al. 1974]*

	Atriplex canescens			*Ceratoides lanata*			*Artemisia tridentata*		
Treatment[a]	Yield[b] (g/plant)	N[c] (%)	N[d] (mg/plant)	Yield (g/plant)	N (%)	N (mg/plant)	Yield (g/plant)	N (%)	N (mg/plant)
Control	55.6	0.60	297	24.9	0.52	180	14.3	0.88	125
100 kg N/ha	79.3	0.52	405	45.8	0.70	401	27.6	1.09	356
300 kg N/ha	79.3	0.94	608	51.0	1.84	889	33.1	2.25	741
F value	11.9	13.1	17.2	25.9	104.6	117.6	29.1	50.9	164.8
$LSD_{.05}$	13.7	0.21	104	9.4	0.24	86	10.8	0.36	63
$LSD_{.01}$	20.7	0.32	149	14.2	0.37	122	16.3	0.54	89

[a]Ammonium nitrate blended with potted soil.

[b]Total plant tops; means of three replicates.

[c]Nitrogen concentrations of leaf tissue.

[d]Nitrogen contents of plant tops.

An important question concerning the biology of desert plants is the extent of their utilization of fixed ammonium in soil. If the relatively large store of exchangeable ammonium N were available to the plants in a significant quantity, which in one case amounts to about 600 kg N/ha in the soil depth of 45 cm (Nishita and Haug 1973), that source may easily supply the nitrogen needs for a given season which, in terms of new uptake by plants, may be around 20 kg N/ha (Wallace et al. 1974).

Studies of exchangeable ammonium N uptake by plants under nonnitrifying conditions are certainly needed. A chemical inhibitor of nitrification, such as 2 chloro-6 (trichloromethyl) pyridine (N-serve), should simplify experimentation. The amount of exchangeable ammonium N in desert soil, however, is insufficient to supply the N needs of perennial desert plants when only 3 kg soil/pot are used for glasshouse studies (Table 9-2). It will be pointed out later, however, that the organic acid composition of desert perennial plants more closely resembles that of plants receiving nitrate nutrition than ammonium N (Kirkby 1969).

Glasshouse Pot Culture Experiments

Some results from pot culture tests using three different species of desert shrubs are given in Table 9-2. Significant increases in both yields and nitrogen contents of foliage were obtained for each of the shrub species investigated, supporting observations from field studies that desert shrubs will respond to nitrogen if other factors are not limiting. A matter of importance is that the low nitrogen values in the plants are much lower than for similar plants from the field, especially for the leaves of control plants grown in potted soil. (Some field data are shown in Table 9-3.) The

Table 9-3 *Nitrogen Concentrations in Leaves of Desert Plants from Rock Valley and Other Locations at the Nevada Test Site*

Species	No. of samples	Mean N (% dry wt)	Standard deviation (% N)	Range of % N	Percent samples over 3.50%
Larrea tridentata	50	2.35	0.285	1.69-3.11	0
Sphaeralcea ambigua	9	2.51	0.518	1.60-3.01	0
Krameria parvifolia	17	2.18	0.386	1.47-3.00	0
Lycium shockleyi	3	3.09	1.542	2.16-4.87	33
Lycium andersonii	40	3.34	0.788	1.65-5.34	45
Ceratoides lanata	32	2.99	0.597	1.40-4.12	32
Grayia spinosa	30	2.70	0.807	1.62-5.29	10
Dalea fremontii	9	2.73	0.725	2.02-4.24	22
Lycium pallidum	14	2.80	1.015	1.12-4.32	36
Stanleya pinnata	9	3.14	0.770	1.96-4.20	44
Lepidium fremontii	4	3.23	0.994	1.83-4.08	50
Mirabilis pudica	12	3.38	1.115	1.42-5.54	33
Atriplex canescens	13	2.47	0.401	1.86-3.00	0
Artemisia spinescens	2	2.72	0.042	2.69-2.75	0
Acamptopappus shockleyi	32	2.48	0.617	1.48-3.87	3
Atriplex confertifolia	21	2.41	0.845	1.34-3.90	14
Coleogyne ramosissima	14	1.81	0.278	1.35-2.21	0
Ephedra funera	5	2.84	0.684	1.79-3.38	0
E. nevadensis	28	2.40	0.822	1.60-4.76	11
Menodora spinescens	12	2.37	0.515	1.82-3.21	0
Ambrosia dumosa	43	3.22	0.871	1.75-5.11	40
Hymenoclea salsola	3	4.17	0.329	3.87-4.52	100
Hilaria rigida	1	2.07	---	---	0
Oryzopsis hymenoides	30	1.16	0.441	0.37-2.32	0
Tetradymia glabrata	1	1.42	---	---	0
Kochia americana	1	4.14	---	---	100
Mirabilis bigelovii	1	2.28	---	---	0
Haplopappus cooperi	3	2.80	0.666	2.34-3.55	33

total volume of soil therefore was limiting in terms of nitrogen supply. This experiment further demonstrated some beneficial effects to desert shrubs that could be derived from root zone contact with fertilizer supplements. In such cases it might be practical to achieve deep placement of fertilizers. Response depends largely upon the amounts assimilated.

Responses of Desert Plants to Nitrogen

Few studies have been conducted of nitrogen effects on desert plants (Romney et al., Chapter 16 of this volume; James and Jurinak, Chapter 15 of this volume). Some studies indicate that little response is obtained for nitrogen fertilization under desert conditions because water is usually a greater limiting factor. When sufficient water is supplied or when plants and soils are moved to controlled conditions, growth responses to nitrogen fertilizers and better transpiration:dry matter production ratios are obtained (Slatyer 1960; Cowling 1969).

At the Nevada Test Site in the northern Mohave Desert, some species of annual plants which often grow in the bare areas between shrub clumps continued to respond favorably to nitrogen fertilizer (Wallace and Romney 1972d) even 6 years after amendments were applied. Such increased growth was associated with uptake of the fertilizer nitrogen (Table 9-4), supporting the idea that the most infertile part of the desert is in the area between shrub clumps (Cowling 1969; Romney et al. 1974; Charley and West 1975). Three of the 12 species sampled did not show growth response on the 6th year after treatment, but even they continued to show elevated nitrogen concentrations in shoot tissues. The apparent lack of yield response of some species to nitrogen is partly a statistical problem with plant distribution. The competition phenomenon is also significant and should be evaluated; increased growth of some species due to nitrogen or other factors can result in crowding out of other species.

The total nitrogen concentration in shoots of annual plants (Table 9-4) was much lower than in leaves of perennial plants (Table 9-5). The shrubs grow in fertile "islands" which probably have supported mixtures of shrubs, annuals and fauna for centuries (Wallace et al. 1972c). Shrubs do seem to be able to rapidly mobilize considerable internal nitrogen (from stems and roots) during spring growth flushes. The nature of this ability must be subjected to future intensive studies. Annual plants responded favorably to nitrogen fertilizer which had been watered into the soil (Table 9-6); however, this marked growth response to supplemental moisture usually resulted in lower nitrogen concentrations in shoot tissues through carbohydrate dilution. Although not indicated in the table, growth response to

Table 9-4 *Influence of Nitrogen Fertilizer on Yield and Nitrogen Concentrations of Annual Plants in Rock Valley, Nevada, May 1973 [Romney et al. 1974]*

Plant species	Yield, plant shoots			N concentration, plant shoots	
	Control (g/m^2)	+N[a] (g/m^2)		Control (%)	+N[a] (%)
Amsinckia tessellata	0.18	1.29		0.87	1.35
Astragalus didymocarpus	0.47	0.24		2.07	2.70
Bromus rubens	1.18	1.01		0.80	0.97
Caulanthus cooperi	0.69	0.40		1.00	1.38
Chaenactis fremontii	7.00	9.08		0.87	0.81
Cryptantha nevadensis	0.30	2.13		0.75	0.85
C. pterocarya	0.72	2.26		0.72	0.72
Malacothrix glabrata	0.59	2.73		0.74	1.00
Mentzelia obscura	1.79	2.60		0.98	1.32
Phacelia vallis-mortae	0.24	0.65		0.78	0.98
Streptanthella longirostris	2.16	4.07		0.43	0.88
Vulpia octoflora	0.16	0.05		0.79	0.82
Totals	15.71[b]	27.37[b]	Means	0.90[c]	1.11[c]

[a]Nitrogen application of 100 kg N/ha (NH_4NO_3) was made in April 1967.

[b]For plant yield, t = 8.48 (significantly different at 0.01 level).

[c]For N concentration, t = 9.95 (significantly different at 0.01 level).

Table 9-5 *Influence of Nitrogen Fertilizer on Leaf:Stem Ratios and Nitrogen Concentrations of Leaves of Shrubs from Rock Valley from Areas With and Without Restrictive Hardpan [Romney et al. 1973]*

Plant species	Leaf:stem		N concentration	
	Control (ratio)	+N[a] (ratio)	Control (%)	+N[a] (%)
		No underlying hardpan		
Ambrosia dumosa	1.78	1.66	3.81	4.18
Ceratoides lanata	1.29	2.10	2.18	2.56
Grayia spinosa	1.19	2.37	1.45	2.64
Larrea tridentata	3.37	5.00	2.10	2.35
Lycium pallidum	0.94	1.17	2.54	3.19
Means	1.71[b]	2.46[b]	2.42[c]	2.98[c]
		Restrictive hardpan at 50-cm depth		
Ambrosia dumosa	0.77	1.51	3.85	3.96
Ceratoides lanata	1.52	2.97	1.98	2.49
Grayia spinosa	1.54	3.10	1.56	2.09
Larrea tridentata	3.54	3.70	2.22	2.70
Lycium andersonii	2.00	2.50	3.15	3.83
L. pallidum	1.00	1.17	2.87	3.64
Means	1.73[d]	2.49[d]	2.61[e]	3.12[e]

[a]Nitrogen equivalent to 100 kg N/ha (NH_4NO_3) applied in April 1967. Plant samples were taken in March 1973.

[b]For leaf:stem ratio and no hardpan, t = 4.07 (significantly different at 0.01 level).

[c]For N concentrations and no hardpan, t = 3.65 (significantly different at 0.02 level).

[d]For leaf:stem ratios and hardpan, t = 2.34 (significantly different at 0.1 level).

[e]For N concentrations and hardpan, t = 6.21 (significantly different at 0.01 level).

irrigation usually was obtained from both annual and perennial plants. In 2 years of irrigation the biomass of perennial plants was increased, as measured by weights of 50 clipped new shoots per species (Wallace et al. 1972e). The nitrogen concentration also was increased by both moisture and nitrogen applications (Table 9-7). The data obtained are not conclusive, however, and it appears that critical work must yet be done to demonstrate that nitrogen is a limiting factor to perennial desert plants only if water is virtually eliminated as a limiting factor (Romney et al, Chapter 16 of this volume).

Partitioning of Nitrogen in Parts of Desert Plants

Nitrogen taken up by plants is distributed among roots, stems, leaves, flowers and fruits. Redistribution of previously accumulated nitrogen occurs in response to phenological events (Table 9-8). Prior to leaf abscission of senescent leaves some

Table 9-6 *Influence of Supplemental Nitrogen and Moisture on Yields and Nitrogen Concentrations of Means of Seven Annual Plant Species in Mercury Valley, Nevada, May 1973 [adapted from Romney et al. 1973; species include* Bromus rubens, Caulanthus cooperi, Cryptantha nevadensis, Descurainia pinnata, Phacelia fremontii, Streptanthella longirostris *and* Vulpia octoflora)

Treatment	Plant yield (shoots)		N in shoots	
	Control (g/m^2)	+N[a]	Control (% dry weight)	+N[a]
Natural desert conditions	29.79	50.86	1.09	1.08
Supplemental moisture[b]	28.13	98.80	0.85	0.98

[a]Application of NH_4NO_3 equivalent to 200 kg N/ha in October 1970.

[b]Supplemental irrigation was applied just after the fertilizer application so that the response was to nitrogen moved into the soil rather than to water per se.

30-60 percent or more of the leaf nitrogen is returned to the plant stems. In the process of senescence, leaf proteins are hydrolyzed to amino acids for transport. This nitrogen then becomes available for rapid shoot extension whenever dormancy is broken on deciduous plants. On evergreen plants such as *Larrea tridentata* the nitrogen would move from old leaves to new growing shoots. As indicated in Table 9-8, nitrogen is not the only element being conserved by retranslocation. This response is in agreement with that observed in many other desert shrubs in Australia, Asia and North America (West, in press b).

Nitrogen appears to be taken up from soil by roots of some species even when shoots are dormant; therefore, roots become a storage site for redistribution later in the season (Table 9-9). More data are needed concerning seasonal trends in root concentrations of nitrogen, so that variations such as those seen in Table 9-9 can be evaluated statistically. There is some evidence that a circulation system is involved in both uptake and redistribution of nitrogen (Lips et al. 1971; Dijkshoorn 1971). Information about factors regulating partitioning and redistribution of nitrogen within desert plants is also urgently needed, since it will contribute to an understanding of nitrogen uptake.

Nitrate in Plant Parts and Nitrate Reductase in Desert Plants

Even though ammonium N generally is readily available to desert plants, indications are that nitrate N may be the form most commonly taken up. There is not enough information, however, concerning the uptake of ammonium that has been adsorbed on clay or soil organic matter colloids to eliminate it as an important source of nitrogen under desert conditions.

Nitrate is generally reduced to ammonium in leaves, but often the reduction occurs in roots. In some species, notably grasses, nitrate accumulates in high

Table 9-7 *Mean Weights of 50 New Shoots and Nitrogen Status of Leaves from Seven Perennial Shrubs from Mercury Valley, Nevada, 2 Years after Fertilization [100 kg N/ha once] and Addition of Supplemental Moisture Annually [Species include* Acamptopappus shockleyi, Ambrosia dumosa, Ceratoides lanata, Grayia spinosa, Krameria parvifolia, Larrea tridentata *and* Lycium andersonii]

Treatment	Control			100 kg N/ha			Percent increase for added nitrogen		
	Dry wt/shoot (g)	N (%)	N (mg/shoot)	Dry wt/shoot (g)	N (%)	N (mg/shoot)	Dry wt/shoot (g)	N (%)	N (mg/shoot)
Natural desert conditions	5.4	2.5	135	5.8	2.9	168	7.4	16	24
Supplemental moisture	6.7	2.5	168	7.4	2.9	215	10.4	16	28
Percent increase for supplemental moisture	24	0	24	28	0	28	---	---	---

Table 9-8 *Nutrient Status of Leaves Before and After Leaf Abscission [10 May 1972]*

Species	Abscised (A) or green (G)	No. of leaves	Dry wt (mg/leaf)	N (%)	N (μg/leaf)	K (%)	K (μg/leaf)	Ca (%)	Ca (μg/leaf)	Mg (%)	Mg (μg/leaf)	P (%)	P (μg/leaf)
Atriplex confertifolia	A	200	7.8	1.07	83	2.21	173	3.58	280	0.78	61	0.029	2
	G	200	8.7	1.75	152	2.29	200	1.92	168	0.62	54	0.082	7
Ephedra nevadensis	A	10	108.0	0.83	896	0.20	216	3.64	3,931	0.13	140	0.024	25
	G	10	290.0	1.71	4,959	0.29	841	2.79	8,091	0.38	1,102	0.076	220
Larrea tridentata	A	200	1.8	0.86	15	2.16	37	1.63	28	0.24	4	0.061	1
	G	200	2.5	1.86	46	2.20	55	0.68	17	0.13	3	0.147	4
Grayia spinosa	A	100	2.3	1.51	34	2.76	63	3.03	69	1.11	25	0.075	2
	G	100	4.6	1.67	76	4.29	184	1.67	76	1.19	54	0.052	2

Table 9-9 *Nitrogen Concentrations in Small Roots of Perennial Desert Plants [means of 4 or 5] in 1972*

Species	N concentration (% dry wt)		
	15 Feb	21 Mar	8 May
Lycium pallidum	2.68 (176[a])	2.12 (13)	2.53 (55)
Ambrosia dumosa	1.17 (45)	0.72 (31)	1.09 (9)
Ephedra nevadensis	0.74 (37)	0.69 (3)	1.18 (35)
Atriplex canescens	1.97 (18)	-b-	1.33 (29)
Larrea tridentata	1.34 (6)	1.48 (20)	1.40 (119)
Lycium andersonii	1.60 (60)	1.88 (38)	2.04 (20)

Note: 15 Feb was before plants broke dormancy; 21 Mar was in the period of vegetative growth and 8 May was when plants were starting to decline.

[a]Weight of small roots per plant (g).

[b]No sample on this date.

concentrations without reduction. One deterrent to heavy nitrogen fertilization of deserts and grasslands is the subsequent accumulation of high nitrate concentrations in forage, resulting in livestock health problems. Information is needed on the site of activity and quantity of the enzyme, nitrate reductase, in some species of desert plants. It is present only if nitrate has been absorbed, so that its presence is significant. Accumulated concentrations of nitrate and activity of the reductase both have regulating roles on nitrate uptake by plants.

The presence of nitrate in plants is evidence that this ion was taken up by the plants, since higher plants possess no mechanisms by which ammonium nitrogen can be oxidized to nitrate. On the other hand, nitrate reductase which reduces nitrate stepwise to ammonium N is ubiquitous, but only after an induction period of a few hours exposure to nitrate in the rooting medium. When nitrates are present in plants in high quantities, the uptake rate is in excess of the rate of nitrate reduction (as when the external supply is high). Usually nitrate reductase is present in both roots and leaves.

Plants seem to develop a capacity for nitrate uptake as it is supplied, probably as the result of an uptake mechanism induced by the presence of nitrate. Perhaps a more important phenomenon is the ability of plants to maintain a higher rate of uptake after the nitrogen supply has been mildly depleted, and to transport to shoots much larger quantities of nitrate N than are transported by plants in which the nitrate supply was not previously depleted (Wallace et al. 1967; Ashcroft et al. 1972). With nitrate supplies so low under desert conditions, it would seem that the kinetic affinity of roots of desert plants for nitrate may be much higher (lower K_m) than those of other plants, especially under depleted concentrations. Toetz et al. (1973) have studied half-saturation constants (K_m) for the uptake of nitrate and ammonium in an aquatic system in an attempt to relate this parameter to seasonal nitrogen requirements. The same needs to be done for desert plants.

Some data for nitrate N concentrations in leaves of desert plants (Table 9-10) indicate low levels for three species and a higher amount for *Lycium andersonii*. Data in Table 9-11 show levels of nitrate, nitrate reductase and protein in several

Table 9-10 *Nitrate Nitrogen Concentrations and Total Nitrogen in Leaves of Some Desert Plants [dry-weight basis]*

Species	Nitrate N (%)	Total N (%)
Atriplex confertifolia	0.093	2.08
Lycium pallidum	0.098	3.02
L. andersonii	0.602	2.98
Ambrosia dumosa	0.084	3.08

desert species supplied with additional levels of nitrogen in potted glasshouse tests. Table 9-12 shows a limited number of data for nitrate reductase in leaves of perennial plants in the northern Mohave Desert. The important conclusions to be drawn from the data in Tables 9-10, 9-11 and 9-12 are that 1) nitrate is taken up by desert plants, as evidenced by its accumulation in these plants; 2) the presence and development of the enzyme, nitrate reductase, in field plants indicates the presence of mechanisms for nitrate utilization; and 3) perennial plant species in the northern Mohave Desert can vary in their nitrate utilization processes. Goodman and Caldwell (1971) and Goodman (1973) found similar patterns and used this variation as a basis for detecting ecotypic differences in Great Basin Desert species.

Table 9-11 *Effect of Nitrogen Fertilization on Nitrate Assimilation in Leaves of Three Desert Shrubs Grown in the Glasshouse*

Species	Treatment	Nitrate reductase	Protein	Nitrate N	Chlorophyll
Ambrosia dumosa	Control	0.053	7.28	297	1.04
	0.25	0.141	7.91	445	1.21
	0.50	0.166	13.43	1336	1.19
Atriplex canescens	Control	0.016	6.10	298	0.67
	0.25	0.136	7.37	622	0.67
	0.50	0.230	8.86	1651	0.64
Lycium andersonii	Control	0.014	4.52	313	0.82
	0.25	0.045	4.26	1798	0.78
	0.50	0.047	4.27	4570	0.82

Note: All treatments were duplicated. Plants were growing in 3-kg soil containers and received 0, 0.25 or 0.50 g nitrate N/wk as KNO_3 solution. Nitrate reductase is reported as µmole nitrate N/g fresh wt per hr. Leaf protein and chlorophyll are in mg/g fresh wt, and nitrate N content of leaves in µg/g dry wt.

Nitrate and Ammonium Nitrogen Uptake by Some Desert Plants

Studies were made with several species of desert plants grown in solution culture of nitrate uptake from KNO_3 and NH_4NO_3 solutions and also of ammonium uptake from $(NH_4)_2SO_4$ and NH_4NO_3 solutions. The measurement technique involved use of NO_3^- and NH_4^+ electrodes in the solutions and total analysis of plant parts following completion of the tests. Rooted cuttings of four test species were prepared (Wieland et al. 1971) and pregrown for 40 days in quarter-strength nutrient solution (renewed once) so that the plants would be in a low-nitrogen condition at the beginning of the uptake tests. The species used were *Tamarix ramosissima, Atriplex canescens, Ambrosia dumosa* and *Lycium pallidum.* The cuttings were transferred to 1,700-ml test solutions containing 1 m*M* solution of KNO_3, $(NH_4)_2SO_4$ or NH_4NO_3, which is in the range of Mechanism 2 (Epstein 1972) but which approximates conditions found in soil solutions in the field. Each solution contained 1 m*M* $CaCl_2$; each treatment for each plant was duplicated. At the conclusion of the tests, dry weights of plant parts were determined and the results were reported on a root-weight basis even though whole plants were involved. The rate of nitrogen uptake was determined periodically (0, 4, 8, 24, 48 and 96 hours) by sampling the external solution and determining ammonium and nitrate concentrations using specific ion electrodes (Orion) for the analyses.

The data on uptake by the different species for the two nitrogen sources supplied separately are given in Figure 9-1, while those for the ions supplied together appear in Figure 9-2. The uptake ratios are in Table 9-13. In most cases the uptake of ammonium N exceeded that of nitrate N after 96 hours, but at 4 and 8 hours and some other periods this was not always the case, at least when the sources were applied separately. When the sources were supplied together, there was a tendency

Table 9-12 *Nitrate Reductase Activity in Leaves of Plant Samples Collected from the Field in 1974*

Species	Collection date (1974)	NR[a]	Protein[b]	NO_3^--N (µg/g dry wt)	NH_4^--N (µg/g dry wt)	Total N (% dry wt)
Larrea tridentata	27 Mar	0.97	22.1	113	49	2.41
	13 May	1.04	3.4	322	64	2.49
Ambrosia dumosa	27 Mar	13.4	37.2	308	40	4.47
	13 May	18.3	78.8	700	50	2.62
Lycium andersonii	27 Mar	0	12.0	735	31	2.35
	11 May	0	21.5	1001	24	1.54
L. pallidum	1 Apr	3.88	7.6	1120	64	1.92
	7 May	0	38.0	560	31	2.05
Grayia spinosa	1 Apr	0	35.4	385	32	1.94
Ceratoides lanata	1 Apr	2.74	97.6	448	54	2.87

[a]Nitrate reductase (NR) = µmole NO_2^-/g dry wt per hr.

[b]Soluble protein = mg soluble protein/g dry wt.

for the plants to use a substantial quantity of the ammonium N before using much of the nitrate ion. In *T. ramosissima* the ammonium N was essentially depleted from the test solutions during the test. With nitrates supplied alone, there appeared to be an induction period necessary for nitrogen uptake by *T. ramosissima*, whereas for *A. canescens* the same seemed to hold for ammonium uptake.

Further uptake studies were made with small plants of *L. andersonii*, *L. pallidum*, *Ceratoides lanata*, *Artemisia tridentata* and *T. ramosissma*. The experimental design was similar to the preceding except that 1 m*M* $CaCl_2$ was replaced with 1 m*M* $CaSO_4$, to avoid Cl^- interference with the nitrate electrode. In these latter tests only about one-third of the supplied nitrogen was depleted from the test solutions, so that depletion was not a factor in the results.

The results (Figs. 9-3 and 9-4 and Table 9-14) indicate differences in uptake for plant species and for nitrogen sources when the two nitrogen sources were supplied separately rather than together. For all of the species, except *A. tridentata*, the nitrate uptake eventually exceeded that of ammonium N. With this species the ratio of the uptake for the two sources did not change with time whether applied together or singly. When supplied as NH_4NO_3, the ammonium uptake exceeded that of the nitrate ion. In general, the proportion of nitrate to ammonium N increased with time (except for *L. pallidum*).

The results imply that both nitrate and ammonium N sources are available to the species studied and that the rates of uptake are not limiting factors to utilization of available forms in the field. This type of study should be refined in the future and

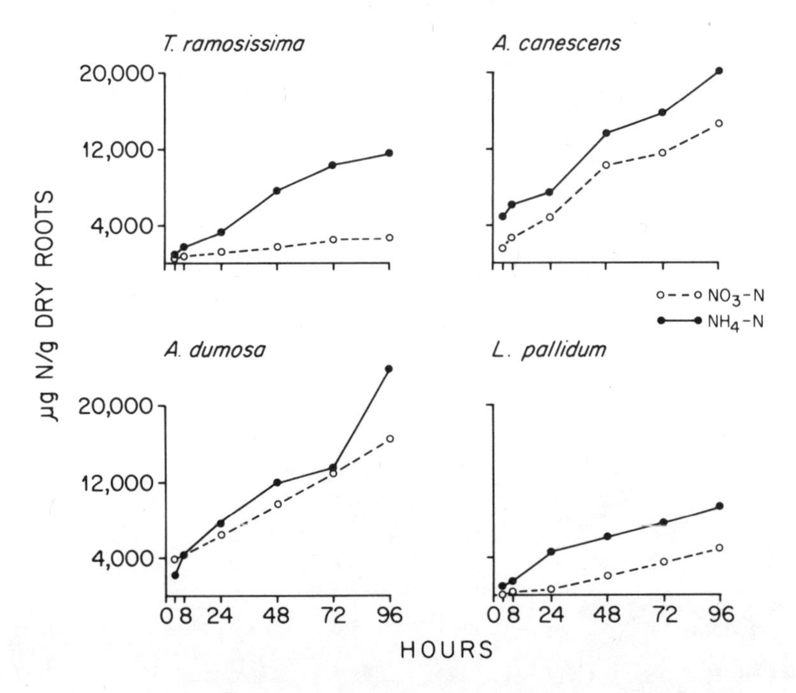

Figure 9-1 *Cumulative nitrogen uptake by four species of desert plants from KNO_3 or $[NH_4]_2SO_4$ applied separately via 10^{-3} M solution.*

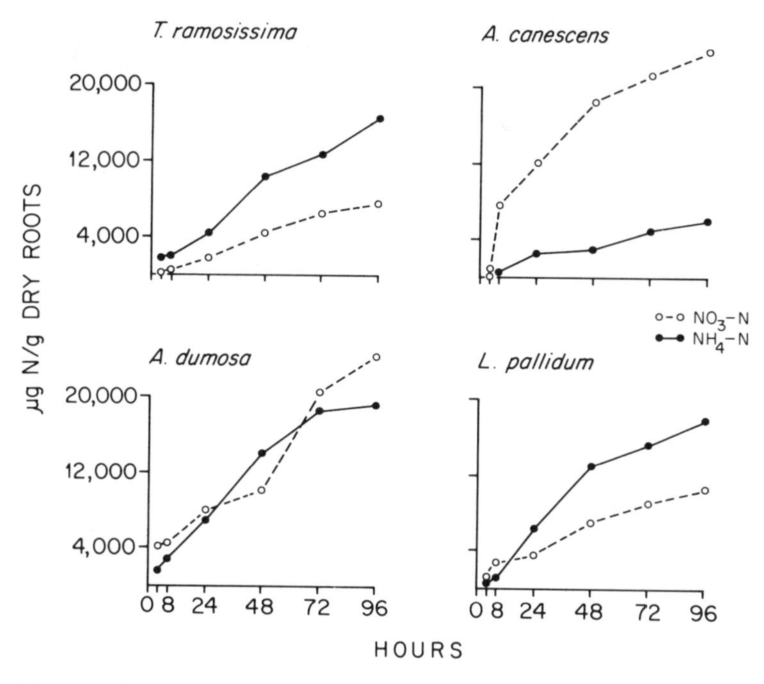

Figure 9-2 *Cumulative nitrogen uptake by four species of desert plants from 10^{-3} M NH_4NO_3 solution.*

Table 9-13 *Ratios of Nitrate N to Ammonium N Uptake by Some Intact Desert Plants*

Species	Cumulative uptake ratios by roots at different time intervals					
	4 hr	8 hr	24 hr	48 hr	72 hr	96 hr
	From KNO_3 and $(NH_4)_2SO_4$ solutions					
Tamarix ramosissima	0.06	0.16	0.43	0.42	0.50	0.58
Atriplex canescens	1.87	14.94	4.73	6.23	4.36	3.93
Ambrosia dumosa	2.34	1.56	1.17	0.72	1.11	0.76
Lycium pallidum	1.95	2.28	0.58	0.53	0.59	0.60
	From NH_4NO_3 solutions					
Tamarix ramosissima	0.40	0.50	0.36	0.24	0.23	0.24
Atriplex canescens	0.30	0.38	0.62	0.73	0.73	0.73
Ambrosia dumosa	0.57	1.00	0.86	0.82	0.86	0.70
Lycium pallidum	---	0.37	0.15	0.31	0.48	0.55

more data obtained. It should be noted that rates for only one concentration for each source were determined, while uptake rates from different external concentrations are needed for determination of K_m values.

Nitrate vs. ammonium uptake by three species of desert plants was studied by Skujins and West (1974) using ^{15}N to label plants removed from the field to be grown in solution culture. *A. tridentata*, *C. lanata* and *Atriplex confertifolia* were grown in solution enriched with either $(^{15}NH_4)_2SO_4$ or $K^{15}NO_3$. Each solution contained an equal quantity of N from an ammonium or nitrate source. Complete aerated Hoagland solutions, containing 210 mg N/liter from either 105 mg ammonium ^{15}N plus 105 mg nitrate N, or 105 mg ammonium N plus 105 mg nitrate ^{15}N, were placed in 1,000-ml Erlenmyer flasks. One plant per flask of each type completed the setup. Samples were taken at 0, 1, 2, 4, 8, 16 and, finally, 32 days when the experiment was

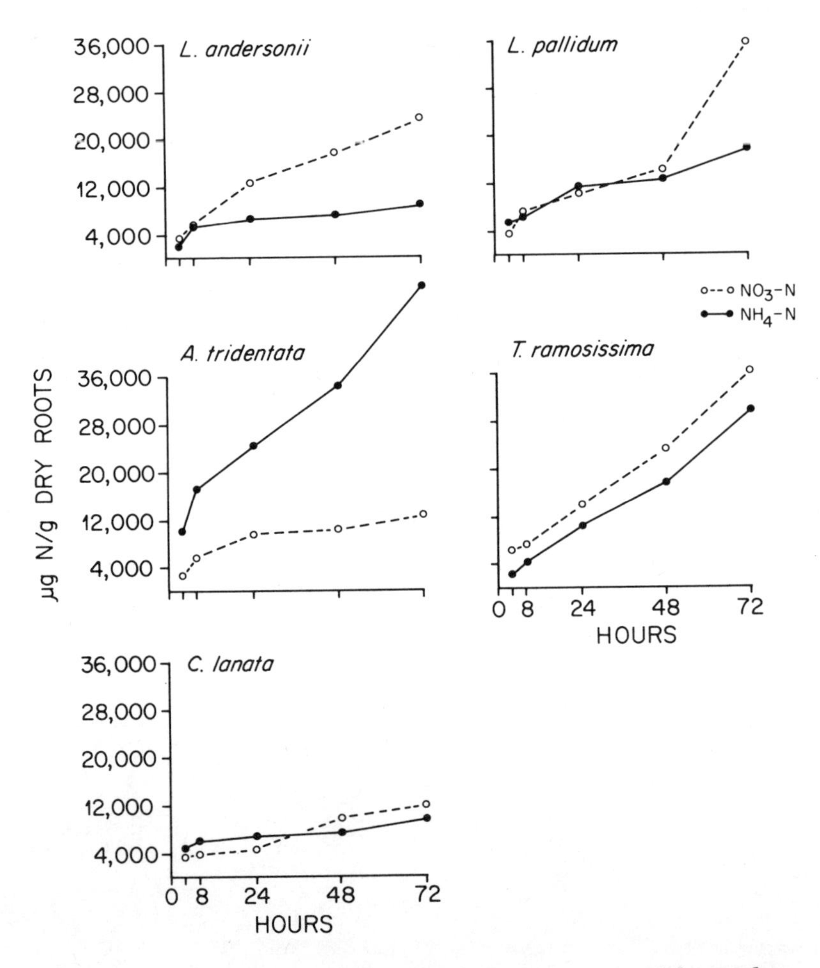

Figure 9-3 *Cumulative nitrogen uptake by five species of desert plants from KNO_3 or $[NH_4]_2SO_4$ applied separately in 10^{-3} M solutions.*

terminated. Plants were separated into foliage and stems plus roots for analysis. A complete balance sheet of ^{15}N uptake and recovery was made.

When *A. tridentata, C. lanata* and *A. confertifolia* were grown in solutions containing both ammonium and nitrate, little difference was found between total uptake of the two nitrogen forms. The pattern of uptake varied somewhat, however, as shown in Table 9-15. Generally there was a lag period before noticeable uptake of nitrate, during which time nitrate reductase enzyme production was being initiated. In soils where nitrate was present, this enzyme would already be active and no lag period would have been evident. These plants were extracted from the desert soil prior to growing in solution culture. If the soils contained nitrate, then nitrate reductase should have been present. The roots could have been extracted during a period of low soil nitrate N.

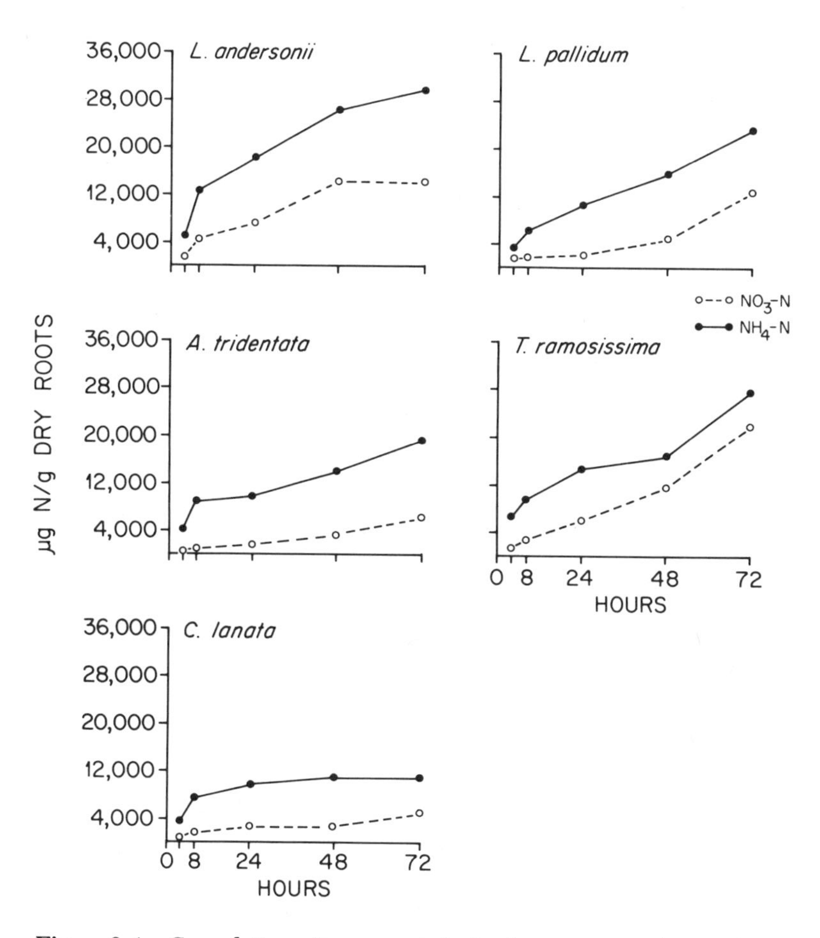

Figure 9-4 *Cumulative nitrogen uptake by five species of desert plants from 10^{-3} M solution of NH_4NO_3.*

Table 9-14 *Ratios of Nitrate N to Ammonium N Taken Up by Roots of Some Desert Plants*

Species	Cumulative uptake ratios by roots at different time intervals				
	4 hr	8 hr	24 hr	48 hr	72 hr
	From KNO_3 and $(NH_4)_2SO_4$ solutions				
Lycium andersonii	0.60	1.03	1.87	2.38	2.66
L. pallidum	0.66	1.12	0.91	1.12	1.97
Artemisia tridentata	0.28	0.34	0.40	0.31	0.25
Tamarix ramosissima	2.91	1.66	1.33	1.30	1.21
Ceratoides lanata	0.76	0.66	0.72	1.26	1.26
	From NH_4NO_3 solution				
Lycium andersonii	0.28	0.34	0.41	0.54	0.49
L. pallidum	0.56	0.30	0.21	0.31	0.54
Artemisia tridentata	0.14	0.09	0.18	0.23	0.33
Tamarix ramosissima	0.20	0.30	0.41	0.70	0.79
Ceratoides lanata	0.23	0.23	0.27	0.25	0.46

Concentration studies are needed to calculate kinetic constants for uptake of both nitrogen sources; temperature, inhibitor and pH studies would help explain the nature of uptake (Jackson et al. 1973; Morgan et al. 1973). All important, in terms of subsequent uptake behavior, is the prior nitrogen status of plants (Wallace and Mueller 1957; Dijkshoorn 1971), which must be fully evaluated.

Table 9-15 *Uptake of Nitrate N and Ammonium N by Three Species of Desert Plants Determined with* ^{15}N *[Skujins and West 1974]*

Days	*Artemisia tridentata*		*Ceratoides lanata*		*Atriplex confertifolia*	
	Nitrate N	Ammonium N	Nitrate N	Ammonium N	Nitrate N	Ammonium N
	Cumulative mg N uptake per plant					
1	6	4	1	1	1	6
4	4	6	2	1	1	3
8	4	7	3	2	3	4
16	11	12	7	4	4	8
32	14	15	13	13	10	12

Ionic Balance in Desert Plants

Desert soils commonly are highly calcareous; many are extremely sodic as well. In consequence, plants adapted to such areas are usually high in cations, which must be balanced by an equivalent amount of anions. Nitrogen concentrations of desert plants are often high relative to agricultural plants (Table 9-3), one reason being the high concentrations of cations taken up and the metabolic consequences which follow. Some leaves of desert shrubs contain 300-500 milliequivalents of cations/100 g dry plant material. Much of the anion concentration necessary for balance comes from oxalates or other organic acids, which arise either from bicarbonate (an anion) or assimilation of nitrate. Unassimilated NO_3^- can contribute to the balance, as can Cl^-, $SO_4^=$, $H_2PO_4^-$ or other mineral anions. The balance is of considerable importance in the regulation of nitrogen uptake.

Nitrogen, whether in the nitrate or ammonium form, is not, of course, the only inorganic ion taken up by plants. The uptake processes are related to the total cation-anion balance; thus this factor must be considered in assessing uptake rates of any single ion. For example, it was observed that the concentrations in bush beans of potassium, calcium and magnesium were changed only slightly when the nitrogen sources were nitrate, ammonium ions, urea or ammonium nitrate. The nitrogen, phosphorus and sulfur concentrations in the plants varied considerably with the source of nitrogen (Wallace and Ashcroft 1956).

There must be an electrostatic balance in the cation and anion relationships in any organism, or an electropotential difference would arise. There are only minor electrical potential differences between membranes; thus ionic balance is normally maintained.

The major anions commonly absorbed by plants include NO_3^-, $H_2PO_4^-$, $SO_4^=$, $SiO_3^=$, HCO_3^- and Cl^-; the cations include NH_4^+, K^+, Na^+, H^+, Ca^{++} and Mg^{++}. The magnitude of uptake of H^+ and HCO_3^- is difficult to measure (Wallace and Mueller 1963). The uptake of both bicarbonate and nitrate with cations leads to synthesis of organic acids because of metabolism of bicarbonate and nitrate (Dijkshoorn 1971; Kirkby 1970). Organic acids derived from nitrate metabolism usually arise in leaves, while those from bicarbonate metabolism usually remain in roots. The biochemical reactions from bicarbonate involve dark carbon dioxide fixation to yield the oxalacetic acid anion. Nitrate is reduced to ammonium, which is then incorporated into amino acids. An organic acid anion arises to balance any cation originally associated with nitrate. The resulting increase in organic acids maintains the cation-anion balance. The uptake and assimilation of ammonium N results in either the production of hydrogen ion within a plant or in the loss of an organic acid anion, assuming that an inorganic anion was taken up with the ammonium.

If nitrate is the only source of nitrogen for plants, and if net hydrogen ion uptake is zero, then the organic acid concentrations in plants can be estimated from the total concentrations of nitrogen, potassium, phosphorus, calcium, magnesium, sodium, silicon and sulfur and the concentration of unassimilated nitrate. The assimilated nitrogen in the plant will be balanced by an equivalent amount of organic acids and, in addition, there will be organic acids approximately equivalent to the differences between $K^+ + Na^+ + Ca^{++} + Mg^{++}$ (all the cations expressed as ionic equivalents) and nitrogen expressed as nitrate, phosphorus, sulfur, chlorine and silicon (each expressed as the ionic form). The role of nitrogen in cation-anion balance in desert plants is illustrated with four species.

From the data in Table 9-16, it can be seen that nitrate assimilation may provide the source of all of the organic acids for the two *Lycium* species and *Ambrosia dumosa* in leaves, and that bicarbonate assimilation is important in *Atriplex confertifolia* both in leaves and in roots. The pattern in general is highly indicative of nitrate uptake and metabolism.

Table 9-16 *Cation-Anion Balance in Whole Plants of Four Desert Perennial Species Collected in the Field* [*in me/100 g dry wt*]

	Atriplex confertifolia	*Ambrosia dumosa*	*Lycium andersonii*	*Lycium pallidum*
		Leaves only		
Cations[a]	438.9	396.6	509.5	468.6
Anions[b]	290.4	525.1	526.5	505.2
Nitrogen[c]	148.5	217.9	210.3	210.1
OA from HCO_3^-[d]	148.5	-128.5	-17.1	-36.6
Total OA[e]	297.0	89.4	193.3	173.5
C-OA[f]	141.9	307.2	316.2	295.1
		Roots only		
Cations[a]	285.3	192.5	105.2	257.3
Anions[b]	133.0	128.7	145.2	197.1
Nitrogen[c]	82.0	65.1	109.6	125.3
OA from HCO_3^-[d]	152.3	63.8	-40.0	60.2
Total OA[e]	234.3	128.8	69.6	185.5
C-OA[f]	51.0	65.5	35.6	71.8
		Whole plant		
Cations[a]	278.6	192.5	153.4	233.3
Anions[b]	134.5	128.7	135.3	177.4
Nitrogen[c]	76.4	65.1	86.7	109.6
OA from HCO_3^-[d]	144.1	63.8	18.1	55.9
Total OA[e]	220.5	128.9	104.8	165.5
C-OA[f]	58.9	65.5	48.6	67.8
g dry wt/plant	126.8	47.2	494.8	106.7

[a] K^{++}, Ca^{++}, Mg^{++}, Na^+ = cation (C).

[b] Assumes all N, P, S, Cl, Si taken up as NO_3^-, $H_2PO_4^-$, $SO_4^=$, Cl^- and $SiO_2^=$, respectively = anion (A).

[c] All nitrogen as NO_3^- (actually one source of organic acids).

[d] OA = organic acids derived from HCO_3^-.

[e] Total OA is the sum of milliequivalents of total nitrogen as (NO_3^-) and (C-A) or the bicarbonate-assimilated OA.

[f] The salinity component of the inorganic constituents (the quantity of the anions which are Cl^-, $H_2PO_4^-$, $SO_4^=$).

Table 9-17 *Cation and Anion Concentrations in Field-Grown* Atriplex confertifolia *from the Nevada Test Site*

Plant part	Height or depth from base (cm)	Dry wt of part (g)	N (%)	P (%)	S (%)	Cl (%)	Si (%)	Anion Σ (me/100 g)	K (%)	Ca (%)	Mg (%)	Na (%)	Cation Σ (me/100 g)	Cation:anion ratio
Large stem	10–20	4.43	0.60	0.028	0.42	0.17	0.05	77	1.09	2.19	0.30	0.07	165	2.14
Small stem	10–20	9.90	0.59	0.043	0.12	0.14	0.07	57	0.44	1.56	0.10	0.01	98	1.72
Leaves	10–20	11.01	2.10	0.083	1.09	2.46	0.07	292	1.90	3.55	0.68	3.87	451	1.54
Large stem	0–10	30.20	0.80	0.019	0.38	0.10	0.26	94	0.65	4.39	0.33	0.53	287	3.09
Small stem	0–10	23.22	0.67	0.024	0.52	0.23	0.09	91	1.17	2.58	0.33	0.51	209	2.30
Leaves	0–10	10.20	2.06	0.080	0.99	2.32	0.35	289	1.77	3.65	0.72	3.28	426	1.47
Large root	0–10	14.57	1.07	0.165	0.30	0.08	0.64	126	0.50	5.06	0.50	0.58	333	2.64
Small root	0–10	1.48	0.16	0.247	0.61	0.24	0.73	162	1.00	2.56	0.66	0.64	236	1.46
Large root	10–20	5.78	1.13	0.235	0.36	0.11	0.61	136	0.43	4.59	0.56	0.57	312	2.29
Small root	10–20	1.91	1.20	0.081	0.38	0.16	0.20	124	0.75	2.78	0.62	0.55	234	1.89
Large root	20–30	3.83	1.15	0.175	0.30	0.24	0.50	131	0.57	3.41	0.66	0.57	265	2.02
Small root	20–30	5.29	1.19	0.142	0.47	0.21	0.38	138	0.90	2.08	0.78	0.78	226	1.64
Large root	30–50	1.25	1.07	0.041	0.35	0.15	0.42	119	0.43	2.95	0.54	0.94	244	2.05
Small root	30–50	3.72	1.40	0.159	0.34	0.19	0.42	147	0.65	2.37	0.74	0.53	220	1.50

Table 9-18 *Cation and Anion Concentrations in* Lycium pallidum *from the Nevada Test Site*

Plant part	Height or depth from base (cm)	Dry wt of plant part (g)	N (%)	P (%)	S (%)	Cl (%)	Si (%)	Anion Σ (me/100 g)	K (%)	Ca (%)	Mg (%)	Na (%)	Cation Σ (me/100 g)	Cation: anion ratio
Large stem	30–42	0.60	0.86	0.071	0.18	0.07	0.03	78	0.45	1.10	0.13	0.02	78	1.00
Small stem	30–42	0.40	0.95	0.064	0.27	0.09	0.04	91	0.64	1.59	0.20	0.02	113	1.14
Leaves	30–42	0.44	3.11	0.198	3.90	1.00	0.10	504	1.38	4.67	1.15	1.25	419	0.83
Large stem	20–30	4.40	1.12	0.084	0.19	0.09	0.03	98	0.42	2.84	0.13	0.02	164	1.67
Small stem	20–30	3.60	1.19	0.076	0.32	0.10	0.06	112	0.62	3.16	0.24	0.02	195	1.74
Leaves	20–30	2.84	2.90	0.178	3.89	1.00	0.13	489	1.69	5.07	1.26	1.39	462	0.94
Large stem	10–20	5.35	1.07	0.071	0.16	0.04	0.11	94	0.33	2.53	0.09	0.01	143	1.52
Small stem	10–20	8.58	1.05	0.061	0.31	0.09	0.13	103	0.56	2.53	0.16	0.02	155	1.50
Leaves	10–20	2.98	3.14	0.130	3.78	1.05	0.14	499	1.69	5.42	1.14	1.28	465	0.93
Large stem	0–10	10.69	1.03	0.276	0.18	0.11	0.83	126	0.54	2.81	0.18	0.07	172	1.37
Small stem	0–10	7.57	1.08	0.057	0.25	0.15	0.05	101	0.32	2.92	0.09	0.01	162	1.60
Leaves	0–10	1.38	2.80	0.187	4.18	2.56	0.20	547	1.95	4.78	1.18	1.42	449	0.82
Large root	0–10	23.56	1.45	0.417	0.32	0.27	1.04	182	0.86	4.96	0.30	0.08	298	1.64
Small root	0–10	3.52	1.93	0.174	0.29	0.33	0.31	182	0.95	4.95	0.24	0.05	294	1.62
Large root	10–20	3.79	1.74	0.236	0.40	0.40	0.56	188	0.30	2.50	0.20	0.03	151	0.80
Small root	10–20	5.70	2.28	0.273	0.46	0.51	0.39	229	0.98	5.11	0.23	0.06	302	1.32
Large root	20–50	9.90	1.97	0.229	0.58	0.50	0.45	215	0.75	2.54	0.23	0.05	168	0.78
Small root	20–50	3.86	2.15	0.267	0.46	0.62	0.47	217	0.90	3.83	0.29	0.06	241	1.11

A. confertifolia probably has the most oxalate of the four species and it has the greatest total organic acids in the whole plant. A high potential for cation uptake appears to exist in many species, which results indirectly in a potential arising from an increase in organic acids, thereby increasing nitrogen uptake by these perennial species in the northern Mohave Desert. Tables 9-3 and 9-16 indicate much higher nitrogen concentrations in leaves than for many agricultural crops. For example, alfalfa (a legume) has nitrogen concentrations of usually 2.5-2.7 percent dry weight. This is higher than most agricultural plants but lower than many of the desert plants.

Representative values for the individual cations and anions for two of the plant species summarized in Table 9-16 by plant part are given in Tables 9-17 and 9-18.

SUMMARY

Nitrogen uptake and assimilation by plants are key steps in nitrogen cycling in any ecosystem. Under desert conditions, information is lacking in most areas related to these problems. The forms and amounts of nitrogen which are available on a seasonal basis have not been satisfactorily documented. Nitrate concentrations are obviously low, but they are seasonally high enough to meet much, if not most, of the short-term plant need for nitrogen. Desert plants commonly respond to addition of nitrogen fertilizer, especially when additional soil moisture is added. Mass flow of salts in water moved to the root could bring substantial quantities of nitrate to root surfaces even though the concentration in soils at any one time is low. Nitrogen recycling within the plant is a conservation measure which decreases need for new supplies of nitrogen from the external medium. The two processes just mentioned perhaps are most important in meeting the nitrogen needs of desert plants. The presence of nitrate in the leaves of some plant species implies that they absorb nitrate. Since nitrate reductase is an induced enzyme, its presence in field-grown plants also implies that nitrate is being utilized. There is more to be learned about the status and implication of nitrate reductase concentrations in desert plants. Ammonium and nitrate uptake have been studied for several species of desert plants under glasshouse conditions. The two forms of nitrogen can be taken up and assimilated with equal facility. In some cases, however, there seemed to be a lag in the rate of nitrate uptake since the ratio of nitrate:ammonium uptake increased with time. No critical studies have yet been made of the kinetics of nitrogen uptake by desert plants to indicate the degree of affinity which such plants might have for particular ionic forms of nitrogen. The ionic balance in some field-grown desert plants for which cation-anion values were estimated implies that nitrate is a major form of nitrogen taken up by desert plants. A steady state of nitrogen input and outgo in desert ecosystems may involve relatively small annual rates of each, with a sufficiently large pool of soil organic N which can buffer against high usage or high loss years.

10

ANIMAL CONSUMER ROLES IN DESERT NITROGEN CYCLES

C. S. GIST and P. R. SFERRA

INTRODUCTION

In the first chapter of this volume some brief observations were made on the allocation of nitrogen to the standing crop biomass of desert ecosystems. Subsequent chapters considered the role of plants and decomposers in the various nitrogen-transforming processes taking place in deserts. We would now like to add a discussion of the role of animals in the nitrogen cycles of arid to semiarid environments.

Although the animal part of the community has relatively little of the total ecosystem nitrogen tied up in its biomass at any one time, the apparent unimportance of this small standing crop is deceptive. Animals compose perhaps the most dynamic portion of the biological community. Their consumption, growth and death occur in widely fluctuating degrees, often on short time scales. Consequently they are able to affect the nitrogen cycle by accelerating the flow of material from living to dead states out of proportion to their biomass. Not all of the acceleration of nutrient cycling is affected by ingestion and excretion. Some food materials are ingested but are relatively unaltered by digestive tracts. Other materials are preconditioned for easier attack by decomposers through reduction in particle size. Larger animals accelerate the cycle by trampling live and standing dead materials into litter. Finally, by their death animals inevitably become part of the detritus that has concentrations of nitrogen in easily decomposable form.

With this overview, we will now look at some of the detatils of animal function in the nitrogen cycle in desert contexts. Only in recent years have ecologists begun to appreciate that minimal amounts of the vital elements could be secondary controlling factors of desert productivity. Nitrogen has been found to limit plant production when better-than-average soil moisture was present in deserts of several continents (West, in press b).

GENERAL NITROGEN EXCRETION

Conversion of amino acids to carbohydrates or fats results in the removal of the amino and nonamino nitrogen and often results in the formation of ammonia as an end product. Ammonia is quite toxic to animal tissue and must be diluted and excreted directly (ammoniotelism) or detoxified and excreted as urea (ureotelism) or uric acid (uricolytism). For ammoniotelic fauna, large quantities of water are used to dilute the ammonia, a luxury afforded only by animals with a virtually unlimited water resource such as aquatic animals. In terrestrial animals, where water conservation is of prime importance, the nitrogen is converted into a less toxic form and later excreted. In the formation of urea there is more hydrogen used to detoxify a given atom of nitrogen than in the case of uric acid (Chapman 1969); therefore it would appear that animals which operate under the greatest water stress would be uricolytic.

Urea is the most soluble of the three end products and uric acid the least (Fig. 10-1). Since the solubility of uric acid is very low, it precipitates from the urine

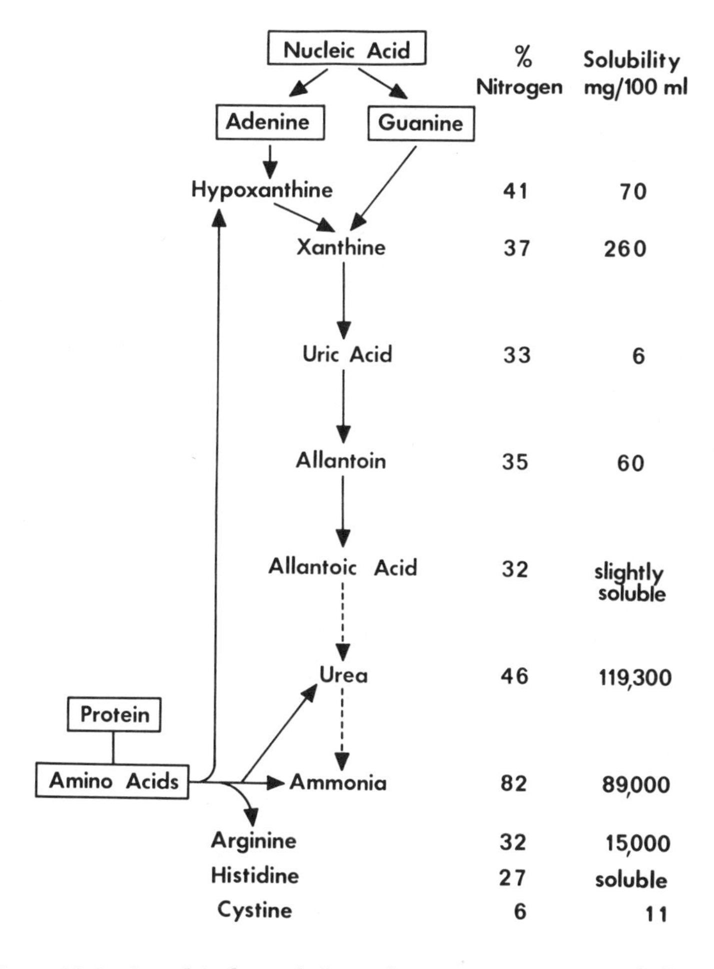

Figure 10-1 *Simplified metabolic pathways of nitrogen metabolism in animals and the relative solubilities of end products* [*Bursell 1967*].

solution in a crystaline form quite easily. As will be shown later, this precipitation proves to be quite useful to animals which cannot concentrate the urine above the isotonic level of plasma. Uric acid is generally formed through the catabolism of both nucleic acids and amino acids in contrast to urea and ammonia, which are formed from amino acid catabolism only (Fig. 10-1). There are certain insects in which the latter does not hold, and nucleic acids as well as amino acids form the ammonium end product.

NITROGEN EXCRETION FOR SPECIFIC ANIMAL GROUPS

Invertebrates

Nitrogen metabolism, because of its diversity in invertebrates, is a complex study. All of the common nitrogen excretory products are found in this group; in addition, a given species may shift from one form of nitrogen excretion to another according to the stage in its life history. With the exception of spiders and isopods, most terrestrial forms are uricolytic and most aquatic forms are ammoniolytic.

Wiser and Schweizer (1970) have shown that certain isopods excrete nitrogen via gaseous ammonia in contrast to the usual urine route. Because special mechanisms are required to volatilize ammonium in the unproteinated form, this type of ammonia excretion should be distinguished from the classical ammoniotelism of aquatic forms in which the ammonia is released in the proteinated form NH_4^+ (Speeg and Campbell 1968). The direct release of ammonia may be more common among terrestrial invertebrates than is now believed (Campbell et al. 1972).

Rao and Gopalakrishnareddy (1962) found that certain arachnids excreted guanine in addition to uric acid and, in certain cases, guanine was the major nitrogen end product. For scorpions and spiders, guanine represents 34-90 percent of the nitrogen excreted.

The picture is further complicated by the fact that terrestrial invertebrates shift from ammoniolytic to uricolytic or uricolytic to ammoniolytic during different stages in their life histories (Chapman 1969), or they may retain the nitrogen excretory products throughout their lives, or they may excrete nitrogen in a form somewhat unique to the animal group. The aquatic life stages tend to be ammoniolytic and the terrestrial life stages uricolytic. A superficial exception to this would be terrestrial animals which have almost a pure protein diet in a moist environment, such as the blow fly and bot fly larvae, which excrete almost pure ammonia. Nitrogen excretory products which are not excreted tend to be stored in various regions of the body according to the invertebrate group in question. In Collembola, *Periplaneta* and *Culex*, uric acid crystals are stored in fat bodies (Chapman 1969). In *Rhodnius*, uric acid accumulates in permanent stores, which contribute to the insect's color pattern (Chapman 1969). Similarly, *Pieris* accumulates 80 percent of the uric acid produced in the pupal instar and it is stored in wing scales.

In many insects subject to water economy, uric acid is excreted rather than ammonia. The uricotelic characteristic, an adaptation to a terrestrial life, was considered to be the common means of excreting end products of nitrogen metabolism in nonaquatic insects. It is now known that in a variety of insects a uricolytic pathway originating with nucleic acids produces at least allantoin and even allantoic acid, which are excreted as such. Other insects metabolize and excrete urea or ammonia from amino acids. Thus a generalization characterizing all insects as uricotelic is invalid (Craig 1960; Bursell 1967; Chapman 1969).

In those insects with complete metamorphosis, metabolic waste products (especially uric acid) accumulate during the pupal stage. When the adult emerges these waste products are discharged in the meconium. There is also some allantoic acid in the waste of the pupae of Lepidoptera and Hymenoptera, while urea has been found in the pupae of the dipteran *Phormia* (Chapman 1969).

In an extensive review by Bursell (1967) on the excretion of nitrogen in insects, summarized data show that excretory metabolism varies within the Class and that the state of knowledge is incomplete. The reactions of adenase and guanase deamination, the oxidative deamination of amino acids, the glutamic dehydrogenase system involved in ammonia formation, the origin of urea and its role in nitrogenous excretion and the pathway of uric acid synthesis are processes not yet fully understood. Much progress has been made recently, however, especially concerning uricolytic enzymes and xanthine dehydrogenase. A summary of Bursell's survey of research by others concerning nitrogenous end products by insect Order follows:

Collembola—Composition of excreta has not been determined; however, three of the uricolytic enzymes, uricase, allantoinase and allantoicase, have been found.

Orthoptera—Many excrete mainly uric acid, but some have been found to excrete mostly allantoic acid, others allantoin. There is no correlation of excretion products with the presence of the corresponding uricolytic enzymes. Uric acid deposited in cells of the fat body accounts for up to 10 percent of the total dry weight of some species and may serve as a nitrogen reserve.

Odonata—In aquatic larvae, ammonia contains most of the excretory nitrogen and is probably the chief end product of protein catabolism, while less than a 10th of the ammonia nitrogen appears as uric acid, which is probably the end product of purine metabolism.

Hemiptera—Allantoin has been found in the excreta of a number of species of plant bugs studied, which correlates well with the presence of a highly active uricase in this group. Amino nitrogen, urea and uric acid alone or in combination have been found in the excreta of various hemipterans studied.

Homoptera—Large quantities of amino acids have been found in the honeydew of aphids and coccids. These substances possibly are in excess of the insects' requirements and are voided without having been involved in any metabolic processes.

Coleoptera and Hymenoptera—In some species, uric acid is the dominant end product; in others, allantoin, allantoic acid or urea may be found and may dominate in some.

Diptera—Many of the adults studied were found to be uricotelic, but allantoic acid and allantoin were found in the excreta of some species.

Lepidoptera—Data on adult and larval excretory metabolism show little correlation. In some species the adults are uricotelic, but the larval excreta contain high amounts of allantoin; in other species the larvae are uricotelic, but the main end product of the adults is allantoic acid.

Reptiles

The excretion of nitrogen by reptiles is associated with both the associated metabolic pathways and the actual functional capacity of the kidney. The reptilian kidney is a simple tubule which may be classified roughly into proximal, medial and distal segments. Reptiles lack the ability to concentrate urine above the osmotic pressure of plasma (Coulson and Hernandez 1970). Therefore, with the precipitation of insoluble uric acid the urine is maintained at near plasma isotonic levels, while passing large quantities of nitrogenous wastes with little water loss.

Reptiles appear to reduce water loss via nitrogen excretion in two ways, production of uric acid and reduction of urine. Coulson and Hernandez (1970) have stated that most terrestrial reptiles are uricolytic, with a semisolid excreta containing 5-90 percent nitrogen. A high degree of water economy is achieved by the precipitation of urine and its excretion as a semisolid mass.

Land reptiles are not strictly uricotelic, as was once thought, and the relative proportions of uric acid and urea may vary greatly. Khalil and Haggag (1955) observed changes from predominantly uric acid to urea metabolism and vice versa for a given individual of *Testudo leithii*. Drilhon and Marcoux (1942) suggested that this shift in nitrogen excretion in land turtles might be a direct function of temperature and hydration; the formation of uric acid increasing when water balance was less favorable. The picture seems to be less clear for snakes (Cloudsley-Thompson 1971) and apparently unknown for lizards.

Birds

Birds appear to excrete all three of the common nitrogen end products, although uric acid may constitue 50-80 percent of the total nitrogen excreted (Sturkie 1965; Brown 1970). Ammonia appears to be the second most common nitrogen end product and may be 15 percent of the total nitrogen excreted (Brown 1970).

Urine production appears to be directly related to water intake. Under "normal" conditions (e.g., an adequate water supply) the urine usually is cream colored, contains thick mucusoid material and is abundant in uric acid (Sturkie 1965). Under water stress the urine becomes much thicker, and the uric acid concentration is increased. Under certain circumstances such as diuresis, the urine may become thin and watery.

Mammals

Mammals are ureotelic. Ureotelism requires a great deal of water in the formation and voiding of metabolic waste nitrogen; therefore mechanisms for water conservation are needed. One such mechanism for water conservation is the concentration of urea in the urine. This concentration is accomplished in the kidney by water reabsorption and results in a 100-fold increase in the urine over the blood plasma. Scheer (1963) has stated that desert mammals are able to concentrate urine to an even greater extent as a mechanism to further conserve water. Prosser et al. (1952) stated that urea concentrations in the blood and urine depend on protein intake and protein catabolism.

General Observations

The preceding information may be summed up in the following generalities: First, with few exceptions, ammonia excretion of waste nitrogen is performed by aquatic animals. Second, urea and uric acid are the main forms of nitrogen excretion in terrestrial animals. The terrestrial approach to nitrogen excretion is expensive from an energy standpoint, since any concentration effort is energy consumptive but does conserve water. Third, the amount of urine and/or nitrogen excretory products produced is a function of the animal's water balance and diet (mainly protein intake). Fourth, certain animals may shift from one nitrogen excretory product to another in response to some environmental stress, namely water or stage-of-life cycle.

CONSUMERS ROLE IN THE NITROGEN CYCLE

The role consumers play in the cycling of nitrogen can be examined from two views: the effect of fauna on the cycling rate of nitrogen and the effect of fauna on nitrogen distribution. This approach may be further subdivided into the faunal effects on nitrogen cycling in the "green food chain," as well as their effects on seeds, the predators' effects and the faunal effects in the detritus food chain.

Green Food Chain

Vertebrates

The "green food chain" is generally the group of most interest to the general public and those involved in resource management. This food chain usually includes cattle, sheep, rabbits, etc.; in other words, the food chain based on the above-ground standing vegetation.

In arid lands, green food chain consumers appear to have a significant impact on the cycling and distribution of nitrogen. This seems especially true for grazed arid range lands. There appears to be a certain enhancement of plant and sheep productivity on grazed *Kochia* and *Atriplex* ranges in northwestern South Australia (Trumble and Woodroffe 1954). Trumble and Woodroffe concluded that grazing increased nitrogen content in new plant shoots, and dung deposits enhanced the nutrient status of the grazed range. It was concluded that grazing accelerated the nutrient cycle and, given adequate soil moisture, increased the deposition rate and decomposition rate of plant litter.

Grazing animals return roughly 60-80 percent of consumed nutrient elements in herbage, and only small proportions of these plant nutrients are removed from the system as animal products (Cowling, 1977). The majority of the nitrogen returned to the soil is present in the form of urine. The excretion of nitrogen in feces is constant per unit of dry matter consumed, with an average near 0.8 g N/100 g forage (Cowling, 1977). Nitrogen in excess of the above amount is excreted mostly as urine, and the nitrogen concentration of the urine is a direct function of the consumed plant material.

Due to the metabolic pathways in the ruminant, nutrient elements assimilated in organic molecules of plant origin are released in feces and urine as smaller molecules or in ionic form. The reduction in molecular complexity tends to enhance decomposition rates of the fecal material. Thus, Barrow (1967) has demonstrated that plant material fed to sheep breaks down more completely than ground plant material which was mixed with soil and incubated. It was concluded that the rumen microflora and enzymes are responsible for more rapid breakdown than the soil microfloral counterparts.

Volatile losses from urine as ammonia result from hydrolysis of urea. Williams (1970) has stated that urea makes up 70-80 percent of the total nitrogen in livestock urine. Simpson (1968) and Watson and Lapins (1969) have shown losses of 60-80 percent of the nitrogen in urine or urea applied to the soil surface. They attribute these losses to urease activity at, but not below, the soil surface.

Vercoe (1962) determined energy and nitrogen losses in urine and feces of Merino wethers grazing on Wimmera ryegrass (*Lolium rigidum*) and subterranean clover (*Trifolium subterraneum*). Daily loss of nitrogen from the urine and feces varied 6.8-25.1 g and 3.5-9.4 g, respectively. However, the validity of these data are questioned by Vercoe because the collection technique may constitute an interference with normal grazing behavior. There seems to be no doubt that nitrogen loss in pastures from urine is considerable. According to Watson and Lapins (1964, 1969), for every 100 kg of herbage nitrogen ingested by sheep, 45 kg or more can be lost mainly from urine nitrogen volatilization. This amounts to a daily loss of nitrogen from urine of as much as 200,000 kg during the winter and 100,000 kg during the summer for 22.5 million sheep in western Australia alone.

The fate of urine in soils is rapid decomposition; ammonia is produced and alkalinity increases in consequence. A 12 percent loss of urine nitrogen as ammonia from the turf was detected by Doak (1952). Simpson (1968) determined the amount of loss to be expected when urea is applied to pastures under intermittently dry conditions; presumably, comparable nitrogen losses would occur from urine under the same conditions. Simpson found that large losses, up to 60 percent of the nitrogen applied, occurred only when the urea was applied to the soil surface. Losses were less under cooler conditions, i.e., under 25 C, or when the urea was applied below the soil surface. Nitrogen losses were reduuced from 60 percent to 46 percent without the top centimeter of soil and to 33 percent loss without the second centimeter of soil. With the accumulation of organic matter in pastures, there is a resultant surface concentration of urease activity which is mainly responsible for the high nitrogen losses on and just below the soil surface.

McGarity and Hoult (1971) found ureolytic activity not only in soil supporting pasture sods, but also in plant and litter components. They suggest that ureolytic phylloplane microorganisms or urease might be present on the plants (viz., Italian ryegrass in these experiments) and may be responsible for ammonia volatilization.

Soil nitrogen may be augmented as a result of dung decomposition. Wood (pers. commun.) as cited in Cowling (1977) has stated that sheep pellets placed in the field increased in nitrogen content over a period of time.

In North American arid lands, the jackrabbit (*Lepus californicus*) is probably one of, if not the most, important native vertebrate grazer. In the Chihuahuan Desert, jackrabbits may annually process roughly 7 percent of the standing nitrogen crop. This calculation is based on 1.34 rabbits/ha (Whitford et al. 1973); 2 kg mean adult live weight (French 1959; Stoddart 1972); mean consumption rate of 40 g (dry)/day per kg of rabbit; assimilation of 0.35 and 10 g of plants cut and dropped/rabbit per day (Shoemaker et al. 1973); and a mean nitrogen concentration of 8.7 percent. In addition, it was calculated that 3.9 percent/year of the total nitrogen standing crop appeared as rabbit feces, and 1 percent is cut and dropped directly into the litter annually. In effect, the jackrabbits are annually "short circuiting" roughly 5 percent of the total nitrogen standing crop to the soil to undergo decomposition and recycling. The cut but uneaten portion will probably decompose at the same rate as the material which undergoes "normal" mortality. The fact that cuttings are involved in the decomposition processes earlier than the other input would enhance the nitrogen cycling rate to a small extent. The 4 percent of the nitrogen standing crop which is deposited as feces will undergo accelerated decomposition and will be available for recycling sooner. Although jackrabbits annually short circuit only 5 percent of the total nitrogen pool in the above-ground portions of primary producers, this may be significant in a nitrogen-limited system.

Herbivorous reptiles in the American deserts are represented by the genera *Sauromalus* and *Dipsosaurus*. *Sauromalus* and *Dipsosaurus* feed primarily on flower parts of desert plants when available and shift to other plant structures or carnivory during the remainder of their active period (Johnson 1965; Norris 1953).

It is difficult to assess the impact of herbivorous lizards on the rate of nitrogen cycling, since very little is known about their feeding rates. However, based on available information from Johnson (1965), *Sauromalus* probably consumes less than 1 percent of the total nitrogen standing crop annually. Possibly *Disposaurus* may consume more, since Norris (1953) has stated it is one of the more abundant reptiles in the southwestern United States. Obviously, the role these animals play in nitrogen cycling and nutrient cycling in general is an area for further research.

Invertebrates

Although few studies have been conducted on the impact of green food chain invertebrate consumers on desert vegetation and the subsequent impact on nitrogen cycling, some inferences may be drawn from work performed on grassland ecosystems. It has been found that the invertebrate green food chain consumers are quite important in grassland systems (Blocker 1969). Grasshoppers consume a large percentage of the vegetation standing crop in local areas (Mulkern et al. 1964; Kelly and Middlekauff 1961). In addition, the Diptera may cause extensive seed damage to certain grass and annual species (Watts and Belloti 1967; Crawford and Harwood 1964; Starks and Thurston 1962; Schwitzbegel and Wilbur 1943). Hemiptera and Homoptera also appear to consume large quantities of the annual grassland vegetation and may be two of the more important consumer groups in grassland systems (Crawford and Harwood 1964; Osborne 1912 and 1939).

Leafhoppers and grasshoppers may forage out across the desert but return to resting places in desert shrubs. They can therefore concentrate frass and thus nitrogen around the more mesic and fertile shrub microsites (Daubenmire 1975).

Granivores

Invertebrates

In arid lands harvester ants may be quite significant in their effect on the cycling rate and distribution of nitrogen. Gentry and Sitritz (1972) found that, in an old field of the southeastern United States, the harvester ant *Pogonomyrmex badius* tended to concentrate nutrients in the mounds around the nest hole. This increase was probably caused by deposition of debris by "spring cleaning" of the nest system in March and April, producing a pulse of waste material deposited on the mound. The resulting soil enrichment accelerated growth and increased the standing crop of plants immediately surrounding the mound. It was estimated that 4,736 individuals were in a colony with 10-15 percent as foragers.

Whitford (1973) observed intense local (< 2 m) foraging of the ants *P. rugosus* and *Novomessor cockerelli*, although there was substantial foraging up to 12 m from the entrance to the nest. He also observed extensive use of the seeds from the plant species preferred as food.

Whitford calculated that 90 percent of the seeds produced by *Eriogonum trichopes* and 71 percent of the *Bouteloua barbata* seeds were removed by ants. This represents about 80 percent of the total seed production of the two species and, likewise, represents 80 percent of their total seed nitrogen pool. The nitrogen flux, based on the above two species alone, represents 3.2 kg N/ha per year, which is equivalent to an annual flux of 5.8 percent of the total nitrogen pool to the harvester ants. This nitrogen-containing material is taken to the nest and reintroduced to the system in pulses due to nest cleaning. The amount of decomposition the seeds undergo in the nest, if any, is unknown; consequently it is difficult to ascertain if harvester ants actually accelerate the nitrogen cycling rate or depress it. However, there can be little doubt that they are a significant influence in nitrogen redistribution.

Vertebrates

Rodents also play a role in the cycling and redistribution of nitrogen. There has been little done in this respect, although some work by Whitford et al. (1973) indicates rodents may be important. The seed-eating rodents annually process about 3 percent of the total nitrogen pool in the Chihuahuan Desert of New Mexico, as compared to 5.8 percent for the harvester ant. Although the rodents annually process less nitrogen, it should not be assumed that their importance in the nitrogen cycle is proportionately reduced. Data are unavailable at this time to calculate the nitrogen turnover rate in rodent populations; however, it may be somewhat slower than the jackrabbit and more rapid than the harvester ant.

The above calculation of 3 percent was based on the following assumptions: mean seed weight used by rodents equals 3.5×10^{-3} g (Franz et al. 1973); mean nitrogen concentration in seeds is 3.05 percent; there are 16.5 rodents per hectare, with a rodent biomass of 715 g/ha (Whitford 1973); 73 g seed are consumed per gram of rodent per year (calculated from Franz et al. 1973).

The role birds play in nitrogen cycling is very difficult to assess; therefore it is not surprising that very little information exists on this subject. Most, if not all, of the information on desert birds in the literature is in terms of breeding pairs per unit area or biomass of same. The total area over which a given breeding pair feeds is difficult to measure and is seldom reported. In addition, the feeding rates of many desert birds are unknown.

It is possible that desert birds may have an influence on the nitrogen cycle and distribution other than feeding. They may also influence the nitrogen dynamics through nesting activities. Desert birds construct their nests from various kinds of desert plant litter which, no doubt, slows down the decomposition rate of this nesting material due to its remoteness from soil decomposers and to a generally drier environment. This would slow down nitrogen reentry into the cycle. Although this effect is probably slight, it is nevertheless an unknown.

Predators

The role predators play in nitrogen cycling and redistribution is probably qualitatively constant for vertebrates and invertebrates alike, although there likely are quantitative differences. The predator's role in nitrogen dynamics is somewhat similar to the herbivore's in that the predators act as a short circuit in the cycle, only in this case the predators are a short circuit of a short circuit. Where the primary consumers cause plant material to enter the decomposition process more quickly through feeding activities as well as by converting the plant material into a more easily decomposed form, the predators do the same thing to the consumer tissue. This activity results in a decrease in the time nitrogen will be bound in the consumer's tissue, therefore further expediting the rate of nitrogen cycling.

As is the case with so many desert animals, it is difficult to quantify the role predators play in nitrogen cycling and redistribution. Based on few available data, predators are considered not as quantitatively significant in nitrogen cycling and redistribution as the herbivores. This does not imply, however, that predators are less important in nitrogen dynamics than herbivores.

Decomposers

How do consumers influence the cycling rate and distribution of nitrogen? In the case of grazing animals such as sheep, 95 percent of the total nitrogen consumed is returned to the system in the form of urine or feces (West, in press b). The returned urine nitrogen in the form of urea is rapidly volatilized or hydrolyzed into ammonium carbonate, a form which is available to the primary producers (Delwiche 1970). The fecal material deposited is decomposed more rapidly than plant litter (West, in press b). In addition, the nitrogen contained in the urine of the consumers is broken down and reenters the cycle more rapidly than the nitrogen of plant litter (Skujins 1974, pers. commmun.).

The rapid decomposition of the fecal material may be due to processes described and subsequently labeled the "external rumen" analogy by Mason and Odum (1969). In the course of passage of the plant material through the digestive tract of the consumer, it is attacked by the rumen bacteria as well as by strong digestive fluids; consequently the vegetal material has undergone a certain amount of decomposition before it is deposited on the desert floor. After initial breakdown, decomposition is enhanced as a result of an increased concentration of the nutrient substrate for the decomposer food chain. In addition, the high volume:surface area ratio of the fecal material allows a greater moisture-holding capacity. Since it has been demonstrated that moisture may be a limiting control on decomposition in arid systems (Comanor and Staffeldt, Chapter 4 of this volume), it would seem that increased water content would, in turn, increase the decomposition rate. The large volume:surface area ratio allows for a relatively constant and often warmer temperature, thus enhancing the decomposition rate in a manner grossly similar to digestion in the rumen.

Faunal communities associated with litter breakdown are difficult to study because of their complex structure and function, and because little is known about their feeding habits and taxonomy. Overall knowledge of the natural history of desert soil and litter invertebrates is woefully lacking. The importance of below-ground litter invertebrates and ants in nitrogen cycling is probably greater than currently thought. What little work has been done demonstrates that this group redistributes large quantities of plant material (Dregne 1968; Went et al. 1972).

There is little information published on the dynamics of litter breakdown by fauna in the deserts. However, some generalities may be made dealing with this subject. First, as with forested soils, the most important contribution of small fauna is probably the fragmentation of litter through mechanical breakdown by passage through the gut. Ingestion also serves to inoculate the litter with microflora and to incorporate some of the litter with mineral soil. These processes have been documented in forested systems (Edwards et al. 1970; Raw 1967; Ghilarov 1967; Kevan 1962; Crossley 1970).

Termites appear to be one of the major animals involved in litter removal and incorporation of fragments into the soil in arid lands. However, it appears that this may retard the rate of nitrogen cycling. Lee and Wood (1971) show that the control by termites over the rate of nitrogen cycling depends on whether or not they build mounds. The mound may make nitrogen unavailable for 100 years or more. The non-mound-building termites appear to enhance cycling to the point that the Dinka tribes in the Sudan depend on it. These Dinka tribesmen follow the practice of felling trees and shrubs, stacking them in fields. Within a few months, the woody material is destroyed by termites and the nutrients are released into the soil, thus enhancing the fertility of poor or depleted land (Lee and Wood 1971). The mound-building termites appear to redistribute and concentrate nitrogen. Lee and Wood (1971) observed that 4.6-5.3 percent of the total nitrogen (in the form or uric acid) was incorporated in these mounds and it was bound for the life of the colony. Moore (1969) observed the recycling of nitrogen in termite colonies through eating of dead, injured or surplus members in a given caste and, in cases of severe nitrogen stress, by quite extensive cannibalism. Kozlava (1951) stated that suitable conditions of temperature and moisture in mounds favored mineralization of nitrogen and that these nitrogen-rich mounds were used as fertilizer in central Asia.

Since the termite feces are used in mound construction, there is virtually no loss of nitrogen while the colony is living (5-100 years). After colony death, however, the mound will erode away and all contained nutrients will be returned to the soil (Lee and Wood 1971).

Termites in the Sonoran Desert consume approximately 7 percent of the total annual woody litter production (Thames et al. 1973; Nutting et al. 1973), and it can be assumed that a like fraction of the nitrogen in woody litter is also consumed. In addition, based on data from Nutting et al. (1973) and McBrayer et al. (1973), it was calculated that the nitrogen pool in termite populations had a turnover of roughly 17 days in a Sonoran Desert community. This does not reflect the turnover of nitrogen, since many species recycle all the organic material either as a food source or a structural material in cast construction.

Internal nitrogen fixation has been proposed as another nitrogen source for termites. Benemann (1973) measured nitrogen fixation rates of 24-566 μg/month per gram (wet weight) of *Kalotermes minor*. Breznak et al. (1973) reported a lower termite fixation rate and postulated that nitrogen-fixing bacteria or their metabolic products may be a direct nitrogen source for termites. However, this nitrogen pathway requires further investigation before its importance can be assessed.

The termite cycling of nitrogen in the Sonoran Desert may be more rapid than in some other arid lands since these termites are not mound builders and cast their fecal material from the nest into the surrounding soil (Haverty 1974, pers. commun.). Naturally the soil decomposers readily attack this material and mobilize the nitrogen.

The dominant environmental parameter controlling activities of soil arthropods appears to be moisture (Gist 1972; Blower 1955; Cole 1946; Maynard 1951; Kunhelt 1955; Hale 1967; Lowrie 1942; Mullin and Hunter 1964). Therefore one would expect soil arthropod numbers to increase during the more moist portion of the year and to be greatly reduced during dry periods, especially in arid lands. Because of this considerable pulsing, it is difficult to assess the reported populations of desert soil arthropods unless the time of year is clearly stated. At any rate, the populations are no doubt lower than those in the temperate forest (Table 10-1).

Since rain comes to deserts in a basically random manner from year to year and in a somewhat cyclic rate within years (Noy-Meir 1973), soil arthropod activity responds in a like manner. One could look at the activities of the soil arthropods in litter decomposition as a broad-peaked spike, in that they probably are active at lower soil moisture than the microbes associated with decomposition (narrow-peaked spike). Because of this broader tolerance and increased mobility as compared to the microbes, soil arthropods can incorporate litter fragments from the dryer soil surface into the lower, more moist, soil regions, thus enhancing the rate of microbial decomposition. Information about the amount of litter decomposition and subsequent impact on nitrogen cycling is presently incomplete; however, the data generally indicate that harvester ants and termites are more important than soil arthropods in the general

Table 10-1 *Comparison of Densities [N/m^2] of Soil Microarthropods in Temperate Forests to Arid Lands*

Community	Source	Collembola	Mesostigmata	Cryptostigmata
Mixed hardwood	Gist 1972	4,492	202	22,433
Liriodendron	Moulder et al. 1970	19,153	4,177	17,153
Beech	Kitazawa 1967	780,000[a]	---	90,000[a]
Beech	Bornebusch 1930	1,383[a]	---	3,049[b]
Oak	Bornebusch 1930	493	---	967
Desert steppe	Wood 1971	1,050	40	320
Mohave Desert	Wallwork 1972	110	20	720

[a]Includes mineral and litter horizons.

[b]All mite groups.

nitrogen cycle in arid lands. Hopefully, further studies will firmly establish the importance of soil fauna in nutrient cycling.

SUMMARY

It appears that consumers generally decrease the cycling time of nitrogen through their feeding activities. This decrease is accomplished in two ways; early introduction of plant material into the decomposition process through feeding and transformation of the consumed material into a more easily decomposed form. In the case of consumers in the decomposition process, nitrogen cycling is enhanced by mechanical breakdown of litter fragments, inoculation of these fragments with microflora and processes referred to as the external rumen. Based on data presented here, it appears that the enhancement of the nitrogen cycle may not be significant to the desert community. This is especially true in nitrogen-limited arid lands.

However, there is a price for the enhancement of the nitrogen cycle through the loss of nitrogen via volatilization. Nitrogen loss from volatilization appears to be heaviest where vertebrates are concerned and may represent as much as 84 percent of the nitrogen consumed.

11

PHYSICAL INPUTS OF NITROGEN TO DESERT ECOSYSTEMS

N. E. WEST

INTRODUCTION

Elaboration of the complete nitrogen cycle requires consideration of the amounts of nitrogen in the atmosphere, the regolith, the biota in or on the earth's surface, and fluxes to and from these compartments. Ecologists have stressed the biological aspects and have generally overlooked the importance of physical contributions in mineral cycling, especially atmospheric inputs (Bormann 1969). Ingham (1950) points out that 98-99 percent of the biomass of terrestrial green plants is ultimately derived from the air. Growing sophistication of chemical analysis techniques has allowed more attention to atmospheric sources of both vital elements and pollutants. Since nitrogen can fit both categories, it behooves us to broaden our understanding of the cycling of this element in all kinds of systems, including deserts.

All nitrogen in biomass was originally in the atmosphere. Thus, in considering the nitrogen cycle, we think first of gaseous nitrogen and its fixation by various specialized microorganisms. This topic has, of course, received considerable attention. For desert contexts both free fixation (Rychert et al., Chapter 3 of this volume) and symbiotic fixation (Farnsworth et al., Chapter 2 of this volume) have been reviewed. Much less attention has been directed to nitrogen contributions by the purely physical processes of precipitation, dry fallout and leaching from plant surfaces.

Liebig (1855) believed that the nitrogen in soils was derived from the small amount of ammonia in the air and rain. Much data in apparent support of this contention were accumulated in the latter half of the 19th century, but reinterpretation became necessary when the new science of microbiology came into being. Henceforth, it has been assumed that additions of nitrogen to the soil were primarily due to biological and not to physical causes (Ingham 1950). Let us now take another look at these physical inputs.

TYPES OF INPUT

Precipitation

Precipitation is here considered to be all liquid and solid materials that are pulled by gravity from the atmosphere to the earth's surface. The scrubbing action of wet precipitation in the forms of rain, snow, sleet, hail, grapple, fog drip, rime, frost and dew, scavenges various aerosolic particles, larger particles and gases from the atmosphere (Eriksson 1952). Categories involved in wet removal are rainout (the collection of aerosols in the formation of rain and snow); washout (the capture of aerosols by falling rain and snow); and chemical reactions between aerosols and hydrometeors (Hidy 1970). Hermann and Gorham (1957) contend that rainwater contains more nitrogen than snowfall, whereas Barica and Armstrong (1971) state that snow is a more efficient scavenger of nitrogen than rain. Regardless of the resolution of this argument, one is led to the conclusion that the relative amounts of the different types of precipitation can be important.

Dry Fallout

Dry fallout (also known as dry precipitation or atmospheric sedimentation) is also an important source of nutrient input, especially in arid regions (Drover and Barrett-Lennard 1953; Junge 1958). Dry processes include sedimentation, inertial deposition (impaction), diffusional deposition on obstacles in the air stream, collision with other particles and chemical reactions on existing particles.

Bulk Precipitation

Deposition of particulates contributed by rainout, washout and sedimentation is often measured by capture in open funnels or pans and is commonly lumped under the term "bulk precipitation." The recent work of Schlesinger and Reiners (1974) is an example. Deposition of particulates on obstacles in the airstream involves a number of microphysical processes, including electrostatic attraction, diffusophoresis, thermophoresis and impaction. The relative importance of each process is controlled by several variables, especially particle size. Because data on particle size are usually lacking, "interception" is used as a general term for capture of aerosols from the airstream by obstacles. The term "deposition" can be used to describe the combined capture of water and dissolved material through bulk precipitation being intercepted (Schlesinger and Reiners 1974).

Leaching

Leaching is here used to describe the removal of chemicals by water flowing over plant surfaces. The material removed may be either particulate or dissolved, and while it is possible to filter out the particulate fraction, without isotopic tracers it is impossible to determine whether the soluble forms of nitrogen encountered are derived from the intact plant or from the dust and organic debris collected on the surface of the plant.

PROCESSES

Precipitation can scavenge a number of forms of atmospheric nitrogen produced by various natural and man-directed processes. Electrochemical fixation reactions are one possibility, with lightning providing the energy for oxygen to unite molecular nitrogen with oxygen to form various nitrogen oxides. Popular opinion once held lightning to be the greatest source of fixed nitrogen obtained from the atmosphere, but this has been shown to be doubtful (Uman 1971). The modern view is that only about 8 percent of the total nitrogen fixed around the world each year is due to electrochemical fixation in the atmosphere (Delwiche 1970). Much more nitrogen is probably fixed by photochemical production of NO, N_2O and NO_2, while the related process of photo-oxidation of NH_3 may also contribute minor amounts. Other minor but original sources of nitrogen in scavengable forms are produced by cosmic radiation, volcanic activity and meteor trails.

A fourth and increasingly important source of nitrogen, which may contribute about 30 percent of the nitrogen fixed over the earth, is due to industrial activity (Delwiche 1970). Most of this is from intentional fertilizer manufacture, but some is due to volatilization of various nitrogen-containing gases from industrial stacks and engine exhausts. Monitoring remains to be done to see if these sources are important in deserts. Although fertilization of desert rangelands is rare, and great distances exist between most deserts and urban areas where industrial pollutants are concentrated, there may still be some movement of nitrogen-containing pollutants into deserts (e.g., the Mohave Desert in southern California). Recent construction of coal-fired electric generating plants in U.S. desert regions may well be augmenting the sources of input.

Dust is an important and perhaps increasing source of nitrogen in desert environments. Aerosols and larger particulate matter can be moved long distances. It is not known whether the net input exceeds or equals the net output in desert regions, since what is picked up at one place must be deposited elsewhere. Because of the generally greater proportion of bare surface area in these regions, one would expect that deserts are a source rather than a sink for nitrogen and nitrogen may be transported to other environments even though local input appears to be an addition.

Rainfall absorbs nitrogen, particularly ammonia. Similarly atmospheric ammonia can be absorbed by wet plants, soils, snow, frost, fog or dew, organic matter and clay minerals (Ingham 1950). Plants have recently been shown to take in ammonia through stomata and metabolically capture some directly (Hutchinson et al. 1972).

INPUTS IN DESERT ECOSYSTEMS

Precipitation and Dry Fallout

Very little work has been done on the quantitative significance of these various physical inputs to desert ecosystems; the bulk of the foregoing information having been derived from more general studies in urban or agricultural contexts in forest and grassland regions. What little information there is pertaining to deserts is usually found in supporting data for studies of the biological aspects of nitrogen cycling. Few workers have analyzed the amounts of various forms of nitrogen falling in precipitation over desert regions.

Eriksson (1952) reported that 0.6 mg N/liter of rainfall came as ammonium and 0.1 mg/liter as nitrate at Tashkent, USSR. Yaalon (1964) found that the concentration of ammonia in rainfall at desert stations in Israel varied between 0.16 epm (equivalents per million) in the cold winter months and 0.68 epm in the warmer

Table 11-1 *Seasonal Inorganic Nitrogen Concentrations [ppm] in Rainfall at Four Stations in Desert Regions, 1960-1966 [Lodge et al. 1968]*

Season	Stations				Average
	Pocatello, Idaho	Ely, Nevada	Grand Junction, Colorado	Winslow, Arizona	
Dec-Feb	0.21	0.07	0.21	0.38	0.22
Mar-May	0.27	0.00	0.13	0.03	0.11
Jun-Aug	0.17	0.73	0.36	0.32	0.40
Sep-Nov	0.17	0.27	0.21	0.05	0.18

spring months. He suggested that most of the ammonia evolved from high pH soils, particularly when the warmer temperatures stimulate the ammonification processes. Nitrate concentrations in rainfall were lower and less variable than ammonia; the annual average being 0.04 epm. Junge (1958) found much more ammonia coming from regions of alkaline soils over the United States, while high proportions of nitrate were exclusive to regions of acid soils. Lodge et al. (1968) also found that total inorganic nitrogen in rainfall collected at several stations in the semiarid western United States varied somewhat between seasons (Table 11-1).

All of the aforementioned studies are based only on rainfall, and thus are probably underestimations due to omissions of the dry fallout component (Chapin and Uttormark 1972). Junge (1958) estimated that 70 percent of the atmospheric nitrogen contribution in arid climates comes from dry fallout. Since nitrogen content

Table 11-2 *Summary of Estimates of Nitrogen Inputs in Precipitation at Stations in Desert Regions [kg N/ha per year]*

Station	Nitrogen input				Source of data
	NH_3	NO_3^-	NO_2^-	Total N	
Tashkent, USSR	2.5[a]	0.4[a]	---	2.9[a]	Eriksson (1972)
Salt Lake City, Utah	5.56[a]	0.40[a]	---	6.05[a]	Eriksson (1972)
Tbilisi, USSR	2.37[a]	0.80[a]	0.30[a]	4.62[a]	Bobritskaya (1962)
Grand Junction, Colorado	0.6[a]	1.3[a]	---	---	Junge (1958)
Ely, Nevada	1.0[a]	0.7[a]	---	---	Junge (1958)
Snowville, Utah	0.5[a]	2.5[a]	tr.	11.9[b]	Skujins and West (1974)

Note: --- indicates data not available.

[a]Precipitation only, excluding dry fallout.

[b]Includes dry fallout.

is highest at the start of a storm, if the collection is started a little late it will intensify the underestimation (Chapin and Uttormark 1972).

Skujins and West (1974) found that about 75 percent of the total nitrogen was in particulate matter collected in automatic wet and dry gauges at Curlew Valley, Utah. The heaviest input was as nitrate N after spring dust storms. Drover and Barrett-Lennard (1953) also implicated the heavy input of particulate matter in western Australia.

Past workers have taken rainfall data and multiplied by the nitrogen concentrations to make estimates of yearly inputs. Some examples of this include Eriksson's (1952) review of European data and that of Chapin and Uttormark (1972) for the United States. Table 11-2 summarizes the data for stations in United States desert regions and adds Tbilisi, USSR, from Bobritskaya's (1962) data.

Skujins and West (1974) calculated the total nitrogen input via precipitation and dry fallout over a shrub steppe region in northern Utah to be about 12 kg N/ha per year. Of this total input about 21 percent was nitrate N, 4 percent as ammonia N, trace amounts of nitrite N and the remainder as particulate matter. If as much dust is assumed to blow off an area as blows on, then only about 4 kg N/ha per year could be assumed to be the true input. On the other hand, if more blows off than on, a net loss could result.

Leaching from Plants

No prior work on leaching of nitrogen from desert plants or litter could be found. The only relevant observations are apparently those of Skujins and West (1974), who used simulated rainfall on two species of cool desert shrubs in Utah. Leaching, although significant for the experimental conditions used, is probably insignificant in nature because of the low frequency of storms intense enough to produce leaf drip and stem flow. These conditions did not occur on average-sized, mature plants until after 4 minutes of a simulated 5 cm/hour intensity rain. Particulate matter contained almost all of the nitrogen released. Thus, if there is no net input of dust, leaching cannot be considered an important nitrogen input to desert soils. This pathway cannot be an absolute input, but only another local circulation pathway.

CONCLUSIONS

Net input from precipitation accounts for 4 to 6 kg/ha per year of new nitrogen coming into desert ecosystems. This input is low compared to the precipitation input to other systems (Stevenson 1965; Chapin and Uttormark 1972); however, the uptake of nitrogen into new desert plant growth is also low on a per-unit-area basis (Wallace et al., Chapter 9 of this volume). The relative importance of physical input is thus comparatively high. Depending on how active free or symbiotic fixing microorganisms might be (Farnsworth et al. Chapter 2 of this volume; Rychert et al., Chapter 3 of this volume), these physical inputs could account for between 20 and 50 percent of the total input of new nitrogen to desert ecosystems. Since losses are also high due to wind and water erosion (Fletcher et al., Chapter 12 of this volume) and gaseous transformations (Westerman and Tucker, Chapter 7 of this volume; Klubek et al., Chapter 8 of this volume), net gains are unlikely. This budgeting also depends on where the boundaries of the system are drawn. Plant uptake and leaching could be considered merely circulation processes within a larger system. Regardless of system size, leaching of nitrogen from plants appears to be of negligible importance.

Physical input processes are collectively important in the nitrogen cycle of deserts because they are "free" work nature does to replenish the nitrogen in the system. Although losses may be quantitatively small, man cannot economically supplement the generally small export of nitrogen in animal protein and fiber taken from grazing over vast areas of arid to semiarid landscape. Any atmospheric input goes toward countering these losses, plus erosional and gaseous losses.

12

EROSIONAL TRANSFER OF NITROGEN IN DESERT ECOSYSTEMS

J. E. FLETCHER, D. L. SORENSEN and D. B. PORCELLA

INTRODUCTION

Erosive Factors

The erosive action of wind and of water is recognizable wherever deserts are found. Wind is considered the most important means of erosion in arid areas, whereas water erosion is the more important form in semiarid areas (Marshall 1973). Erosion affects all aspects of soil structure in desert ecosystems (Dregne 1968) and plays a significant role in the translocation of nitrogen, an element which is an important limiting factor to production in deserts.

Because soil particles seem easily translocated by the actions of wind and water, it is logical that nitrogen and other nutrients (especially as part of biological material which is less dense than the inorganic soil particles) are significantly affected by erosional processes in desert ecosystems. The question of whether nitrogen is actually lost from desert ecosystems by erosional transfers, rather than merely translocated, depends on interpretation of the boundaries of the ecosystem and the magnitude of the erosional event. Wind may move soil and its associated nitrogen from the desert floor to the nearest mountain range kilometers away, or only a few centimeters to the nearest hummock. Water may erode soil and thus nitrogen, and carry it to the nearest lake or river, or it may be deposited in the nearest rill, gully or playa. Human uses and ecological management procedures can result in significant changes in desert erosion whether by wind or water (Branson et al. 1972; Sturges 1975).

This chapter attempts to address the problem of erosional nitrogen transfer in deserts, using a general approach that hopefully can be applied to most world deserts. Data presented are the results of studies conducted by the authors in desert areas of the western United States.

Methods used by the authors consisted of field measurements and calculations with models. Field measurements were performed using steel-frame plot boundaries around selected sites with a known area (>1 m^2) having a catchment basin. Runoff generated by natural precipitation or with a rainfall simulator using deionized water ("rainulator," see Dogan 1975) was studied. Sediments carried in runoff were measured by dry weight of the filtered sediments and nitrogen was measured by Kjeldahl or nitrogen analyzer techniques (Sorensen et al. 1974).

Soil Surface Accumulation of Nitrogen

It is interesting that water, the most growth-limiting factor in the desert, should be a vehicle for transporting and removing vital nitrogen from the ecosystem. Microbial nitrogen fixation at or near the surface of many desert soils is dependent on soil moisture, and the fixation results in an accumulation of nitrogen in surface materials (Cameron and Fuller 1960; Mayland et al. 1966; Dregne 1968; Rychert and Skujins 1974b; and Sorensen et al. 1974).

Charley and Cowling (1968) have shown that although the concentration of nitrogen remains comparatively low, there is a greater proportionate accumulation of nitrogen and organic matter in the surface soils of arid areas as compared with subhumid and humid type soils. The organic carbon and nitrogen concentratioons in arid soils rapidly diminish with increasing depth. Arid area surface soils have carbon:nitrogen ratios usually between 6 and 14, with lower soil horizons having ratios of 2 or less (Porcella et al. 1973; Balph et al. 1973). The surface horizons in forested soils in humid to superhumid areas frequently have carbon:nitrogen ratios in excess of 50 with little vertical change, except in the deepest horizons. Grassland soils have carbon:nitrogen ratios between 14 and 20 with very little change into the lower horizons. In forested (humid to superhumid) zones, mineral nitrogen is so leached from the soil as to be rarely detectable, except in surface litter. In grasslands, where the precipitation effectiveness (PE) index drops below 60, the mineral nitrogen accumulates in the A horizon of the soil with all other horizons remaining near zero (Martin and Fletcher 1943). This is also true of arid soils. Data presented by Balph et al. (1973) show that nitrate concentrations are usually not greater than 2 ppm in Curlew Valley soils below 20-50 cm, but concentrations of nitrate may be as high as 30 ppm at or near the surface.

This accumulation of organic and mineral nitrogen at the surface of arid area soils apparently results from litter and microbial accumulation of nutrients as plants translocate nutrients from deeper horizons of the soil to the surface and as microbial nitrogen fixation proceeds. Erosional processes at the surface could result in a net loss of biologically important nutrients from the desert soil system. In undisturbed soil systems, the nitrogen fixed by microorganisms could be mineralized, nitrified and leached into the root zone and become available as a plant nutrient. In contrast, the nitrogen in arid system surface soils (probably mostly biological material or mineralized nitrogen closely associated with the soil particles in the form of ions on exchange sites) may be selectively carried off in runoff or by winds.

WATER EROSION OF NITROGEN

The forces associated with falling rain and/or flowing water may dislodge the surface soil, algal crust and litter, and transport it and its associated nitrogen away from sites where the nitrogen would be available for plant growth. Besides providing forces for erosion, water also carries certain amounts of dissolved nitrogenous compounds as it flows across the surface.

Water erosion phenomena in deserts may be thought of primarily in terms of the upland phase of the sediment yield process as described by Bennett (1974). Thus, the erosional process is inseparably related to individual precipitation or snowmelt events. Sheet and rill erosion compose the first and second stages of upland erosion, respectively, and are primarily responsible for nitrogen transport since they cause suspension of the surface materials. This transport of surface materials and nitrogen is accomplished by detachment of particles by raindrop impact and runoff water, and transport by raindrop splash and runoff. The erosion by snowmelt is accomplished by the detachment and transport of surface material by runoff water only, and may be more or less severe. However, because the total magnitude of runoff waters associated with snowmelt may be greater than other runoff events (at least in the Great Basin Desert), the total sediment load from snowmelt runoff may also be greater.

Probably of greatest importance in soil and thus nitrogen loss in arid to semiarid environments is water erosion associated with the extreme hydrologic events resulting from thunderstorms. Comparatively small storms can remove considerable amounts of nitrogen-rich organic matter (Gifford and Busby 1973). The accumulation of such materials on desert playas is evidence of these transfers. More widespread extreme storms can cut arroyos and drop water tables, thereby triggering desertification of heavily grazed grasslands (Hastings and Turner 1965).

If bare soils are subjected to long-duration, uniform-intensity rain, the erosion rate rises to a very high level as runoff proceeds and then decreases gradually as runoff continues. The high surge represents the material that was unattached plus the material loosened by the beating action of the rain. As the surface detention increases, the surface of the soil is covered with a deeper cushion of water and the splash detachment of soil decreases. A typical curve for a simulated 1-hour duration, 7.6 cm/hour and a similar curve for a 3.8 cm/hour simulated rain are shown in Figure 12-1. The two density curves of Figure 12-1 are shown in a cumulative form in Figure 12-2. Similar curves were obtained using Curlew Valley runoff plots and measuring runoff from a "rainulator" (Dogan 1975). By determining such erosivity relationships it is possible to develop models for estimating sediment eroded in relation to other factors.

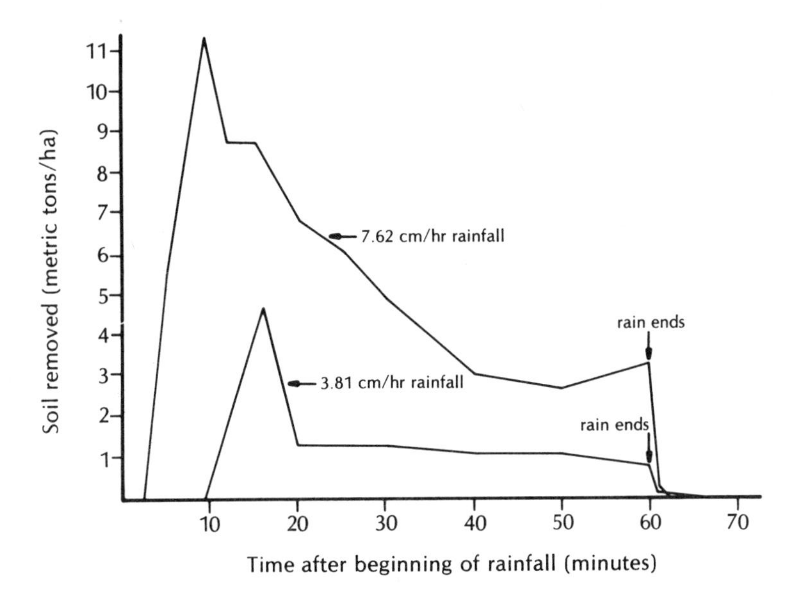

Figure 12-1 *Rate of erosion in Curlew Valley, Utah, as a function of time after rain begins* [*7.62 and 3.81 cm/hour rainfall*].

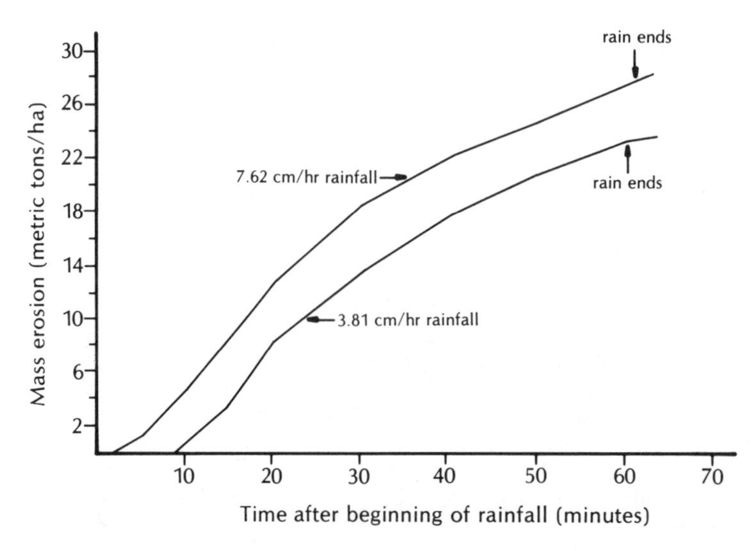

Figure 12-2 *The relation between mass erosion and time in Curlew Valley, Utah, after beginning of rain* [*7.62 and 3.81 cm/hour rainfall*].

The Universal Soil Loss Equation

Over the past 25 years Wischmeier and associates (Zingg 1940; Smith 1941; Musgrave 1947; Smith and Whitt 1947; Wischmeier 1959; U. S. Department of Agriculture 1961; Wischmeier and Smith 1965) have been studying factors involved in the erosion process. As a result of these investigations an equation, now known as the *Universal Soil Loss Equation*, was developed. The equation for the annual soil loss by water is as follows:

$$A = RKLSCP \qquad (12\text{-}1)$$

where

A = the annual average soil loss in metric tons/ha. It may be computed on some probability level, but R must be at the same level or probability of recurrence.

R = the rainfall erosion factor calculated on an annual return period only (the 0.5 probability of return is 1 year). It is composed of the local rainfall-erosion index for a given return period, EI, where E is the rainfall kinetic evergy in metric ton meters/ha per cm, and I is the 30-minute rainfall intensity at whatever probability level the A value is desired. The probability level is defined as the frequency with which a specific year's rainfall re-occurs.

K = the soil erodibility or soil factor, expressed as metric tons or soil loss/ha per unit of rainfall EI for a slope of 9 percent and 22.13 m long. It expresses the soil lost from a nonvegetated area.

L = the slope length factor as a ratio of the soil loss to a similar one from a 22.13-m slope length or

$$L = (\lambda/22.13)^{0.5} \quad (12.2)$$

where

λ = the length of desired slope in compatible units

S = the steepness of the slope factor:

$$S = 0.520 + 0.363s = 0.052s^2 \quad (12\text{-}3)$$

where

s = the percent slope

C = cover factor. It is the ratio of the soil loss from a nonvegetated, tilled area to an area with a cover or in undisturbed condition.

P = practices factor. On farms this means engineering practices for erosion control such as contouring, terracing, etc. On grasslands, deserts and forests this means engineering treatments of any type.

Soil erodibility (K in Equation 12-1 above) is dependent on such factors as temperature (especially freeze-thaw cycles) and soil properties; namely, texture, structure, permeability, compaction and infiltration capacity. According to Wischmeier (1961) and his associates (Klingebiel 1961; Olson 1961), erodibility decreases with an increase of grade and size of soil structure; with coarseness of texture; with the amount and size of coarse fragments; with organic matter content; and with increased transmissivity of the subsoil materials.

In addition to the above factors, soil crust algae in arid to semiarid environments (Booth 1941; Fletcher and Martin 1948; Fletcher 1961) have a profound influence in both the K factor and runoff. This influence, however, is transitory. For example, if the algal-stabilized crust is intact and moist, the K factor is reduced to near zero and runoff is increased to nearly 100 percent of the rainfall. When this same algal layer is dried and incised, it still shows a very low K or C factor, but the runoff may be reduced to a very low value. Last, if the algal layer is disturbed, the soil essentially reverts to the state it was in before being stabilized, except that it will have higher carbon and nitrogen levels to the depth of disturbance. The erosion will be near but slightly less than its original value, as will the runoff.

Cover such as desert algal crusts influences erosion by binding soil crumbs with secreted mucilaginous material to protect them against detachment and to enhance infiltration. Higher plants retard erosion by dissipating some of the energy of raindrops before they strike the soil, by retarding runoff flow and by improving permeability through root effects (Bennett 1974).

Erosive Transfers of Nitrogen by Water in Western U.S. Deserts

Wischmeier did not extend the application of the soil loss equation west of the eastern front of the Rocky Mountains. With the advent of the western precipitation intensity maps by Miller et al. (1973) and the high correlation at measured sites between *EI* with the 2-year 6-hour precipitation, we have extended the isoerodent map to the areas of the Sonoran and Great Basin deserts which we studied. The two isoerodent (*R* or annual *EI*) maps are shown in Figures 12-3 and 12-4.

Using the criteria of Fletcher and Beutner (1941), the soils of the western deserts would be expected to be among the most erodible in the United States. On the other hand, they are subject to smaller erosive forces, as evidenced by the lower values for the *R* factor shown in Figures 12-3 and 12-4.

The northern Great Basin Desert, with cooler temperatures and lower precipitation intensities, has even smaller *R* values. The Arizona-New Mexico *R* values are greater than 70, while the Curlew Valley *R* values are less than 25. This difference is a major factor in the erosional movement of nitrogen in the two areas.

Nitrogen transfers in the Arizona study area were estimated using the universal soil loss equation to normalize the data to mean annual values and data collected from runoff plots (Fletcher 1961). Nitrogen losses ranged from 0.24-24.7 kg/ha per year, with a mean value of 7.2 kg/ha per year in this area. These results are similar to those derived by Renard et al. (1974). In limited portions of this desert area that are severely disturbed by plowing or other activity, nitrogen losses due to water erosion can be as high as 18 metric tons/ha per year. These large losses are due to the complete washing away of the soil particles.

Nitrogen transfers in the Curlew Valley, Utah-Idaho, study area were measured using the rainulator and runoff plots (Dogan 1975). Losses of nitrogen by water erosion range from 0.24-3.1 kg/ha per year and average 1.62 kg/ha per year.

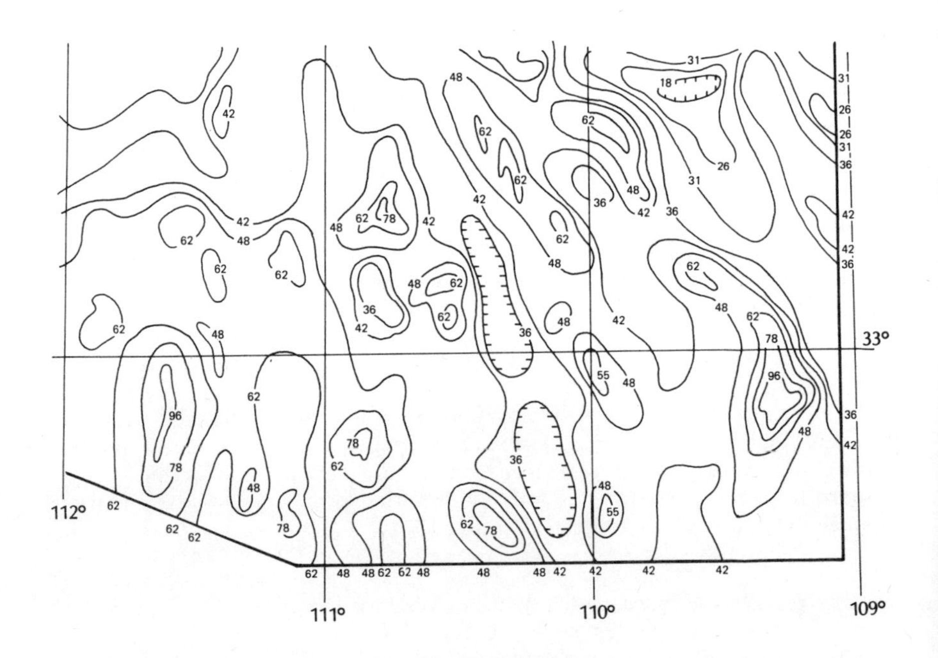

Figure 12-3 *Isoerodent* [R *or annual* EI] *map of southern Arizona.*

Measurements of runoff nitrogen from natural events summarized in Table 12-1 indicate similar values for losses of nitrogen over a 10-month period from seven 1-m^2 plots located throughout Curlew Valley (Sorensen et al. 1974). With the exception of site 4, all plots showing lower values for nitrogen loss were covered with a well-defined soil-algae crust having a high incidence of soil lichens. The site 4 plot was located in a semidesert area artificially converted to a rather heavy crested wheatgrass (*Agropyron desertorum*) cover.

The ratio of nitrogen content of the suspended soil and organic material in the runoff to nitrogen content of the soil surface to 1 cm depth ranged from about 2 to 10 in both study areas. In Curlew Valley the ratio ranged from nearly 17 to less than 2, with a mean of 5 and standard deviation of 3. It is thought that a large contributor to this condition is the fact that litter and organic matter, high in nitrogen content, are much lower in density than mineral soil particles, and so are much easier to remove by wind or water. This is particularly true in the losses from the more minor rainstorms.

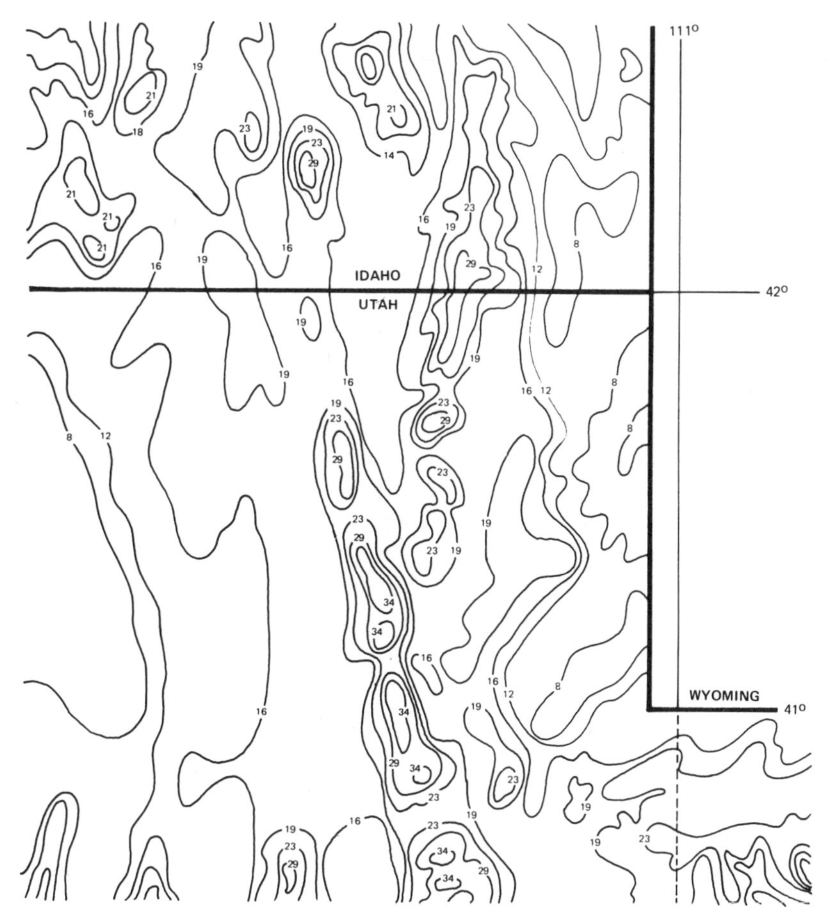

Figure 12-4 *Isoerodent* [R *or annual* EI] *map of Curlew Valley area, Utah, and vicinity.*

Table 12-1 *Summary of Runoff Water and Runoff Nitrogen Removals [25 November 1972 to 27 September 1973] in Curlew Valley [from Sorensen et al. 1974]*

Site	Total rainfall during period (cm)	Total runoff flow (liters/m^2)	Ratio of runoff:rainfall	Total N carried in runoff water (kg/ha)
1	18.8	16.8	0.89	1.50
2	20.4	16.9	0.83	.52
3	23.4	90.5	3.87	1.50
4	27.4	7.1	0.26	.20
5	25.1	86.2	3.44	1.47
6	23.5	83.1	3.54	2.56
7	24.4	53.7	2.20	1.63

WIND EROSION OF NITROGEN

As is the case with sediment transported by water, soils transported by wind carry with them their associated nitrogen and may contribute to the loss of this element from the ecosystem. It is probable that nitrogen concentrations in wind-removed sediments would also vary signficantly from that of the soil itself, due to the lower mass of litter and organic matter.

Chepil (1945a-c, 1946a,b, 1953, 1954, 1955a-c, 1957a-c, 1958), Chepil and Woodruff (1957), Chepil et al. (1955) and Skidmore and Woodruff (1968) developed technology for evaluating the potential for wind erosion by some relatively simple criteria. Variations of weather are the most distinctive features of wind-eroded regions (Zingg 1953). These climates are characterized by 1) extremely variable and often high wind velocities, 2) generally low and variable precipitation, 3) high frequency of drought, 4) rapid, extreme changes in temperature and 5) high evaporation rates. All of these conditions could be substituted as a description of desert environments.

The conditions under which wind erosion occurs are few and obvious. Whenever the soil surface is finely divided, loose, light and dry, the surface is smooth and bare, and the fetch area is unsheltered and wide in the windward direction, erosion may be expected.

Primary conditions that influence wind erosion may be tabulated as follows:

Primary Conditions		*Equivalent conditions*	
Percent nonerodible soil particles (particles > 0.84 mm)	A		
		Soil erodibility	I
Stability of aggregates	F		
Surface crust (transient)	T		
Wind velocity	U		
		Climatic factor	C
Surface soil moisture	M		
Soil surface roughness	K		

Quantity of vegetal cover	R		
Kind of vegetal cover	K_o		
Distance of wind fetch	D	Field width	L
Distance shielded by barrier	D_b		

Wind erosion E (mass per area) can then be expressed as a function:

$$E = f(ICKVL) \tag{12-4}$$

where

I = soil erodibility (soil loss from a large, isolated field with a bare, smooth surface in the vicinity of Garden City, Kansas)
C = local wind erosion climatic factor
K = soil surface roughness
V = vegetal cover
L = equivalent width of field

For details and use of Equation 12-4 the reader is referred to Woodruff and Siddoway (1965). The map of the C factor (Fig. 12-5) extrapolates Chepil's work to western U.S. desert areas.

Nitrogen data have been handled in a similar fashion to those in the case of water erosion, where wind deposits or siltation products are analyzed for nitrogen and the annual movement is computed from the relative value of the C factor during the period sampled and the annual C factor.

By using the wind soil loss equation, and the nitrogen composition of a limited number of hummock and interhummock depositions to estimate saltation drifts and interdunal nitrogen content, an estimate of wind translocation of nitrogen was obtained (Fletcher, unpubl. data; also see Shields et al. 1957). The nitrogen loss by wind in Arizona was estimated to range between 0.28-5.1 kg/ha per year, with an average value of 1.7 kg/ha per year.

The corresponding nitrogen losses from wind erosion calculated for Curlew Valley ranged between 1.0-5.6 kg/ha per year, with an average value of 3.4 kg/ha per year.

OTHER PHYSICAL TRANSFERS OF NITROGEN

Without question, the major transfers of nitrogen from arid to semiarid landscapes by physical means come through water and wind erosion. A few other physical processes do exist which may transport or remove minor quantities of nitrogen. Burning is the only other process considered here.

Fire would seem a likely candidate for freeing nitrogen immobilized in plant material so that it could be removed from the ecosystem. Few data are available on which to base estimates of nitrogen losses from arid or semiarid environments due to fire (Humphrey 1974). The most likely mechanism for nitrogen loss in fire is the conversion of organic nitrogen to gaseous oxides of nitrogen. However, as Komarek (1970) points out, the requirements for extremely high temperatures and/or high pressures of combustion to form these gases are not met in fires typical of those encountered in drier environments. This factor, combined with the requirement for sufficient plant cover to carry a fire, leads us to believe that fire is negligible as a vehicle for nitrogen loss in arid environments. In semiarid environments the loss or redistribution of nitrogen can be considerable. For instance, Murray (1975) estimates that burning of *Artemisia tridentata*-dominated areas in southern Idaho volatilizes about 70 percent of the nitrogen in above-ground biomass. Some of this nitrogen may come down in adjacent ecosystems (Clayton 1976). There is a short-term "green manuring" effect of the decaying roots left after a fire (Murray

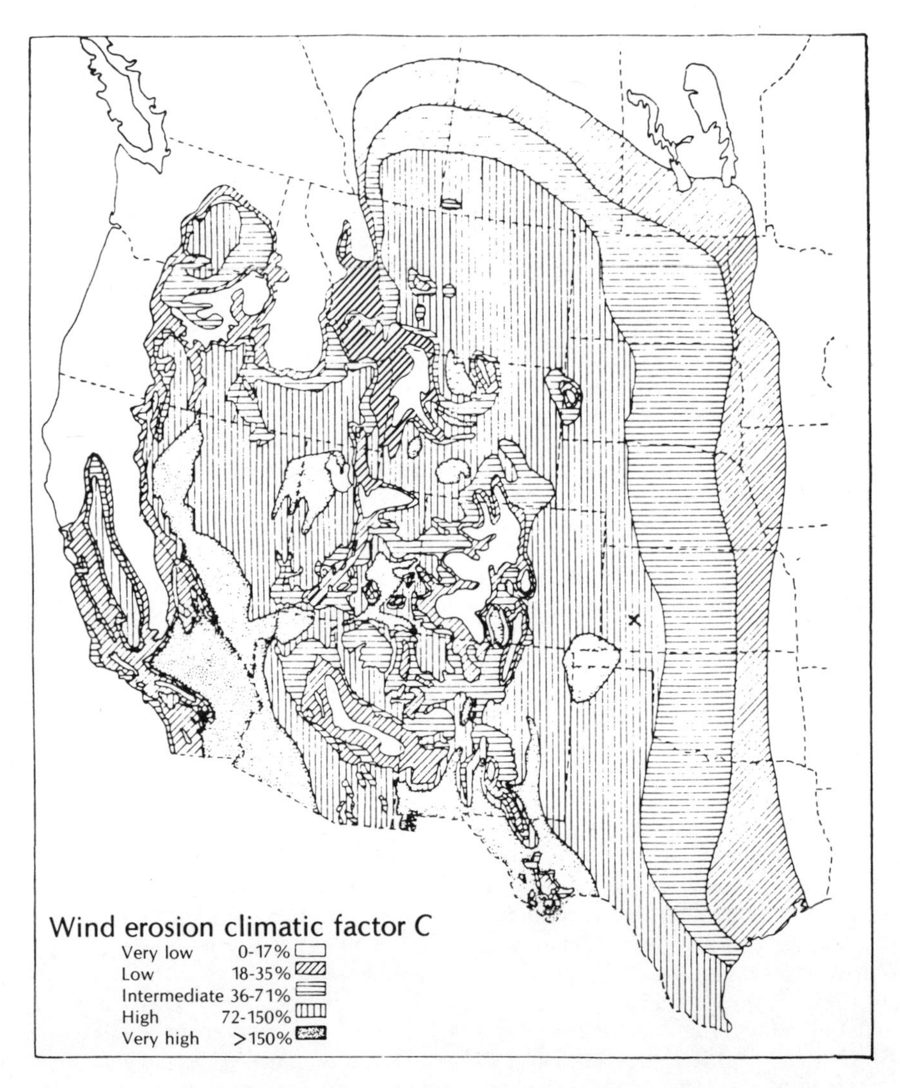

Figure 12-5 *Wind erosion climatic factor expressing percent soil loss by wind in the vicinity of Garden City, Kansas, marked by* X [*Chepil, pers. commun.*].

1975). Fire could be important in mineralizing organic nitrogen in litter and living plant material, but it would likely be quickly reutilized or removed by normal erosive processes and not have great impact except in the case of large-scale fires, which might produce nitrogen losses on a scale such as at Hubbard Brook (Likens et al. 1970).

SUMMARY

Erosion by water and wind accounts for practically all of the physical transfers of nitrogen from deserts in the solid state. Results from studies in two desert areas of the western United States, one in Arizona and one in northern Utah, show total erosive losses of nitrogen as follows:

	Arizona	*Curlew Valley, Utah*
Range:	0.5-30 kg/ha per year	1-9 kg/ha per year
Mean:	9 kg/ha per year	6 kg/ha per year

It is notable that nitrogen, because of its concentration in lighter surface materials, is selectively eroded from a soil by water (and possibly by wind), with the result that total nitrogen content of eroded material is several times greater than the original soil.

Soil profile truncation by severe erosion may result in a lowered concentration of the amount of nitrogen removed by erosive processes, since nitrogen is much less concentrated in subsurface soil horizons in arid areas. However, acute erosion that removes the soil profile more rapidly than natural biological processes can replace the surface nitrogen would probably result in an increase in total mass of nitrogen removed, since a large increase in mass of eroded soil would be expected.

In addition, the proper microflora can repopulate these truncated areas at an accelerated rate and thus rapidly restore the nitrogen content of the new surface to the original level. As the microflora invades, it restabilizes the surface of the truncated soil, thus preventing further losses by wind and water.

The role of erosion in nitrogen dynamics of desert systems cannot be clear until ecosystem boundaries are defined. On a small scale, erosive nitrogen losses from specific sites or additions to other sites may have significant impact on production. Whether output is greater than input over a greater landscape is not known. Further analysis of the net impacts of such nitrogen translocations on overall desert ecosystem dynamics is needed.

13

A COMPUTER SIMULATION MODEL OF NITROGEN DYNAMICS IN A GREAT BASIN DESERT ECOSYSTEM

C. S. GIST, N. E. WEST and M. McKEE

INTRODUCTION

The nitrogen cycle has long been modeled schematically. Nearly every introductory textbook in soils, botany, microbiology and ecology contains at least one figure with the biogeochemical cycle of nitrogen outlined. In some, pool sizes and average flux rates are given. Most such schematic models draw their data largely from, and point to, the relevance of the nitrogen cycle for intensive agriculture. For example, Delwiche (1970) estimated the size of the components of the global nitrogen cycle. Fewer schematic models have dealt with nutrient cycles in wildlands. The nitrogen cycle of deserts has rarely been specifically addressed. As can be seen in the foregoing chapters, many of the processes of the nitrogen cycle that are important in agronomic and other wildland ecosystems are minor in deserts, whereas other processes are uniquely important in arid land contexts.

One way to synthesize the foregoing information is to produce schematic models for deserts. For example, Figure 13-1 presents an annual budget of the nitrogen cycle in an *Atriplex confertifolia*-dominated ecosystem in Curlew Valley, Utah, a representative area in the Great Basin Desert. Although this diagram gives some overview of the various processes previously discussed in this volume, it gives little insight into the within-year dynamics of the total cycle, nor does it allow us to project what might happen with various perturbations of the system. It also ignores animals and their functions in the nitrogen cycle. We therefore developed a digital computer model that would stress what data might be directly available or derivable by difference. Our goal was to simulate nitrogen cycle dynamics in a cold winter desert system, as it is currently understood, and to predict system responses to perturbations involving this nutrient.

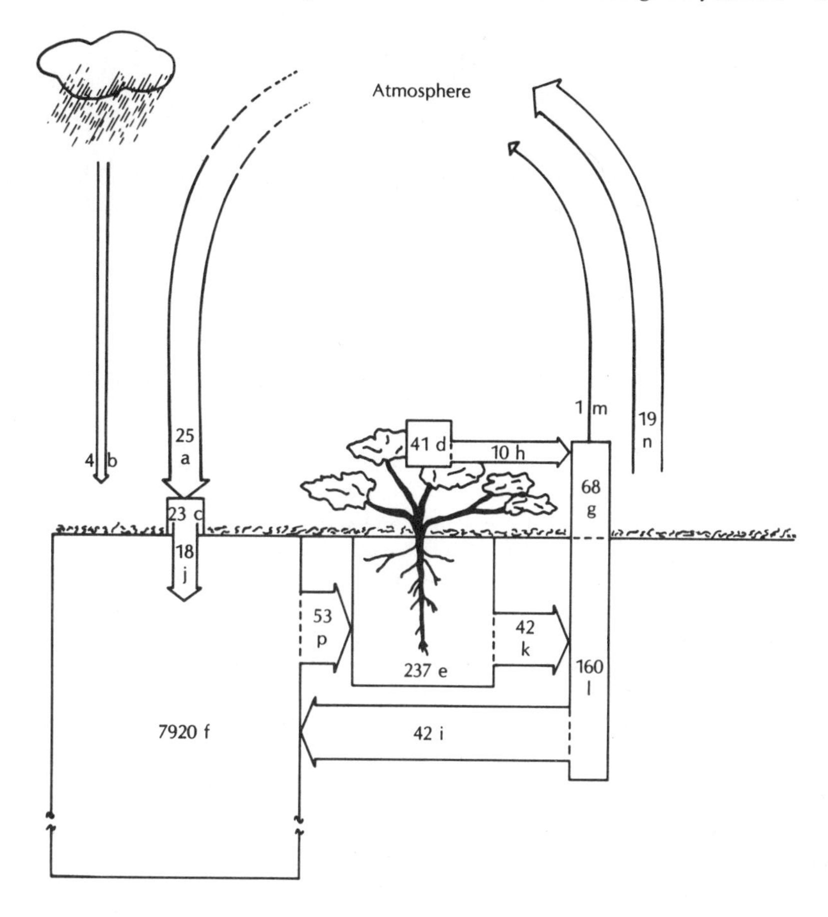

Figure 13-1 *Schematic representation of the main aspects of the annual nitrogen balance for an* Atriplex confertifolia-*dominated ecosystem in Curlew Valley, Utah. Arrows are fluxes [kg N/ha per year]. Width of arrows approximates the magnitudes given in numbers inside arrows. Boxes represent the highest annual values of the components [kg N/ha]. Average of 3 or more years' data. Sources of data itemized in West and Skujins [1977]. a = biological fixation, mainly by blue-green algae in cryptogamic crust; b = input in wet and dry precipitation [including dust]; c = total N in cryptogamic crust; d = total N content of above-ground higher plant biomass; e = total N content of living below-ground higher plant biomass; f = total N mineral content of soil, for up to 90 cm [rooting depth]; g = total N in above-ground phanerogamic plant litter; h = total N in above-ground higher plant litter production; i = mineralization rates of N in higher plant litter; j = mineralization of cryptogamic plant litter; k = below-ground litter production; l = total N in below-ground litter standing crop; m = volatilization; n = denitrification; p = plant uptake.*

MODEL STRUCTURE AND RATIONALE

Compartments

The quantified variables and the sources of data may be found in Tables 13-1 to 13-4. The Boolean matrix associated with the schematic compartmental and flow model shown in Figure 13-2 may be found in Table 13-5.

X_1—*Fixers*

All organisms that fix nitrogen are lumped into one compartment because it is difficult, under field conditions, to separate the groups involved. In most terrestrial ecosystems, symbiotic bacteria associated with legumes are the major producers of fixed nitrogen. However, it is the cryptogamic crusts common to cold winter semideserts that account for most of the biological input to the system (Rychert et al., Chapter 3 of this volume). The biomass of the cryptogamic crust is estimated to be 72.5 g/m^2 (Lynn and Cameron 1973). Assuming a ground cover of 25 percent with 15 percent nitrogen content in organisms (Lynn and Cameron 1973), one would expect about 3.3 g N/m^2 for the average annual standing crop in this compartment.

Field measurements of nitrogen fixation by Rychert and Skujins (1974b) using acetylene reduction methods have demonstrated up to 450 mg N/m^2 per hour fixed under optimal conditions. Weather records would indicate that optimal conditions exist between 10 and 25 days in the late spring and sometimes early fall; ^{15}N studies of diurnal patterns have shown that fixation is higher in the early morning and late

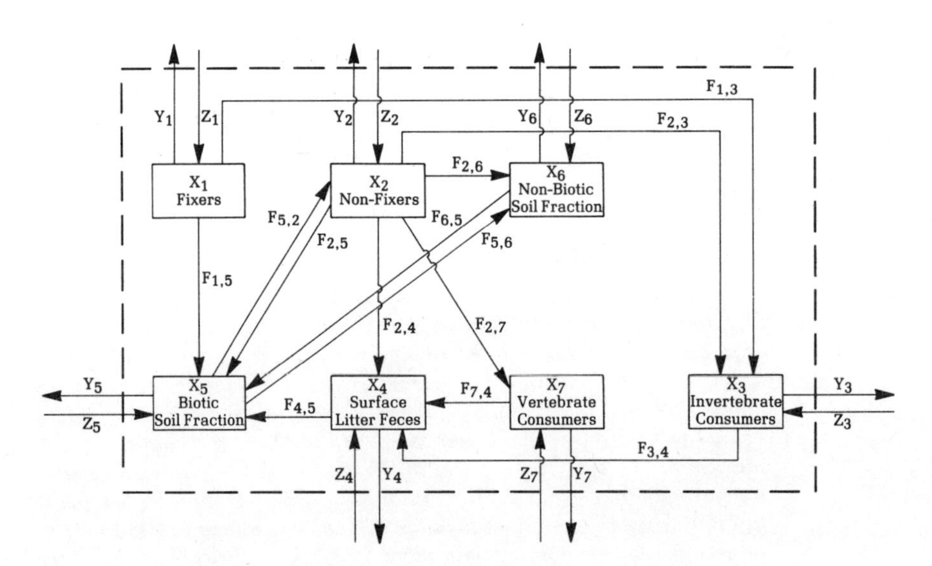

Figure 13-2 *Schematic compartmental and flow model of nitrogen in a generalized Curlew Valley ecosystem.*

Table 13-1 *State Variables and Ranges and Average Annual Standing Crops of Nitrogen in the Curlew Nitrogen Model* [*Original, Pretuned Data*]

Compartment	Value (g N/m^2)	Source
X_1: Fixers	3.3	Lynn and Cameron (1973)
X_2: Nonfixer vegetation	4.3	Bjerregaard (1971) Balph et al. (1973) West (1972)
X_3: Invertebrate consumers	3.0	Balph et al. (1973)
X_4: Surface litter and feces	10.0	Bjerregaard (1971) West and Fareed (1973) Westoby (1973) Intuition
X_5: Biotic soil fraction	800	Bjerregaard (1971) West (1972) Skujins and Eberhardt (1973)
X_6: Inorganic soil fraction	3.0	Bjerregaard (1971) West (1972) Skujins and Eberhardt (1973)
X_7: Vertebrate consumers	0.000091	Smith (1973) Malechek and Smith (1972) Gross et al. (1974)

evening (Skujins and Eberhardt 1973). It is likely that only 2-3 hours/day comprise the effective periods. By multiplying an average rate of 160 mg N/m^2 per hour by a 2.1-hour day of effectiveness by 20 days when climatic and soil moisture conditions permit over 25 percent of the landscape yields an average value of 16 kg N/ha per year for the total flow of fixed nitrogen from X_1 to other compartments. Appropriate adjustments are made for simulations under varying environmental conditions.

X_2—*Nonfixer Vegetation*

This compartment contains all of the above-ground parts of vascular plants in our cool desert situation. Using the data of Bjerregaard (1971) and Balph et al. (1973) we obtained a range of 30-50 kg N/ha in above-ground biomass of *Atriplex* and *Ceratoides*-dominated communities in midsummer following the peak period of growth. West (1972) has shown peak standing crops of 85 kg N/ha in the drier types of *Artemisia*-dominated vegetation. The species dominating these communities have an overall average of 5.05 percent nitrogen content. By multiplying 850 g/m^2 of biomass by 5.05 percent nitrogen content we obtain an average value of 43 g N/m^2 as the above-ground standing crop of X_2 (Table 13-1).

The dynamics of above-ground growth of the dominant shrubs is given in Table 13-2. Growth is initiated in late spring with the advent of warmer temperatures. Soil moisture content is highest during this season. Growth should continue until the middle of June. The resumption of growth could occur in the fall with adequate rainfall before temperatures become too cool. If a short period of favorable conditions should occur, then only flowers and a little vegetative tissue are produced.

Table 13-2 *Compartment Dynamics and Annual Patterns of Change from Curlew Nitrogen Model [Original, Pretuned Data]*

Compartment	Initial condition (g/m^2)		Curve shape (1 Jan to 31 Dec)	Source
	Maximum	Minimum		
X_1	450	0		Rychert and Skujins (1973)
X_2	607	2.1		West (1972) West and Fareed (1973)
X_3	2.8	0.1		Mack (1971) Balph et al. (1973) Gist (pers. commun.)
X_4	15.0	5.6		West and Gunn (1974) Sferra (pers. commun.)
X_5	820	780		Bjerregaard (1971) West (1972) Skujins (1972)
X_6	10	0.5		Bjerregaard (1971) West (1972) Skujins and Eberhardt (1973)
X_7	0.000091	0.00000009		Gross et al. (1974) Malechek and Smith (1972) Smith (1973)

Besides N_2- and CO_2-fixing blue-greens there are some non-N_2-fixing algae which act as additional primary producers and may provide substrate for denitrification and other heterotrophic processes. They are not considered here because of the lack of adequate data.

X_3—*Above-ground Invertebrate Consumers*

The invertebrate consumer compartment includes all nonsaprophytic invertebrate organisms in the system of interest. Soil-inhabiting consumers or soil-inhabiting life stages of certain organisms are especially important.

Data from Curlew Valley on a variety of invertebrates were available (Table 13-2) but are generally skimpy. Many dynamic aspects of the organisms' interaction with both the biotic and the abiotic environment had to be estimated. Much more is known about the biology of the vertebrates, even though they are probably less important in ecosystem function and control.

Data on invertebrate biomass, population densities and seasonal gross energy consumption were obtained from Balph et al. (1973). Guidelines for converting biomass values to amount of nitrogen per unit area were found in Bowen (1966). The value for the average nitrogen content in invertebrate consumer biomass (X_3) was calculated to be 3.5 g N/m^2 (Balph et al. 1973).

A total compartment value of 3.5 g N/m^2 was used after consideration of other compartment values, critical analysis of reviewers, weighing of questionable field techniques and the overall behavior and life history tendencies of the invertebrate consumer organisms.

X_4—*Surface Litter and Feces*

Compartment X_4 contains organic detritus, such as leaves and jackrabbit pellets. All evidence indicates that plant litter is the major source of nitrogen in this compartment. Bjerregaard (1971) reported midsummer standing crops in *Ceratoides* and *Atriplex* communities which average approximately 6.3 g N/m^2. This calculation compares well with the data of Mack (1971) for *Artemisia*.

A separate estimate of litter production is possible from the studies of West and Fareed (1973) and West and Gunn (1974). They found that the bulk of plant litter is produced in midsummer after the ephemeral leaves are lost as soil moisture stress develops. By multiplying an average of 4 years of plant litter production by appropriate cover values, they estimated annual plant litter fluxes of 285 g/m^2 per year. This litter had an average nitrogen content of 2.2 percent. Multiplying these factors gave an average input of 6.3 g N/m^2 per day.

In addition, an attempt was made to obtain data on the amount of feces present. The only information available deals exclusively with jackrabbit pellets. Westoby (1973) reported a density of pellets of approximately 143 pellets/m^2. From the data of Arnold and Reynolds (1943), Westoby's counts amount to approximately 44 g/m^2. Skujins (pers. commun.) indicated that the nitrogen concentration of field-exposed rabbit pellets is approximately <1 to 2 percent nitrogen by dry weight. This provides an estimated average standing crop of approximately 0.7 g/m^2 of nitrogen due to rabbit feces. A conservative value of 3.0 g N/m^2 was thought due to other animals' feces, based on reasoning that the remaining animal community, including invertebrates, consumed much more than jackrabbits alone.

The seasonal dynamics of plant litter in this compartment were taken from Mack (1971), West and Gunn (1974) and West and Fareed (1973). It was assumed that feces production followed animal population trends. With the first frost, a substantial number of insect carcasses would be added to this compartment (estimated nitrogen contribution of 3 g/m^2; from P. R. Sferra, pers. commun.). Plant litter was maximal during late summer. With the spring thaw and the initiation of decomposer activity, the amount of nitrogen would decrease to approximately 5.6 g/m^2, initially dropping rapidly and then tapering off in the form of an upwardly concave curve. Beginning in late June with leaf fall there would be an increase in this part of the compartment to approximately 1.1 g N/m^2 which probably would decline after mid-July.

X_5—*Biotic Soil Fraction*

Compartment X_5 consists of nitrogen in organic constituents below the soil surface in the upper meter of the profile where all plant roots occur. This compartment also includes bacteria, fungi and other microbes, as well as the nitrogen-containing organic residues found in the soil. For a Curlew Valley desert soil there was an estimated 800 g N/m^2 (Bjerregaard 1971; West 1972); this represents the largest nitrogen-containing compartment within the model. However, much of the nitrogen is almost permanently fixed to clay and humic fractions of the soil and is thus practically unavailable for plant growth or any biological function.

There is little seasonal variation in this total. The zone of maximum fluctuation, the upper 5 cm, drops 0.05 percent in total organic nitrogen content during the spring-summer-fall growing season, which is 25 percent of total nitrogen concentration (Skujins 1972; Rychert and Skujins 1973).

X_6—*Inorganic Soil Fraction*

The compartment "nonbiotic soil nitrogen," labeled X_6, consists of all the nitrogen in below-ground, nonbiological material. The data were obtained primarily from Bjerregaard (1971), West (1972) and Skujins and Eberhardt (1973). In these studies, inorganic soil nitrogen was sampled at intervals during the year. Measurements were taken for at least four different depths in the soil profiles. Using these data, an average value of nitrogen content was found to be 4.0 g/m^2. These studies also described seasonal variation in X_6.

X_7—*Vertebrates*

Vertebrates are the most conspicuous, but not necessarily the most important, consumers in this system. Jackrabbits are present year-round following seasonal and longer-term fluctuations in their populations (Westoby 1973; Gross et al. 1974).

Cattle represent a special problem since they are usually allowed to graze for only several months each winter. During this time they exhibit a gross energy consumption of about 315,000 kcal/ha (Malechek and Smith 1972). Of this, 35 percent is returned to the desert ecosystem as feces and urine (0.22 g N/m^2). The remainder, amounting to 0.4 g N/m^2, is lost as export when the cattle are removed for grazing elsewhere or are harvested (Y_7).

Although the total contribution by cattle to nitrogen cycle dynamics in this system is very small, they were included to investigate the effects of grazing management practices on the nitrogen cycle in Curlew Valley. Cattle were the only vertebrates which we could extensively manipulate. In contrast, jackrabbits are uncontrollable and have highly variable population trends. The invertebrates collectively dominated the faunal above-ground nitrogen standing crop by at least an order of magnitude.

Fluxes

Let us now consider nitrogen fluxes among the compartments. These are outlined in Table 13-3. A discussion of the definitions, values and their derivation follows.

The movement from fixers (X_1) to other compartments in our model includes the following:

Table 13-3 *Fluxes in Curlew Nitrogen Model* [*Original, Pretuned Data*]

Symbol	Parameter	Average fluxes (g N/m^2 per yr)	Source
$F_{1,3}$	Cryptogamic grazing	0.3	Gist (pers. commun.)
$F_{1,5}$	Cryptogamic decomposition	3.0	By subtraction
$F_{2,4}$	Plant death (litter formation)	6.3	West and Gunn (1974)
$F_{2,3}$	Invertebrate grazing	1.3	Intuition
$F_{2,5}$	Plant leaching of organic nitrogen	0.0001	Skujins and West (1974)
$F_{2,6}$	Plant leaching of inorganic nitrogen	0.0000001	Skujins and West (1974)
$F_{3,4}$	Invertebrate death and waste	0.63	Intuition
$F_{4,5}$	Litter decomposition	4.1	Comanor and Prusso (1974)
$F_{5,6}$	Replacement of nitrogen used by plant growth (10% of X_5 per yr)	0.1	Skujins and West (1974)
$F_{6,5}$	Nitrogen uptake by roots from nonbiotic fraction	.00073	Fernandez (1974)
$F_{5,2}$	Nitrogen uptake by above-ground plant parts	9.7	West (1972)
$F_{2,7}$	Vertebrate herbivory	0.025	Westoby (1973) Malechek and Smith (1972)
$F_{2,3}$	Invertebrate herbivory	1.452	Gist (pers. commun.)
$F_{7,4}$	Vertebrate death and waste	0.0268	Malechek and Smith (1972)

$F_{1,3}$—*Consumption of Cryptogamic Crusts*

This process is mostly by invertebrates. Collembolans, termites and small beetles are the major consumers of cryptogamic crusts. We estimate that about 10 percent of the standing crop is annually grazed by these organisms. This flux would account for an average of 0.3 g N/m^2 per year. These organisms are most active during cool, moist weather.

$F_{1,5}$—*Decomposition of Cryptogamic Crust*

Since the biomass of X_1 seems to remain approximately constant and since as much algal material must die as is produced to maintain equilibrium, the overall loss should be equal to input. By difference from flux $F_{1,3}$, about 3.0 g N/m^2 per year should go into the soil fraction through decomposition.

About one-half of the 73 g/m^2 of cryptogamic biomass is estimated to turn over each year; it has a 15 percent nitrogen content, yielding 5.5 g N/m^2 ($F_{1,4}$). The combined flux to litter is about 12 g N/m^2 per year (Lynn and Cameron 1973).

$F_{2,4}$—*Nonfixer Plant Litter Production*

Recent data of Mack (1971), West and Fareed (1973) and West and Gunn (1974) have shown that cool desert shrubs such as *Artemisia, Atriplex* and *Ceratoides* produce from 27 to 30 g litter/m^2 per year. The nitrogen content of this litter ranges between 1.5 and 2.8 percent. By taking a median of both ranges and multiplying, an approximate flux was calculated as 6.3 g N/m^2 per year moving into plant litter.

This flux was controlled and/or affected daily through the year by moisture and temperature parameters which pulse the production of fixer and non-nitrogen-fixing vegetation sources. Meteorological conditions also affect the growth and death rates of the consumers.

$F_{2,5}$—*Plant Leaching of Organic Nitrogen*

Skujins and West (1973 and 1974) found that negligible amounts of organic nitrogen move directly into the soil from leachates carried by leaf drip and stem flow. Most of the nitrogen carried by precipitation is in particulate form. Few storms in the desert are intense enough to cause leaf drip and particularly stem flow (West and Gifford 1976). The fluxes are not zero, but nearly so. Accordingly, we have arrived at an average rate of 0.001 g/m^2 per year. The peak of such activity would be during the spring and especially during the summer when convectional storms can occur.

$F_{2,3}$—*Invertebrate Grazing*

Because the invertebrate consumer compartment is being functionally generalized in the extreme, we might expect great variance in both intensity and timing of this flux. For lack of reliable data, we guess the movement to be 30 percent of the nitrogen in X_2 each year. This is tempered by the realization that the current year's plant growth is the richest in nitrogen (West 1972) and also is probably being foraged the most. The intensity of grazing by all classes of invertebrates is probably greatest in the spring-early summer period. Invertebrate herbivory probably varies between nearly zero and up to 5 g N/m² during some periods. Our estimate is for an average transfer rate of 1.3 g N/m² per year.

$F_{2,6}$—*Plant Leaching of Inorganic Nitrogen*

As mentioned in the discussion of $F_{2,5}$, this process is very minor. Skujins and West (1974) found almost negligible leaching of inorganic forms of nitrogen from cool desert shrubs. The intensity of rainfall required to get much leaf drop or stem flow has a very low probability of occurrence (West and Gifford 1976). Late spring and summer would be the times when this process might occur.

$F_{3,4}$—*Invertebrate Death and Waste*

This flux covers a great many unknowns. The population dynamics of all the invertebrate native fauna would need to be known to obtain specific data for this transfer function. The most difficult aspect of describing fluxes from the consumer compartment occurred in attempting to determine the amount of tissue carried over during the winter months. Because invertebrate data for Curlew Valley were inadequate, only a rough estimate of the turnover rate of consumers could be made. No reliable estimate could be made of the number of invertebrates, especially insects, overwintering as eggs, larvae, pupae, nymphs or adults. The short life span of most arthropods was a criterion for the decision to set a total turnover of consumers on a per-year basis. The transfer rate from X_3 to X_4 was controlled mainly by temperature.

The greatest flux of invertebrate death probably occurs during the first fall freeze. Feces and urine production probably parallels the dynamics of X_3.

$F_{4,5}$—*Litter Decomposition*

Bjerregaard (1971) reports that the average turnover rate of plant litter from some cool desert shrubs is approximately 6 years. Holmgren and Brewster (1972) estimate that if woody tissue is considered, a figure of around 14 years is more likely in salt deserts. This compares well with the report of Mack (1971). Westoby (1973) estimated the turnover time of rabbit pellets to be approximately 2 years, which may be close to that of cow feces. Combining these data, one derives an average turnover rate for nitrogen in litter of approximately 4.1 g/m² per year.

The dynamics probably follow a spring peak, a fall resurge, if rain comes, and a distinct summer trough.

$F_{5,6}$—*Replacement of Nitrogen Used in Plant Growth*

For lack of a better alternative, we assumed a balance in the system and equated plant uptake with nitrogen appearance in new plant growth. Presumably, this is replaced over the long term by deamination, ammonification and nitrification processes. Although much lab work exists, these processes have not been directly measured in the field.

Using the following sources and the above rationale, a flux averaging 0.1 g N/m^2 per year was calculated (West 1972; Skujins and West 1974). The peak rates would occur during spring when soil moisture, root growth, proteolysis, ammonification and nitrification rates are optimum. A small fall peak may occur, if rains come early.

$F_{5,2}$—*Nitrogen Uptake by Above-ground Plant Parts*

This process, like $F_{5,6}$, is believed to be coordinated with plant growth. Nitrate might accumulate if a drier than average year prevents plant uptake, volatilization and denitrification.

$F_{6,5}$—*Nitrogen Uptake by Root and Bacterial Growth*

Immobilization in microbial biomass is probably negligible. Since up to 85 percent of the plant biomass in these systems can be underground (Bjerregaard 1971; Fernandez 1974), a significant immobilization by plant growth occurs. Caldwell and Camp (1974) estimate plant root productivities of 186 g/m^2 per year for *Ceratoides* to 443 g/m^2 per year for *Atriplex confertifolia*. Taking the higher value and combining these values with an average nitrogen content of 1.7 percent for roots of *Atriplex* (West 1972) gives an average flux of 7.5 g N/m^2 per year. Most of the root growth is in spring and summer (Fernandez 1974).

$F_{2,7}$—*Consumption of Nonfixers by Vertebrates*

The function generated to estimate the consumption of above-ground vegetation by cattle followed the work by Malechek and Smith (1972) and Smith (1973). In these studies it was found that the feeding activities of cows were reduced when the temperature was much below freezing, and that this was compounded if the weather was wet and/or windy. However, the dominant factor was temperature. Consequently, the flux to cattle from the vegetation compartment was considered a function of temperature only for the purposes of this model.

$F_{7,4}$—*Cattle Excreta*

Based on the information contained in Malechek and Smith (1972), we concluded that the assimilation efficiency of the cattle was approximately 20 percent. The $F_{7,4}$ parameter was calculated as a constant fraction of the food intake (.80 x $F_{2,7}$).

Forcings and Outputs

The two sources of nitrogen inputs into the system assumed in the model were fixation by the cryptogamic crust and nitrogen in precipitation. Other sources of inputs such as erosional runon were considered to be of negligible importance, but could be included in a future model; if a nearly level midvalley site is chosen, then runoff should equal runon. There are probably small net exports of nitrogen due to livestock grazing (Cowling, 1977). Leaching to a water table does not occur in these arid locations since soil moisture changes are rarely noted below 1-m depths and permanent water tables exist at more than 100 m below the surface for the bulk of the vegetated portions of cold winter desert valleys. Phreatophytic zones, such as those dominated by *Sarcobatus*, *Allenrolfea* or *Salicornia*, are not considered here.

Each compartment in the model contained both external forcings and losses (often referred to as sources and sinks). The source for X_1 was considered atmospheric nitrogen fixation and nitrogen input due to precipitation. The nitrogen content of rainfall was measured in Snowville, Utah, in 1972 and 1974 (Skujins and West 1973, 1974). Although there were some seasonal patterns detected, the overall yearly mean was approximately 0.16 g N/m^2 per cm precipitation. In an average year, the rainfall in this valley is about 25 cm. Thus, the nitrogen flux in precipitation averages about 0.4 g N/m^2 annually.

Nitrogen input due to fixers was assumed to be an internal function of the fixers themselves. The average annual input of biologically fixed nitrogen was estimated to be 1.6 g/m^2. Time variations were first considered to be an on and off function of rain and snowmelt. The function was turned on at a constant rate when there was snowmelt or precipitation. As an approximation, the fixation rate was assumed to be in the form $Z_1 = kM_ST_S$, where k is a constant, M_S is the soil moisture at the time of interest, and T_S is the soil temperature (°F) at the time of interest. It was subsequently felt that the process was in reality a recipient controlled phenomenon; therefore, the final function appeared as $Z_1 = kM_ST_SX_1$.

Table 13-4 *Forcings and Outputs [Sources and Sinks] in Curlew Nitrogen Model*

Symbol	Parameter	Average annual flux	Source
Z_1	Atmospheric fixation by X_1	1.6 g N/m^2 per yr	Skujins and West (1973,74)
Z_6	N in precipitation	.4 g N/m^2	Skujins and West (1973,74)
Z_7	Turn in of livestock	0.00009 g N/m^2	Malechek and Smith (1972)
Y_4	Ammonia volatilization from litter, feces and urine (75% of animal material)	27.85% of standing crop in X_4	Cowling (1977) Intuition
Y_5	Ammonia volatilization and denitrification from soil organic fraction	12.6% of the input to X_5	Skujins and West (1973,74)
Y_6	Volatilization and denitrification from soil solution $Z_6 + F_{2,6} + F_{5,6} - F_{6,5}$	99.7% of the input to X_6	Skujins and West (1973,74)
Y_7	Removal of livestock	Entire standing crop at time of removable variable	Malechek and Smith (1972)

Table 13-5 *An A Matrix for the Linear Equation* $X_i = \Sigma Z_i + \Sigma a_{i,j}X_i - \Sigma Y_{i,o}$ *Where the* $a_{i,j}$ *Elements are Rate Constants for Transfers Between the* i *and* j *Compartments*

Recipient \ Donor	X_1	X_2	X_3	X_4	X_5	X_6	X_7	Z
X_1	A_{11}	0	0	0	0	0	0	Z_1
X_2	0	A_{22}	0	0	0	0	0	Z_2
X_3	A_{13}	A_{23}	A_{33}	0	0	0	0	Z_3
X_4	0	A_{24}	A_{34}	A_{44}	0	0	A_{74}	Z_4
X_5	A_{15}	A_{25}	0	A_{45}	A_{55}	A_{65}	0	Z_5
X_6	0	A_{26}	0	0	A_{56}	A_{66}	0	Z_6
X_7	0	A_{27}	0	0	0	0	A_{77}	Z_7
Y	Y_1	Y_2	Y_3	Y_4	Y_5	Y_6	Y_7	

The source for the vegetation compartment (Z_2) was considered to represent nitrogen contained in precipitation that entered the system through foliar absorption. This was considered to be negligible in the actual simulations.

The forcing for the invertebrate consumers (Z_3) is considered to be immigration into the area. This function does not enter into the dynamics of the simulation model at present because it is assumed that immigration equalled emigration. However, if it becomes desirable to ask questions about the consequences of a sudden influx of invertebrate consumers on nitrogen cycling, the model might provide some meaningful insights.

Precipitation input, foliar drip and leaching represent the forcing on X_4. At present this input is of little consequence in this system (West, Chapter 11 of this volume); however, Z_4 may represent a substantial input if the area of interest is near a large agricultural region in which large quantities of nitrogen fertilizers are used.

The portions of the external nitrogen input which pass through the litter in the organic (Z_5) or inorganic form (Z_6) represent the forcing functions for X_5 and X_6, respectively.

The import of cattle due to seasonal grazing practices represents the forcing (Z_7) on the vertebrate compartment (X_7). This function is simply an all or none condition, in that the cattle are either brought in on a given date or they are not present. Therefore, one may look upon Z as a step function.

The losses associated with a given compartment (Y_i) are often a direct function of the forcings on the same compartment. The loss from the fixer compartment (Y_i) represents the nitrogen loss associated with denitrification and volatilization and at present is considered as a constant fraction of Z_1. For the Curlew Valley simulation this fraction was suggested as 0.83 (Skujins, pers. commun.). The high loss of fixed nitrogen for Curlew Valley is associated with seasonal drying-wetting periods and alkaline soils (Skujins and West 1973).

The loss from the vegetation compartment (Y_2) is a constant fraction of the forcing on this compartment (Z_2). Since Z_2 was considered negligible in the Curlew Valley simulation, Y_2 was also considered insignificant.

For the invertebrate consumer compartment the external loss (Y_3) was considered to consist primarily of emigration from the study site. For the current simulation it was assumed that $Y_3 = Z_3$.

The loss from the litter and feces compartment (Y_4) represents mostly ammonia volatilization. This loss is a function of both mean daily and maximum daily temperatures, a function of soil surface moisture and a function of animal activity. The volatilization associated with the animal activities is a constant fraction of the excretory products deposited. For the insects this fraction represents roughly 0.02 since the nitrogen excretory products are not volatile to any large extent (Gist and Sferra, Chapter 10 of this volume). In contrast, the excretory products of cattle contain large quantities of volatile nitrogen. Williams (1970) estimated that a fraction of roughly 0.8 of the urine and feces is volatilized in semiarid environments. Based on this information, the model calculates nitrogen losses from cow excretory products to be 0.81 of that deposited. The loss of nitrogen from the litter and feces compartment is also a direct linear function of litter production between the above-ground vegetation compartment vs. the soil temperature and soil moisture. However, this linear relationship holds only between certain critical bounds, as may be seen in Table 13-6.

The loss from the system associated with the biotic soil fraction (X_5) has been described as denitrification and volatilization of nitrogen from the organic soil compartment. This loss has been calculated in the model as 0.126 (Skujins, pers. commun.) of the total amount coming into the compartment. Skujins feels that virtually all the mobile nitrogen found in the nonbiotic soil compartment is denitrified or volatilized. Based on this statement by Skujins, Y_6 was calculated as 0.997 of the total input into compartment X_6.

Movement of herds from one grazing area to another in accordance with range management practices, as well as harvesting, represents the loss from the cattle compartment (X_7). This loss function (Y_7) is simply a function of the size of the X_7 compartment at the time of removal. At the time of removal of the cattle, $Y_7 = 1.0 \cdot X_7$.

It is realized that the use of a constant fraction of the input for the calculations of Y_1, Y_4, Y_5 and Y_6 is somewhat artificial. These fractions will probably fluctuate somewhat according to the ambient weather conditions; however, these data are lacking at present, and therefore the constant values were used. If the needed information is available in the future, then it can be incorporated into the model with very little effort.

The mathematical representation of the Y_i's is shown in Table 13-6.

IMPLEMENTATION OF THE MODEL

The model is a piecewise, linear model in ordinary differential equation form (Tables 13-5 and 13-7). Over the whole regime of simulation the model is nonlinear because of discontinuities in the β functions which control the fluxes from one compartment to another. However, within a given set of β values the model is linear. The form of the β involved may be seen in Table 13-7.

The model was executed in PL I, using a Burroughs 6700 computer. The model was written in a general form such that any computer containing a PL I compiler could implement the model with few changes. The coding has also been translated into FORTRAN IV for more universal application.

The physical structure of the model is composed of five parts:

1. Declarations.
   Initializations.
   Changes in default values of parameters.
   Establishment of values of Alpha coefficients (*aij* in Table 13-5).
2. Begins daily calculations, updates counters, writes page headings. If changes in state variables are required, reads in net changes, records these on an output file.
3. If a Monte Carlo simulation is requested, then reads in the daily probability of precipitation, and the mean and standard deviation for precipitation, maximum and minimum daily temperature; calculates precipitation and maximum and minimum daily temperatures assuming a normal distribution; calculates snowfall, snowpack, snowmelt; calculates ground and surface temperatures, ground and surface moisture. Reads in daily

Table 13-6 *Equation System for Forcings and Losses Associated with Each of the Compartments in the Model. Constant Values for the Curlew Valley Simulation*

Where

S_m = soil moisture (%)

S_t = soil temperature (°F)

D_p = daily precipitation (inches)

$Z_1 = .0099\ (S_m)\ (S_t - 32)X_1$

$Z_2 \simeq 0$

$Z_3 - Y_3 = 0$

$Z_4 \simeq 0$

$Z_5 \simeq 0$

$Z_6 = 0.4064\ (Dp)$

$Z_7 = 1.17$

$Y_1 = 0.835\ (Z_1)$

$Y_2 \simeq 0$

$Y_3 - Z_3 = 0$

$Y_4 = 0.2785\ (F_{24} + F_{34} + F_{47} + Z_4)$

$Y_5 = 0.1264\ (F_{15} + F_{25} + F_{45} + F_{65})$

$Y_6 = 0.9973\ (Z_6 + F_{26} + F_{56})$

$Y_7 = 1.0\ (X_7)$ upon removal

otherwise

$Y_7 = 0$

Table 13-7 *Generalized System Equations for Nitrogen Simulation Model for Curlew Valley*

Where

S_m = soil moisture (%)
S_t = soil temperature (°F)
G_m = soil surface moisture (%)
G_t = soil surface temperature (°F)
A_t = air temperature (°F)
Ppt = precipitation (inches)

$\beta_1 = \{e^{.08G_t} - 1.7\}/5.55$	when $32 < G_t < 120$
$\beta_1 = 0$	when $32 > G_t > 120$
$\beta_2 = .5(1 + \sin\{(3.1415927[G_m - 10]/20) - 1.5707963\})$	when $10 < G_m < 30$
$\beta_2 = 1$	when $G_m > 30$
$\beta_2 = 0$	when $G_m < 10$
$\beta_3 = 1$	when $37 < G_t < 120$
$\beta_3 = 1 - \{(G_t - 120)/6\}$	when $120 < G_t < 126$
$\beta_3 = 0$	when $31 > G_t > 126$
$\beta_4 = G_m$	
$\beta_5 = G_t - 32$	when $G_t > 32$
$\beta_5 = 0$	when $G_t < 32$
$\beta_6 = 1 - \{(S_m - 25)/.1\}$	when $S_m > 25$
$\beta_6 = 10S_m$	when $.1 < S_m < 25$
$\beta_6 = 0$	when $S_m > 25.1$
$\beta_7 = .0172A_t$	when $32 < A_t < 90$
$\beta_7 = 1$	when $A_t > 90$
$\beta_7 = 0$	when $32 > A_t$

precipitation, maximum and minimum daily temperatures; calculates ground and surface temperatures and surface moisture by interpolating between data points.

4. Calculates Beta coefficients; calculates inputs and outputs; if inputs are to be held constant, then set

$$Z(1) = 0.054795$$
$$Z(6) = 0.001096$$

Updates counters on any nonzero impulses to state variables; computes daily fluxes and updates the A matrix.

5. Computes changes in the state variables; updates state variables using Euler integration with $\Delta t = 1$ day; calculates values for maxima and minima of all important variables; prints out values of state variables and all important system parameters; returns to Part 2 to continue the daily simulations.

The model may be driven by either deterministic or stochastic processes for precipitation. In the deterministic version of the model, the daily precipitation input will be entered for a specific year of interest, most often from weather station data from the nearest station provided by the U.S. Department of Commerce. This source is also adequate for air temperature data. One could use Desert Biome study site data, if sufficient data are available.

For the Curlew Valley study a Monte Carlo process generator was constructed to generate the precipitation input. This generator was constructed in such a manner that if the probability of rain for a given day of the year was greater than a probability generated by the Monte Carlo process, there was a rain input into the model. For desert areas, the decision whether the system will receive rain or not is necessary but not sufficient; therefore, storms of different sizes were also generated. The storm size was simulated in the same manner as the decision to generate rain. If a Monte Carlo-generated probability was less than the probability of a storm of a given size, then the storm of the proper intensity was used as precipitation input to the model for that day. The probabilities associated with the storms and their intensities for a given day were based on 12 years of data from the Snowville weather station.

For soil surface moisture in the deterministic case, the soil moisture at any time within the simulation period is extrapolated from the known relationships with the weather data (Hanks and Nimah 1973). If the simulation step is shorter than the time spread of the input data, then a linear interpolation of the data is implemented. If the precipitation information is generated stochastically, then soil surface moisture is a function of the air temperature, field capacity, water penetrability and precipitation.

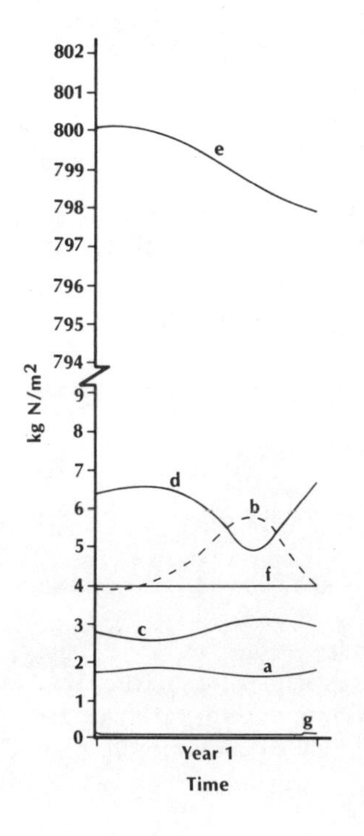

Figure 13-3 *Baseline simulation showing the seasonal dynamics of the nitrogen content [g/m²] of each compartment over 1 calendar year. Deterministic weather generator employed.*

a = nitrogen-fixing organisms; b = non-nitrogen-fixing primary producers; c = invertebrate consumers; d = surface litter; e = biotic soil fraction; f = non-biotic soil fraction; g = vertebrate consumers.

RESULTS

Simulation results are presented in Figures 13-3 to 13-12. Initial simulations were performed to determine if the model was stable and performed in an acceptable manner. Subsequent simulations were performed to address specific questions.

Deterministic baseline simulations were made with climatic information for Snowville in 1972 (Fig. 13-3). Stochastic baseline simulations were made with 1961-72 Snowville weather data. Normal distributions were assumed for daily maximum and minimum temperatures and precipitation. In comparing Figures 13-3 and 13-4, one notes considerable differences between the simulations using stochastic (Fig. 13-3) and deterministic weather generators (Fig. 13-4). The reason for these differences is that rate constants were calibrated for 1972 weather data, but the Monte Carlo weather generator produced somewhat greater than average long-term precipitation and higher temperatures than happened to occur in 1972. Since precipitation and temperatures are major driving variables of the system, these differences in amplitude caused different fluxes between components. Nitrogen fixation and nitrogen content of nonfixers are responses affected by these major driving forces. We simply cannot choose between these two alternatives because subsequent simulations require one or the other of the data bases. We needed the

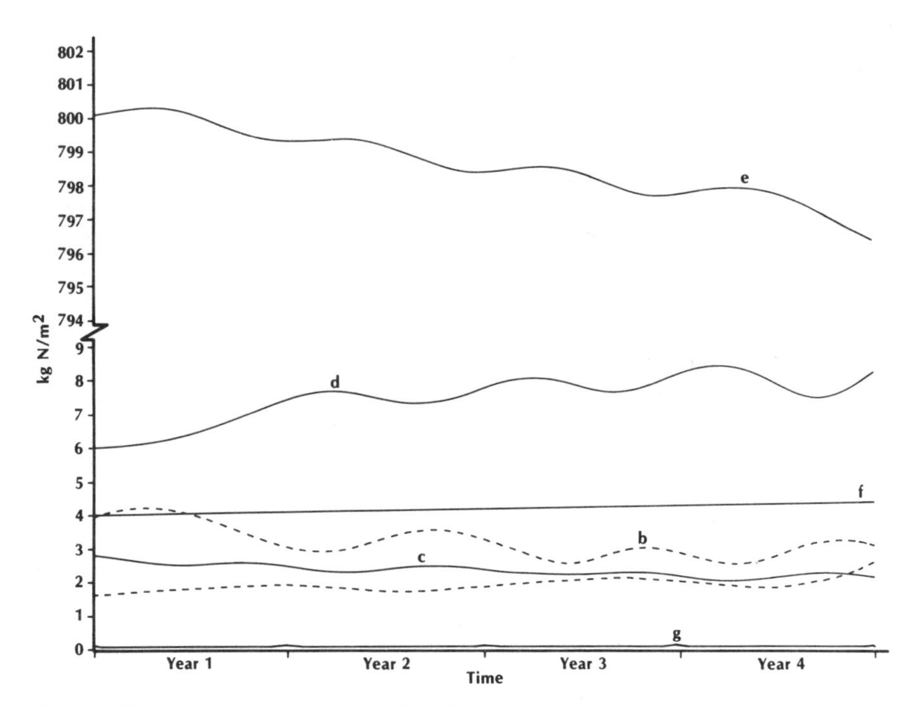

Figure 13-4 *Baseline simulation showing the seasonal dynamics of the nitrogen content [g/m²] of each compartment over 4 calendar years. Stochastic weather generator employed. See Figure 13-3 for identification of curves.*

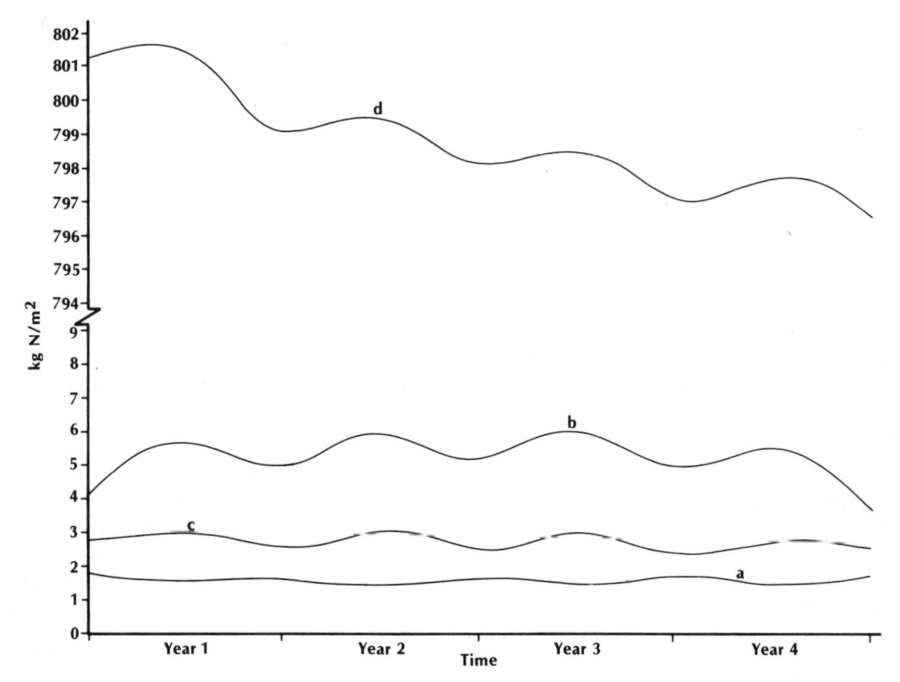

Figure 13-5 *Simulation showing the effect of doubling the number of rainy days on the nitrogen content [g/m^2] of selected compartments. Stochastic weather generator employed over 4 calendar years.*

a = nitrogen-fixing organisms; b = non-nitrogen-fixing primary producers; c = invertebrate consumers; d = biotic soil fractions.

Monte Carlo weather base for anything requiring simulations greater than 1 year long.

Following the preliminary simulations, several questions were asked of the model. These questions were as follows: What would be the effects on nitrogen in the total system if: 1) the total annual rainfall were doubled by doubling the number of rainy days and holding the intensity to "normal"? 2) The total annual rainfall were doubled by doubling the intensity of the storms but holding the number of rainy days to "normal"? 3) Cattle grazing were doubled or tripled? 4) Invertebrate grazing were doubled? 5) Cattle grazing occurred during the summer instead of the winter? 6) The consumers (vertebrate and invertebrate) were removed?

The question of doubling the rainfall was initiated by the work of Trumble and Woodroffe (1954), who suggested that when primary production drops on Australian saltbush ranges over successive years of adequate rainfall, it is due to the limitations of available nitrogen and other nutrients. The question was subsequently expanded to embrace not only the quantity of rainfall but also the "quality" of rainfall (quality here meaning the increase in storm intensity). A third simulation dealing with precipitation consisted of doubling the number of rainy days and doubling the storm intensity, thus in fact quadrupling the total annual rainfall.

Figure 13-5 shows the result of doubling the number of rainy days. Nitrogen fixers (X_1) appear to decrease over time. This decrease may be due to the release of the moisture stress followed by enhanced production and a subsequent resource limitation. The potential limiting resource allowing enhanced production may be organic carbon. Wallace and Romney (1972a) and Skujins and West (1974) have shown that enhanced moisture and nitrogen amendments lead to a limitation of desert plant productivity through organic carbon. Through potential limitation of organic carbon and the associated decline in the nitrogen fixers (X_1), there is a resultant decrease in the flux of nitrogen into the soil organic nitrogen compartment (X_5). The primary producers (X_2) are also released from moisture stress and begin to enhance production, which entails an increased flow from the soil organic nitrogen compartment (X_5) into the primary producer compartment (X_2). The enhanced flow to to the primary producers (X_2), coupled with a decline in the input from the fixers (X_1), results in a net loss in the soil organic nitrogen (X_5) compartment.

The second simulation explored the consequences of doubling annual rainfall through the doubling of storm intensity (Fig. 13-6). It will be noted in comparing Figure 13-6 with Figure 13-4 that the same basic trends hold for each compartment. However, the effects on a given compartment seem to be amplified. The fixer compartment (X_1) exhibits a deeper decline and has less recuperative power; the primary producer compartment (X_2) exhibits lower peaks and deeper troughs; the soil

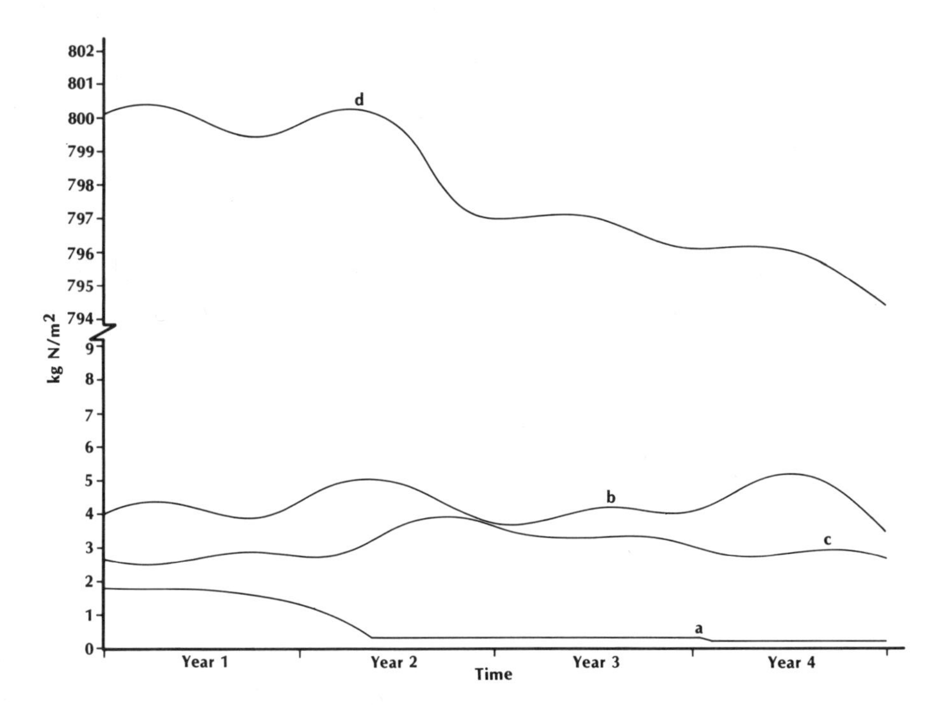

Figure 13-6 *Simulation showing the effect of doubling the average storm intensity on the nitrogen content [g/m²] of selected compartments over 4 calendar years. Stochastic weather generator employed. See Figure 13-5 for identification of curves.*

organic nitrogen compartment (X_5) exhibits a steeper and more pronounced decline to an ultimately lower level than in Figure 13-4. The biological rationale for the results demonstrated in this simulation are difficult to decipher; however, it may be possible to postulate some reasons for the behavior in Figure 13-6. It will be noted that the peak in the nitrogen fixers (X_1) is higher in Figure 13-6 than in Figure 13-5; this would result in an increased utilization of the limited organic carbon resource, thus causing a rapid decline in the standing crop of nitrogen fixers. The resulting decline would in turn cause a general decline in the overall nitrogen stature of the system. Once the fixers have declined to a very low level, the system must then rely for the most part on the inadequate supply of nitrogen input due to atmospheric scrubbing by precipitation. Rainfall sources for nitrogen will not prevent the ultimate decline of the system, but might slow the process. The invertebrate consumers (X_3) in Figure 13-6 demonstrate dynamics similar to that in Figure 13-5.

The enhancement of initial production for both nitrogen fixers (X_1) and primary producers (X_2) associated with the increase in storm size as compared to that of the increase in the number of storms may be related to evaporation. Curlew Valley was on the bottom of pluvial Lake Bonneville and is consequently quite flat. Because of the limited relief in the area there is little if any runon or runoff in the system; therefore, the resulting precipitation tends, for the most part, to remain in place. With little or no change in the movement of water onto or off the area, the heavier storms would tend to saturate the soils to a greater extent than the lighter ones; consequently there would be more moisture available to the fixers and producers for a given amount of precipitation.

In another simulation both the storm size and the number of rainy days were doubled (Fig. 13-7). The results appeared to be a hybrid of the two previous storm treatments. In the fixers (X_1) the initial peak equalled that of the doubling of the number of rainy days; however, the pattern of decline followed that of the increased storm size (Fig. 13-6). The nitrogen content of primary producers (X_2) showed a peak somewhere midway between the peaks associated with the increase in storm size and the increase in the number of rainy days. The general dynamics and peaks associated with the invertebrate consumers (X_3) seemed to follow the enhanced-storm-size simulations most closely.

The biological implications of the results presented in Figure 13-7 are obscure and not decipherable at this point. If arid and semiarid systems have evolved to respond to just so much moisture, this simulation deviated so far from the real world as to yield unreliable results. In addition, the precipitation inputs were so far from the neighborhood of steady state that the simulations associated with a linear model may be considered unrealistic.

The second group of questions for simulation dealt with manipulations of consumer population and activities. First, the effect of doubling insect consumption was examined. The biological realities of such a question may be associated with an insect population explosion, such as the variegated army worm or the black grass bug. During a major outbreak, a given insect species may more than double its population and therefore its consumption. This simulation was restricted to the doubling because of the realities of a linear model. The results of the simulation associated with the doubling of the insect consumption may be seen in Figure 13-8. This simulation was based on a 1-year simulation and shows that doubling the insect population has only a slight effect on the primary producers (X_2), other than depressing the maximum peak. On the whole the consumers (X_3) had little effect on

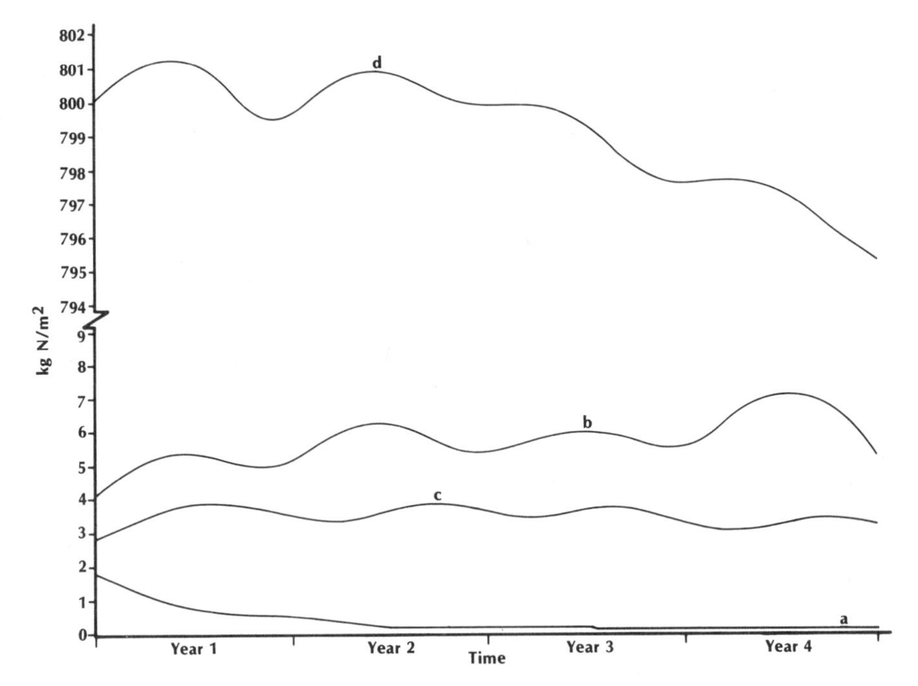

Figure 13-7 *Simulation showing the effect of doubling both the number of rainy days and the average storm intensity on the nitrogen content [g/m²] of selected compartments over 4 calendar years. Stochastic weather generator employed. See Figure 13-5 for identification of curves.*

the producer nitrogen curve form. The invertebrate consumers follow the curves of nitrogen content in both the nonfixers and the nitrogen fixing plants (X_1) quite closely, exhibiting a time lag associated with the peak in X_3. In the real world, invertebrate consumers queue on their food sources to some extent and will increase their populations when food is plentiful. Many species have evolved a life history to maximize population size when food is maximum. In other words, many invertebrate life histories are synchronized with the food-plant's phenology. The close tracking of the invertebrate consumers to the nitrogen fixers may be attributed to the soil surface invertebrates who feed on the cryptogamic crust. In this case, the similarity of the two curves is no doubt based on the knowledge that the two groups require approximately the same environmental conditions for their activity. There is little or no evidence that the invertebrates queue on the nitrogen fixers per se.

To bring practical management questions into the investigation of the model, some questions were asked with respect to the livestock grazing regime at Curlew Valley. The first question dealt with altering the intensity of cattle grazing, such as doubling and/or tripling. These simulation results are presented in Figures 13-10 and 13-11. The nitrogen in primary producers (X_2) appeared not to be very sensitive to the changes in grazing pressure, in that there was little change between doubling and tripling the grazing pressure. Thus, one may conclude that the total nitrogen content of the vegetation remains similar under increased grazing pressure, but

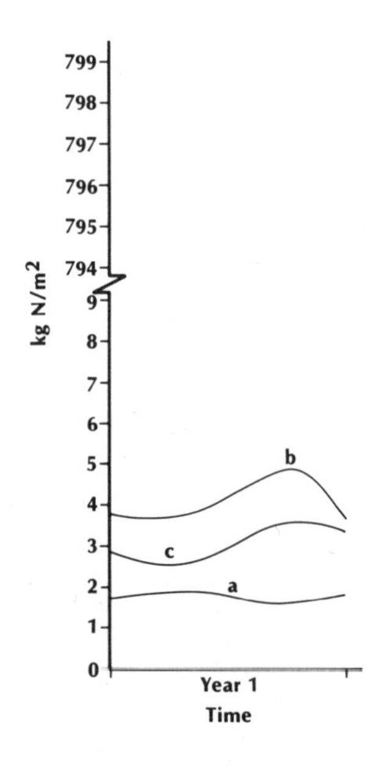

Figure 13-8 *Simulation showing the effect of doubling the grazing by insects on the nitrogen content [g/m²] of selected compartments over 1 year. Deterministic weather generator employed. See Figure 13-5 for identification of curves.*

changes within the vegetation to perhaps different components of different species (Cook and Stoddart 1953; Cook et al. 1954).

The second question associated with grazing was related to the period of use. If the cattle graze during the summer instead of the winter, would the nitrogen in net primary production be depressed? One would suspect that there is indeed such a suppression. However, as seen in Figure 13-11, once again the nitrogen content of the primary producer compartment is not very sensitive to the change in grazing strategy. The results are counterintuitive, since one would suspect extensive suppression of primary production if the system was grazed during the height of the growing season (Cook 1971).

Upon examination of the simulation results associated with the cattle grazing regime, one tends to suspect the potential conclusions. However, here is a classic case where questions were asked of the model which it was not specifically designed to handle. In this case only the nitrogen content of the primary producers and every other compartment is considered. Thus the model was not designed specifically to answer questions about cattle grazing per se; this came as an afterthought. The primary producer compartment is composed of all plant species found in the area, most of which consist of species where the entire plant or a substantial portion thereof is not palatable to the cow. Consequently, the enhancement of grazing may have a severe impact on the palatable species but have little effect on the overall primary producer compartment. In order to bring the model onto a more realistic footing, one should partition the primary producer compartment into those species grazed by cows and those not grazed by cows. Under the aforementioned conditions, the system would undoubtedly be more sensitive to changes in grazing pressure and in grazing regime. In addition, one may need to consider relative palatability of the different plant species in a manner similar to Wilkin (1973).

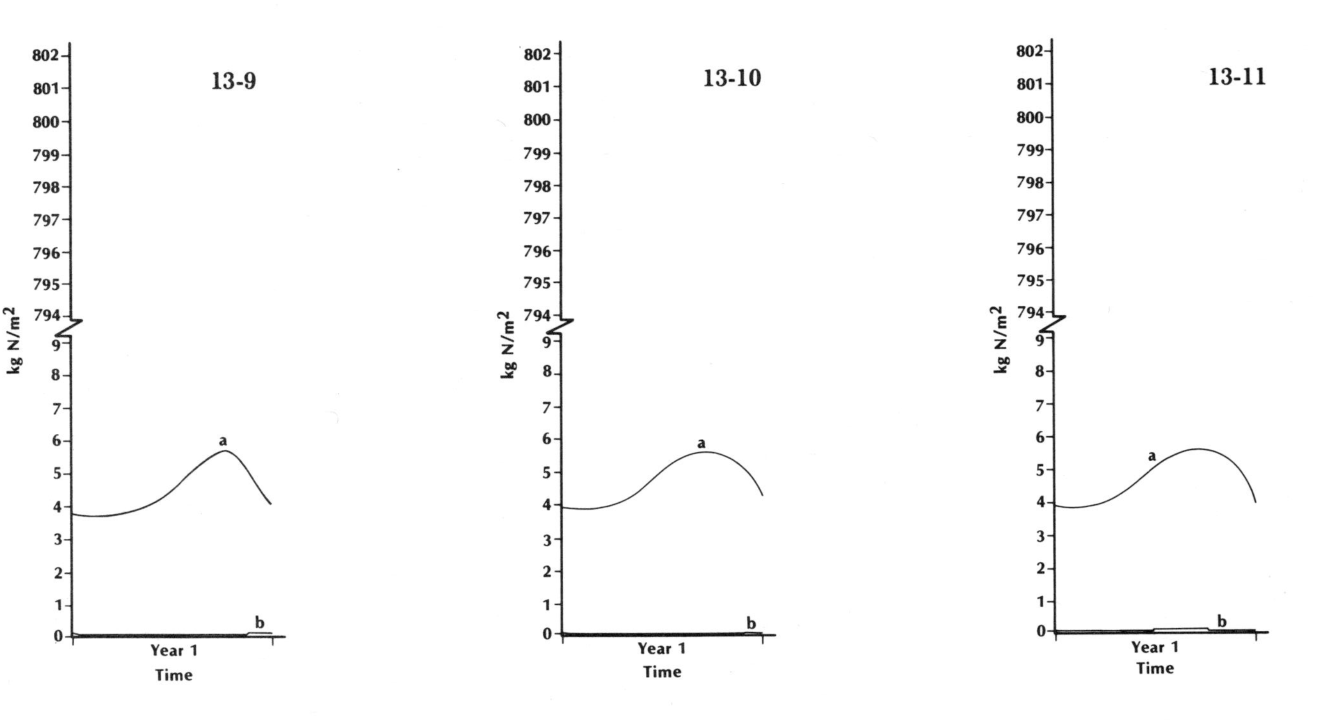

Figures 13-9 to 13-11 *Simulation showing the effects of different cattle grazing treatments on nitrogen content [g/m²] of selected compartments over 1 year.* **Fig. 13-9**, *cattle grazing doubled;* **Fig. 13-10**, *cattle grazing increased threefold;* **Fig. 13-11**, *cattle grazed in summer instead of winter.*

a = non-nitrogen-fixing primary producers; b = vertebrate consumers.

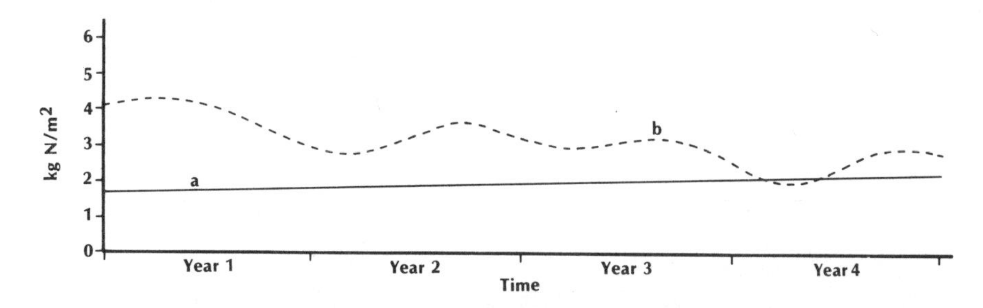

Figure 13-12 *Simulation showing the effect of eliminating herbivory on the nitrogen content* [g/m^2] *of selected compartments over 4 years. Stochastic weather generator employed.*

a = nitrogen-fixing organisms; b = nitrogen-fixing primary producers.

Since most manipulations of herbivory had minor or questionable impact, we decided to simulate the removal of all herbivores. This could be done by fencing out ungulates, rodents and lagomorphs and by killing invertebrates with a systemic such as Temik. These simulations (Fig. 13-12) show little increase in the nonfixer plant components over that of the Monte Carlo baseline (Fig. 13-4). The fixers in the cryptogamic crust, on the other hand, increase about 25 percent over 4 years. Perhaps animal control of community structure and function is more important for the nonvascular than for vascular plants in these ecological contexts.

CONCLUSIONS

The model-building process is a way of examining what is known or needs to be known to simulate the nitrogen cycle of a representative Great Basin Desert area. The paucity of information on many ecologically important compartments and fluxes was readily apparent. Research projects were funded within the Desert Biome to gather data where prior knowledge was scant or nonexistent. This research helped, but did not fill all the gaps. We used the available data and reasonable extrapolations to provide numbers necessary for simulations with a 7-compartment, linear, donor-controlled differential equation model.

Simulations of normal responses seemed within expected bounds. Several experimental simulations with altered precipitation regimes made ecological sense. Simulations with changed intensity and timing of herbivory illustrated the greater relative importance of invertebrate over vertebrate herbivory in nitrogen cycling. More sophisticated conclusions were thwarted by lack of better data and by the simplicity of present model structure.

Models such as the one illustrated here are useful heuristic constructs helping us to understand ecosystem function. Attempts to ask sophisticated questions will require increasingly sophisticated models. This progress cannot be made without more detailed data on compartments and processes that are now understood only in very vague or general ways. The modeling process can serve as a guide to ranking research priorities. As a result of this activity, we can now better enumerate the most critical deficiencies for a better understanding of the nitrogen cycle of Great Basin Desert ecosystems.

14

NITROGEN CYCLE IN THE NORTHERN MOHAVE DESERT: IMPLICATIONS AND PREDICTIONS

A. WALLACE, E. M. ROMNEY and R. B. HUNTER

INTRODUCTION

The nitrogen cycle in desert ecosystems is generally believed to exercise a limitation on the rate of primary production since very low soil organic matter levels associated with low nitrogen cycle activity are usually found in deserts (Jenny et al. 1949; Kononova 1961; Fuller 1974, 1975a, 1975b). These workers report relatively rapid decay of organic residues in desert soils and intense mineralization rates. Information on the nigrogen cycle in deserts, particularly relating to nitrogen fixation, is very meager (Garcia-Moya and McKell 1970; Bjerregaard 1971). Until recently when acetylene reduction assay became a widely used tool (Hardy et al. 1971; Burris 1975), except for nitrogen fixation with legumes and nonsymbiotic nitrogen fixation, mechanisms for input of newly fixed nitrogen into desert ecosystems were not understood (Fuller et al. 1960; Shields and Drouet 1962; Shields et al. 1957; and Tchan and Beadle 1955). This technique and the use of ^{15}N have permitted a fresh look at nitrogen cycles under many conditions (Ness et al. 1963; Stewart 1970; Smith and Legg 1971; Burris 1975). The newer methods have helped considerably in the understanding of nitrogen cycling phenomena in the desert biome (Tucker and Westerman 1974).

The IBP studies of nitrogen dynamics in the Great Basin Desert of Utah, Nevada and Idaho have indicated that there are relatively large annual inputs of nitrogen (10-100 kg/ha per year; see Rychert and Skujins 1974b, Sorensen and Porcella 1974; West and Skujins 1977) by lichen crusts and algae and by precipitation (~12 kg/ha). Nearly all appeared to be fairly quickly lost, however, primarily by denitrification (Skujins 1975; Skujins and West 1973, 1974). Using changes in total nitrogen concentrations in soil from Great Basin Desert areas, Bjerregaard (1971) found the equivalent of from 2 to 3 kg N/ha per year fixed by free-living microorganisms. MacGregor and Johnson (1971) reported that algal crust organisms in the Sonoran Desert soils fixed from 3 to 4 g N/hr per ha of crust following rainfall. They estimated algal crust cover at 4 percent of the surface. Mayland et al. (1966) obtained similar results in the Sonoran Desert.

Algae-lichen crust covers up to 80 percent of the Great Basin Desert soil (Skujins and West 1973; Lynn and Vogelsberg 1974). In contrast, in the northern Mohave

Desert, nitrogen-fixing lichen crust covers only 1 percent or less of the soil (Shields 1957; Shields et al. 1957; Nash et al. 1974; Hunter et al. 1976b) and its fixation of nitrogen has been calculated (Hunter et al. 1976b) as less than 100 g/ha per year. Precipitation in the northern Mohave Desert provides $\leqslant 1$ kg N/ha per year (Hunter et al. 1976b). Losses by denitrification are assumed to be low, as nitrate builds up to very high levels ($\geqslant 1{,}000$ ppm nitrate N) in certain soils receiving runoff (Romney et al. 1973; Hunter et al. 1975b). Total soil nitrogen ranges from 0.01 to 0.25 percent (Romney et al. 1973). Annual utilization of nitrogen by plants is usually less than 10 kg/ha (Hunter et al. 1975b) and most is returned to the soil as organic nitrogen in the litter.

In the Sonoran Desert the lichen crust is considered to be a major source of nitrogen input contributing a maximum of about 0.5 kg N/ha per year (Mayland et al. 1966; Mayland and McIntosh 1966; MacGregor and Johnson 1971), assuming a 4 percent soil cover (MacGregor and Johnson 1971; Nash et al. 1974).

Although data from the Chihuahuan Desert are unfortunately scanty (Staffeldt and Vogt 1974), there are indications of fundamental differences between nitrogen dynamics of the cool, and relatively wet, Great Basin Desert and the three regional warm southern deserts (Fig. 14-1). The authors know of no particular differences among the southern deserts.

Our aim here is to summarize our current understanding of the entire nitrogen cycle in the northern Mohave Desert and contrast it with characteristics of the cycle in other situations.

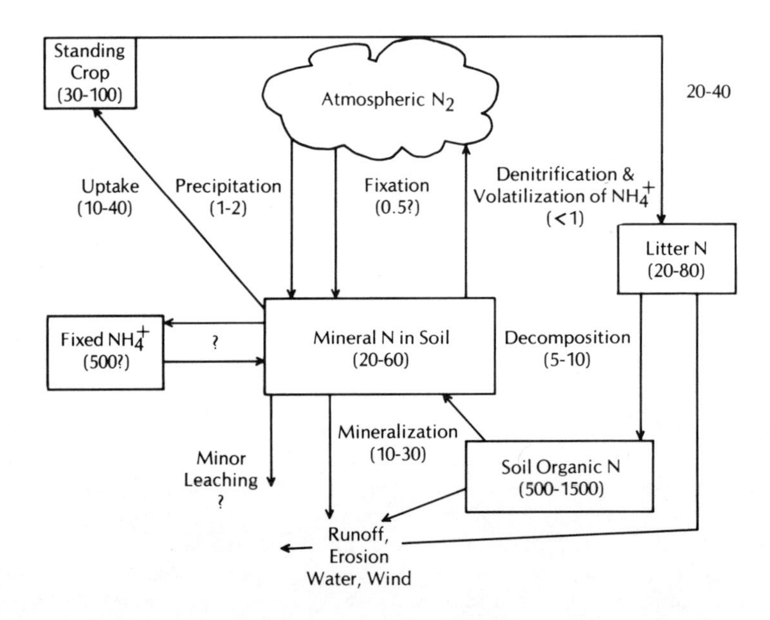

Figure 14-1 *Nitrogen cycle model for northern Mohave Desert; kg N/ha for compartments* [*boxes*] *and kg N/ha per year for fluxes* [*arrows*].

NORTHERN MOHAVE DESERT NITROGEN CYCLE

A generalized outline of the nitrogen cycle of the northern Mohave Desert (Fig. 14-1) indicates a relatively large (for deserts) pool of soil organic nitrogen in equilibrium with a pool of soluble mineral nitrogen. This mineral pool is, ignoring differential distribution problems (Romney et al. Chapter 16 of this volume), more than adequate to meet the needs of seasonal plant productivity in the highest rainfall years (Turner and McBrayer 1974). The soil organic nitrogen is part of soil organic matter which averages about 0.75 percent of the soil with a C:N ratio of about 10 (Romney et al. 1973). Annual inputs and losses of nitrogen in the system are relatively small and much smaller than those suggested for the Great Basin Desert (Skujins 1972; Skujins and West 1973, 1974; West and Skujins 1977). This concept differs greatly from that proposed by Reichle (1975). It parallels the information obtained for short grass prairie, however, where fixation of nitrogen by algae, and bacteria with or without plants, was very small (Woodmansee, in press). A considerable amount of fixed ammonium N exists on the clay of such soils (Nishita and Haug 1973). Its relative availability to plants is probably low. As yet, decomposition rates of litter are not wholly known, but some approximations have been reported (Wallace and Smith 1954; Rodin and Bazilevich 1967). Some leaf material can turn over in less than 1 year (Hunter et al. 1976b), but large amounts of branch and root materials have remained in the system for a decade without much decomposition. The latter, however, contain only a small part of the nitrogen in the system.

Nitrogen in the Plant and Animal Compartments

The question of nitrogen uptake by desert plants has been dealt with elsewhere (Wallace et al., Chapter 9 of this volume). The pool sizes of the major biota (Table

Table 14-1 *Nitrogen Compartment Sizes in Bajada Areas of the Northern Mohave Desert, 0-100 cm Soil Depth [where applicable]*

Compartment	Size range (kg N/ha)
Soil organic matter	500-1500
Undecomposed litter and dead plant parts	20-80
Nonexchangeable fixed ammonium nitrogen	400-800
Biotic nitrogen components	
Perennial plant roots	20-40
Perennial plant branches and stems	4-20
Leaves, flowers and other new growth	4-20
Annual plants	1-10
Animals	0.2-0.3
Total biotic nitrogen	39-90
Soluble mineral nitrogen in soil	20-80

Table 14-2 *Acetylene Reduction by the Rhizosphere from Roots of Desert Species Showing a Positive Test [fresh-weight basis]*

Species	nmole C_2H_4/g per hr	g N/ha per season[a]
Atriplex canescens	0.168	7
A. confertifolia	3.83	156
Coleogyne ramosissima	0.190	4
Ephedra nevadensis	0.004	1
Larrea tridentata	0.003	1
Lycium pallidum	1.378	113
L. shockleyi	0.074	3
Yucca schidigera	0.756	27

[a]The g N/ha per season was estimated on the assumption of activity for 6 mo and an estimate for the root density of the species.

14-1) indicate a need for approximately 5 percent and sometimes up to 10 percent of the total nitrogen in the system to be in the biota at any one time. This compares with nearly 100 percent in some tropical forest systems and with less than 2 percent in some grassland systems. In a deciduous forest, Mitchell et al. (1975) found about 15 percent in biota. This implies a fairly efficient system in the desert, in which the rate of mineralization can be at least equal to that in temperate ecosystems. The efficiency of the desert system is probably related to the relatively trivial amount of leaching in desert conditions, and also to the shrub-clump "fertile island" phenomenon (Garcia-Moya and McKell 1970; Charley 1972; Charley and West 1975; Romney et al., in press).

The amount of nitrogen in the animal compartments is relatively insignificant and no mechanism is known by which either the vertebrate or invertebrate animals in the system can dominate the nitrogen cycle. It is thought that termites may be influential in the Sonoran Desert of North America (Nutting et al. 1975). Termites there may move nitrogen in organic matter below the root zones and, in addition, may fix some nitrogen (Benemann 1973; French et al. 1976).

Problems Related to Input of Nitrogen to the Northern Mohave Desert

Acetylene reduction studies (Wallace and Romney 1972f; Hunter et al. 1975b) indicate only small inputs from semisymbiotic nitrogen fixation; a few hundred g/ha per year in the northern Mohave Desert (Table 14-2.) Roots of a number of species sometimes give reduction (Table 14-3). The irregular features of the data indicate the critical effects of environmental variables on the fixation processes and the difficulty of studies in this area. Lichens and algae may fix a few hundred grams N/ha per year. No one source seems to be major, but several mechanisms seem to additively combine to result in a small input of newly fixed nitrogen sufficient to maintain the system. Each of these mechanisms operates in other ecosystems, but at higher levels, the extent of which depending upon environmental factors. An extrapolation of the curve for acetylene reduction in England (Day et al. 1975; Fig. 14-2) to low soil moisture levels implies that under desert conditions only a very few kg N/ha should be fixed per year ($y = .793 + .134x - .0014x^2$, where $y = \log$ g N fixed/ha per day and x = percent soil moisture; r^2 = .81). At 10 percent soil moisture, 2.6 g N/ha per day would be fixed and 1.2 g at 7 percent soil moisture, according to the curve in Figure 14-2. These values at low soil moisture are less than 1 kg N/ha per year. If the rate at 20 percent soil moisture were maintained for 60 days about 1.3 kg N/ha per year would be fixed. This seems to correspond with values for our acetylene reduction studies. Sprent (1971) suggested that water stress inhibits biological nitrogen fixation from any mechanism.

In extrapolating from the data of Day et al. (1975) to the northern Mohave Desert, the problem of comparative soil texture and soil moisture potential arises, but since they report that most of the nitrogen fixation is with surface blue-green algae, the comparative differences would not be very different for soils from the two places at equal moisture levels.

The very low levels of acetylene reduction encountered in our studies gave us considerable difficulty due to high variability and also because of background endogenous ethylene production. To avoid this problem the method requires many replications and careful evaluation of data.

The number of leguminous plants in the Rock Valley area are as many as 2,500 individuals of perennial *Astragalus* spp. and up to 50 perennial *Psorothamnus fremontii* per hectare in some locations (Wallace and Romney 1972f). These could put about 200 g N/ha per year into the system. Nonleguminous nodulated plants

Table 14-3 *Nonleguminous Northern Mohave Desert Species Whose Roots or Rhizospheres Sometimes Have Been Quantitatively Positive in Acetylene Reduction Assay*

Species	References
Atriplex canescens	Wallace et al. 1974
A. confertifolia	Wallace et al. 1974
Bromus rubens	Wallace et al. 1974 Hunter et al. 1975b
Coleogyne ramosissima	Wallace and Romney 1972d
Hymenoclea salsola	Wallace and Romney 1972d
Krameria parvifolia	Wallace and Romney 1972d
Larrea tridentata	Hunter et al. 1975b
Lichen crust	Wallace and Romney 1972f Wallace et al. 1974 Hunter et al. 1975b
Lycium pallidum	Wallace et al. 1974
L. shockleyi	Wallace et al. 1974
Menodora spinescens	Hunter et al. 1975b
Stipa speciosa	Hunter et al. 1975b
Tetradymia canescens	Wallace and Romney 1972b Hunter et al. 1975b
Thamnosma montana	Hunter et al. 1975b
Yucca schidigera	Wallace et al. 1974 Hunter et al. 1975b

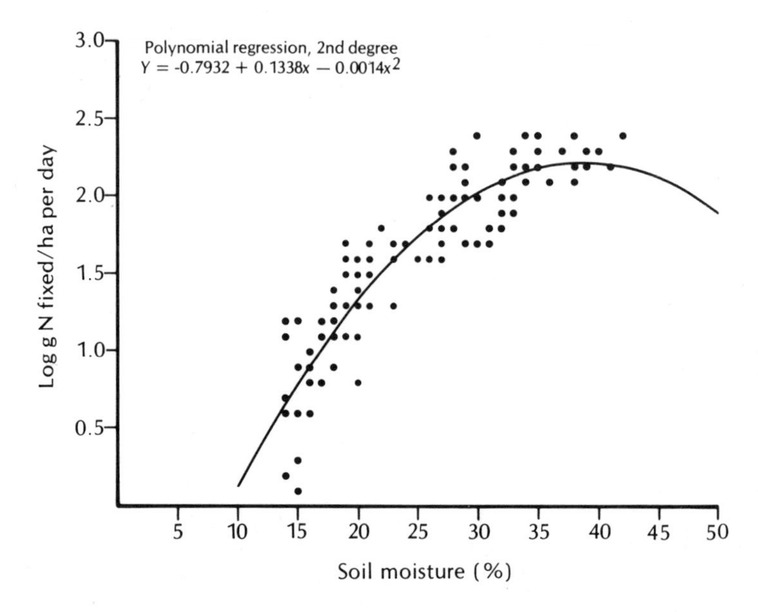

Figure 14-2 *Effect of soil moisture on nitrogen fixation by soil cores with plants [Day et al. 1975; the reference does not give any information on soil moisture percentage vs. soil moisture tension.]*

may be important in some desert ecosystems (Farnsworth et al. 1976), but so far not to our knowledge for the northern Mohave Desert.

The land area covered by lichen crust in the northern Mohave Desert (determined on four 300-m quadrats parallel to the Mercury-Rock Valley road) is sometimes around 1 percent. Such cover could be expected to result in the production of much less than 1 kg N/ha per year (Hunter et al. 1975b), and some of that may be lost after fixation (MacGregor 1972; Skujins 1972; Skujins and West 1973, 1974).

Studies at Mercury, Nevada, indicate that as much as half of the nitrogen input may be as combined nitrogen in rainfall (Fig. 14-1). Although rainfall has always been a source of fixed nitrogen, the exact origin of this nitrogen is not known. The reasons for the fluctuations in this source of combined nitrogen also are not known.

Nitrogen Outgo

Although there are losses of nitrogen from the Mohave Desert system, they are, on the average, relatively smaller than from other ecosystems. Volatilization of ammonium and denitrification are low because the soil is generally dry (Fenn and Kissel 1973, 1976). Leaching losses are low because roots usually occupy the water saturation zone which terminates in a caliche layer in most areas. Large runoff losses of nitrogen can occur in some years when occasional floods are experienced. Soil erosion can occur which results in loss of soil organic matter as well as of mineral forms of nitrogen. In addition, the undecomposed litter can be washed away to

Table 14-4 *Soil Profiles, Some Having Sizable Quantities of Soluble Nitrates [Romney et al. 1973]*

Profile type	Horizon	Depth (cm)	EC 25 (mmhos/cm[a])	Saturation extract nitrate (μg/g soil)
Site 42 - Frenchman Flat; Slope 3% - Physiography Bajada				
Shrub	A1	000-005	1.42	0
Shrub	A2	005-017	0.90	0
Shrub	A3	017-030	1.26	0
Shrub	C1	030-051	1.54	0
Shrub	C2	051- [b]	4.62	0
Bare	A2	000-012	0.57	0
Bare	A3	012-029	0.48	0
Bare	C1	029-054	0.44	0
Bare	C2	054- [b]	4.26	0
Site 43 - Frenchman Flat; Slope 1% - Physiography Bajada				
Shrub	A1	000-028	2.13	0
Shrub	A2	028-041	1.23	0
Shrub	C1	041-063	1.26	0
Shrub	C2	063-097	2.27	0
Shrub	C3	097-153	3.48	4.3
Bare	A1	000-002	0.66	0
Bare	A2	002-012	0.50	0
Bare	C1	012-032	0.41	0
Bare	C2	032-072	0.44	0
Bare	C3	072-128	0.52	0
Site 45 - Frenchman Flat; Slope 0% - Physiography Playa				
Shrub	A1	000-017	1.21	8
Shrub	A2	017-028	0.87	0
Shrub	C1	028-043	0.97	0
Shrub	C2	043-079	1.58	34
Shrub	C3	079-146+	26.08	546
Bare	A2	000-011	0.83	16
Bare	C1	011-024	0.54	0
Bare	C2	024-039	2.61	308
Bare	C3	039-068+	26.08	20
Site 46 - Frenchman Flat; Slope 1% - Physiography Playa				
Shrub	A1	000-010	2.09	13
Shrub	A2	010-027	1.80	0
Shrub	C1	027-042	1.24	0
Shrub	C2	042-061	0.85	0
Shrub	C3	061-087	2.90	9
Shrub	C4	087-125	4.53	171
Bare	A2	000-016	0.78	3
Bare	C1	016-030	0.69	0
Bare	C2	030-049	0.71	0
Bare	C3	049-074	0.71	0
Bare	C4	074-110	4.17	13
Bare	C5	110-135	6.95	300

[a]Of saturation extract.

[b]Caliche layer near.

drainage sinks. Considerable debris following large storms has been observed in the playas. This may be an important source of the nitrate which can accumulate to high levels in these sites (Table 14-4). Such accumulations in playas are indicative of the fact that the fixation processes on the bajadas can be in excess of the need. The levels accumulated in the playas probably represent an input over many centuries from fixation, mineralization and erosion processes occurring on the bajadas above.

There is usually very little visible above-ground litter in the northern Mohave Desert. Fallen leaves are seen usually for a few days only. They could be blown away by wind and broken into small fragments; but if they were, then there should be about as much blown in as blown out. The litter does not seem to be lost by the wind. Another hypothesis is that animals could move much of the litter below ground. This could be of considerable importance, for as much as 10 kg N/ha per year input of above-ground litter cannot be lost, but must be conserved in the system. Validation must await additional studies.

Mineral Pool Dynamics

Nitrate is formed, particularly under shrub clumps, in most months of the year by mineralization and nitrification (Table 14-5). Its level decreases with depth and is higher under shrubs in areas of high fertility (Hunter et al. 1976b) than between shrubs. Ammonium N applied to soil is rapidly nitrified (Table 14-6). Nitrite and soluble ammonium in very small amounts are usually found in the soil solution (Nishita and Haug 1973).

The new growth of the perennial plants requires uptake of somewhere between 5 and 20 kg N/ha per year (but usually less than 10; Table 14-6) during the short growth cycle of springtime, which generally is of 2-3 months duration. Most of this nitrogen must be available in a period of 4 to 6 weeks at the beginning of the growing season. Additional nitrogen is needed for the growth of annual plants. This can vary from less than 1 to 10 or more kg N/ha per year, depending upon rainfall and temperatures (Table 14-1). Thus, totals of 30 kg/ha or more must become available from the various pools in a period of about a month to meet all the nitrogen needs of the biota in the system. In the spring of 1975 we observed up to 70 μg nitrate N/g soil in the top 7.5 cm of soil in the northern Mohave Desert (Table 14-5). This is about 14 kg/ha if the clump soil alone is 20 percent of the area. In the 30-40 cm depth where roots grow, there could be 30 or more kg/ha on the March dates (Table 14-5). In the bare area soils (Table 14-5) mineralization and nitrification also occur and if they represent 80 percent of the soil area, an additional 5 to 10 kg/ha of nitrate N would be available to plants. Since this total of 30-40 kg/ha of nitrate N represents about 2 percent of the total nitrogen in the soil organic matter pool, it may play an important role in nitrogen cycling. Ordinarily under temperate climatic conditions, about 2 percent of the soil organic matter would mineralize in a year. This apparently occurs in about 2 months in the northern Mohave Desert, assuming that total nitrate pool turnover occurs once annually during the growing season.

Table 14-5 *Nitrate Nitrogen Concentrations in Surface Samples Under Five Shrub Clumps and in the Adjacent Bare Areas for 1975 [Each value was determined on a composite of two or three cores. Error estimates are standard errors of the mean. Units are μg nitrate N/g dry weight; ± is standard error of the mean.]*

Depth (cm)[a]	3 Feb	19 Feb	3 Mar	17 Mar	31 Mar	14 Apr	6 May	12 May	27 May	12 Jun	7 Jul	18 Aug	21 Oct	20 Nov	18 Dec
Under five shrub clumps															
0–22.5	2.63	8.96	6.05	17.82	13.52	10.48	7.23	4.09	10.32	5.13	7.84	10.25	8.91	10.53	10.15
	±0.41	±3.24	±2.77	±5.78	±5.08	±2.42	±1.94	±1.01	±1.81	±0.76	±2.23	±2.20	±1.33	±2.28	±1.72
0–7.5	2.66	13.20	11.94	25.06	26.64	16.52	12.94	4.62	16.88	6.86	15.12	15.92	14.74	12.92	11.80
	±0.74	±8.07	±8.11	±12.12	±13.41	±1.87	±4.59	±1.51	±2.84	±1.74	±5.28	±5.02	±1.70	±3.29	±4.49
7.5–15	2.74	7.12	3.28	16.94	10.36	12.16	5.88	4.52	8.20	4.92	6.42	4.92	5.74	12.26	10.86
	±0.88	±4.57	±1.04	±11.33	±4.34	±5.70	±2.17	±1.72	±1.57	±0.85	±1.41	±2.30	±0.94	±5.49	±1.85
15–22.5	2.48	6.56	2.92	11.46	3.56	2.76	2.88	3.12	5.88	3.62	1.98	9.92	6.26	6.40	7.80
	±0.64	±4.21	±0.93	±7.20	±1.27	±0.85	±0.75	±2.26	±2.65	±0.96	±0.34	±2.22	±1.40	±1.70	±2.30
Adjacent bare areas															
0–22.5	1.52	0.78	2.21	2.57	1.30	0.81	1.68	0.78	2.22	1.15	0.80	7.12	1.04	1.49	1.97
	±0.12	±0.06	±1.08	±0.63	±0.24	±0.15	±0.34	±0.17	±0.44	±0.34	±0.06	±4.05	±0.19	±0.25	±0.38
0–7.5	1.54	0.80	1.56	2.44	2.22	0.86	2.80	1.12	2.46	0.98	0.90	4.52	0.62	1.70	2.08
	±0.22	±0.09	±0.48	±0.63	±0.46	±0.37	±0.53	±0.41	±0.55	±0.24	±0.09	±2.12	±0.13	±0.38	±0.66
7.5–15	1.22	0.72	4.24	1.96	0.78	0.62	1.12	0.78	2.90	1.04	0.86	12.88	0.94	1.88	2.64
	±0.16	±0.14	±3.23	±0.47	±0.17	±0.05	±0.46	±0.27	±1.08	±0.49	±0.13	±12.28	±0.29	±0.40	±0.80
15–22.5	1.80	0.82	0.82	3.30	0.90	0.96	1.14	0.44	1.30	1.84	0.64	2.10	1.56	0.90	1.18
	±0.19	±0.12	±0.17	±1.90	±0.21	±0.26	±0.54	±0.10	±0.44	±0.87	±0.07	±1.10	±0.40	±0.48	±0.35

[a]The Al horizon was generally in the first 10 cm with Cl below (Romney et al. 1973).

Table 14-6 *Estimates of Old and New Contents of Nitrogen [kg N/ha] in Above-Ground Plant Parts in the Spring of 1973 in Rock Valley [Hunter et al. 1975a]*

Species	Leaves	New stems	Flowers	Fruit	Total new N	Old stems	Old leaves	Total old N
Ambrosia dumosa	1.0	0.2	1.0	0.7	2.9	0.8	---	0.8
Atriplex confertifolia	0.3	0.1	0.1	---	0.5	0.1	0.2	0.3
Ceratoides lanata	0.2	0.1	0.0	0.1	0.4	0.2	0.1	0.3
Ephedra nevadensis	---	0.9	0.2	0.3	1.4	1.1	---	1.1
Grayia spinosa	1.0	0.3	0.1	0.5	1.9	1.0	---	1.0
Krameria parvifolia	1.5	0.3	0.7	0.3	2.8	1.2	---	1.2
Larrea tridentata	0.7	0.2	0.1	0.2	1.2	2.1	0.7	2.8
Lycium andersonii	2.5	0.6	0.4	0.2	3.7	5.2	---	5.2
L. pallidum	1.1	0.2	0.5	0.3	2.1	1.5	---	1.5
Total					16.9			14.2

Fragile Aspects of the Nitrogen Cycle

The nitrogen cycle of the northern Mohave Desert seems capable of adjusting to whatever vegetational requirements are determined by the amount of rainfall. Nitrogen stress is not usually experienced until extra water is applied by irrigation (Romney et al., Chapter 16 of this volume). Since the input systems result in small amounts (hundreds of grams per hectare) of fixed nitrogen annually, they must be considered fragile because of extreme variation in rainfall. Some input systems very likely do not even operate in some years when there is little need for additional nitrogen or when the environment is not favorable.

The amount of soil organic matter in desert soils is small even though we consider it to be large in comparison to the nitrogen needs of plants in the northern Mohave Desert. Since surface soil is readily lost during floods and sometimes by wind, the status of the nitrogen cycle can be precarious.

In some areas at the Nevada Test Site soil and vegetation were completely changed by disturbance; after 25 years there has been very little tendency for development of a system like the original (Wallace and Romney 1975). The delicate balance among plants, soil organic matter and nitrogen fixation can easily be disrupted.

Loss of the vegetation, if not renewed, can result in loss of the shrub-clump fertile islands (Romney et al., in press). Energy is necessary to maintain the high levels of soil organic matter (and nitrogen) found in such fertile shrub-clump soils under hot desert conditions. Without the continued input of new organic matter, the soil organic matter would be lost, and hence a crucial component of the nitrogen cycle would be out of balance.

Is the Nitrogen Cycle in Equilibrium Under Desert Conditions?

Certainly the nitrogen input and outgo rates per year cannot be constant. A large loss of reserve nitrogen in 1 year via flood erosion may require decades to be restored. Input would have to exceed loss during this period. Then there would necessarily be net-gain and net-loss years; it is postulated that net-loss years could never exceed the net-gain years either in number or in quantity if the system is to reach an equilibrium. It is then possible that the net gain could exceed the net loss, with a continuous buildup of the nitrogen reserve. The imposed limit for such buildup may, on the other hand, be dictated by the moisture regime, which may result in equilibrium. Considerable year-to-year and decade-to-decade fluctuations in rainfall, and thus rates of nitrogen cycle processes and sizes of pools, are to be expected.

Can the Cycle be Induced to Operate at a Higher Capacity?

Irrigation (Romney et al., Chapter 16 of this volume) or a permanent significant increase in rainfall would require that the nitrogen processes operate at higher rates, with some larger pool sizes, to maintain a higher level of productivity. Doubling rainfall could increase the amount of autotrophic nitrogen fixation by a factor of 5 to 10 (Fig. 14-2). The blue-green algae-lichen crust could be more nearly like that in the Great Basin Desert and that alone could achieve this result. The additional organic matter produced with greater rainfall would provide the energy needed to maintain a higher pool of soil organic matter; however, it would require a few, or perhaps many, years to build up to its new level of soil organic matter. Until a new level was reached, there would very likely be a nitrogen stress on the system even though the nitrogen fixation process could be going five times as fast as at the lower moisture level.

What Should Yet be Done?

1. The present data on rates and mean input of nitrogen into the northern Mohave Desert are not yet satisfactory.
2. Why does lichen crust cover so little of the northern Mohave Desert in comparison with the Great Basin Desert?
3. Is the nitrogen in litter really conserved in the system?
4. Are there some years in which no nitrogen is biologically fixed?
5. Is the hypothesis that nitrogen input generally exceeds nitrogen loss a valid one?
6. What are the real limiting factors to nitrogen fixation in the desert?
7. Would the nitrogen cycle truly operate at higher levels, with sufficient time for a new equilibrium to be established, if the water input were increased? How much time would the new equilibrium require?
8. Is the desert nitrogen cycle different now from centuries ago because of the effect of man's industrial activities on input of combined nitrogen into the the atmosphere, and hence in rainfall?

SUMMARY

The nitrogen cycle in the northern Mohave Desert is characterized by a pool of nitrogen in the soil organic matter fraction from about 500 to 1,500 kg/ha, which is in equilibrium with mineral nitrogen fractions similar to those of other biomes and agricultural systems. The input of nitrogen and losses of nitrogen from the system are very small compared to the size of the soil organic nitrogen pool. Even so, the losses via runoff and leaching are sufficient to result in large accumulations of nitrate in the playas of closed basins common to the Mohave Desert. The amount of annual input is perhaps less than 1 kg/ha per year of newly fixed nitrogen and appears to be almost equally divided between that coming via nitrogenase systems and rainfall deposition. The mineralization process occurs at rates sufficiently high (before or near the time of critical phenological events) to create available nitrogen pools large enough to accommodate, without stress, a wide range of annual primary productivities resulting from differences in annual precipitation. If our conclusions about the relatively small annual inputs of newly fixed nitrogen are correct, then hundreds of years were necessary to obtain the pool sizes observed. We are unable to tell if they are at a steady state at the present time.

15

NITROGEN FERTILIZATION OF DOMINANT PLANTS IN THE NORTHEASTERN GREAT BASIN DESERT

D. W. JAMES and J. J. JURINAK

Within the desert rangelands of the Great Basin soil nitrogen availability is generally second only to soil moisture availability as a limiting factor in biomass production. It is apparent that if moisture availability is sufficient for a complete plant growth cycle, then the probability is very high that plants will respond to nitrogen fertilization. Practically speaking, the main question on fertilization of desert rangelands is whether the increase in growth will justify the cost of the fertilizer and its application.

Soil nitrogen-plant relationships were investigated in Curlew Valley adjacent to the IBP Desert Biome's southern intensive study site just south of the Utah-Idaho border. The study involved nitrogen fertilization of *Artemisia tridentata* (big sagebrush), *Atriplex confertifolia* (shadscale) in native vegetation and *Agropyron desertorum* (crested wheatgrass), an introduced species in tilled areas. The purpose of this chapter is to report the effects of nitrogen fertilizer on biomass production and on nitrogen uptake by these plant species.

BACKGROUND

Vallentine (1971, p. 355) and Mayland (1969) reviewed nitrogen soil fertility management of Great Basin Desert areas. The reports covered in these reviews dealt with both native and seeded grass pastures. Significant responses to nitrogen fertilization were summarized for investigations done in Utah, Nevada, Oregon and Washington. A typical result is that given by Cook (1965), who indicated that nitrogen fertilization increased crested wheatgrass herbage, number of heads per plant, basal area of crowns and number of roots. Protein content also increased by about one-third in his studies. Cook indicated that 45 kg N/ha increased herbage up to 1,100 kg/ha. A notable aspect of Cook's results dealt with the longevity of the fertilizer treatments: 60 percent of the total yield occurred the first year, 30 percent the second year and 10 percent the third year. He also pointed out that nitrogen fertilization increased the efficiency of soil moisture utilization.

An important consideration in the Vallentine (1971) and Mayland (1969) reviews was the association between rangeland grass responsiveness to nitrogen fertilization and precipitation. According to Mayland, crested wheatgrass performance is influenced by the total precipitation received in any one year and also by the variable amounts of precipitation that occur from year to year.

Table 15-1 *Form, Rate, Source and Time of Application of Fertilizers Applied to* Agropyron desertorum *Plots*

Treatment no.[a]	Application rate (kg/ha)		Form
	N	P	
Applied Oct 1973			
1	0	0.0	NH_4NO_3
2	56	0.0	"
3	28	11.2	"
4	56	11.2	"
5	224	11.2	"
6	0	22.4	"
7	28	22.4	"
8	56	22.4	"
9	112	22.4	"
10	224	22.4	"
11	56	44.8	"
12	28	67.2	"
13	56	67.2	"
14	224	67.2	"
15	28	22.4	$Ca(NO_3)_2$
16	28	22.4	$(NH_4)_2SO_4$
17	56	22.4	$Ca(NO_3)_2$
18	56	22.4	$(NH_4)_2SO_4$
Applied Apr 1974			
19	28	22.4	$Ca(NO_3)_2$
20	28	22.4	$(NH_4)_2SO_4$
21	56	22.4	$Ca(NO_3)_2$
22	56	22.4	$(NH_4)_2SO_4$
Applied Oct 1974			
23	56	0.0	NH_4NO_3
24	112	0.0	"

Note: All phosphorus was in the form of concentrated superphosphate (0-45-0).
[a]Treatments 1-14 were soil injected and 15-24 were surface broadcast.

Recently Sneva (1973a, b, c) reported on nitrogen fertilization in conjunction with grass clipping treatments, herbicide applications and studies dealing with the time and source of nitrogen fertilization. This work was done in the Great Basin Desert of central east Oregon. Over a 13-year period, 34 kg N/ha gave the following average increases above the control: 83 percent more spring growth, 29 percent more regrowth after clipping and 78 percent more biomass at maturity. Nitrogen fertilization also increased the crude protein content of the mature herbage. There appeared to be no interaction between nitrogen fertilization and clipping of the grass during the growing season (Sneva 1973a).

In a 3-year study involving fall, spring and summer application of nitrogen as urea and ammonium nitrate (Sneva 1973b), mature herbage yields of crested and Siberian wheatgrasses increased with fertilizer application rate, but there was no interaction with application date. In this study Sneva indicated that urea gave a 3 percent higher mean yield than did ammonium nitrate, which is evidently of no practical consequence. In the third study (Sneva 1973c) there was a significant response to nitrogen and an apparent positive interaction between nitrogen and the herbicide, paraquat, in yield and quality of crested wheatgrass.

In an economic analysis of nitrogen fertilization, McCormick and Workman (1975) reported that ammonium nitrate stimulated early spring growth initiation in grass growing on a Utah range. Nitrogen rates of 28 to 34 kg/ha advanced spring range readiness by 11 to 13 days. They indicated that during the first year of the study, nitrogen fertilization of crested wheatgrass pasture would have been a profitable substitute for purchased hay.

Comparatively few studies have been made of fertilizer effects on desert shrubs. Carpenter (1972), working in a Colorado Plateau area similar to the Great Basin, reported that 134 kg N/ha applied to sagebrush yielded an 81 percent increase in total leafy material compared to the nontreated control. Bayoumi (1975) reported on a study done in the Bear River Mountain Range of northern Utah involving nitrogen fertilization of a big-game winter pasture. The nitrogen fertilization of up to 100 kg N/ha linearly increased twig growth of both sagebrush and bitterbrush (*Purshia tridentata*).

EXPERIMENTAL STUDY

In the present field study, one area was used for *Agropyron desertorum*, and two other small areas were used for *Artemisia tridentata* and *Atriplex confertifolia*. The soil at the site is classified as Thiokol silt loam (a typic calciorthid). Some physical and chemical properties of this soil were discussed by Evans and Jurinak (1976).

Crested Wheatgrass

The area selected for the grass study was a tilled seeding made in 1965 (Shinn 1975). It was a uniform and essentially pure stand of *Agropyron*. About 1 ha of the area was prepared in October 1973 by mowing the grass stubble. The individual field plot size was 2.4 x 12.2 m. The fertilizer rates, forms and times of application are shown in Table 15-1. The first 14 treatments of Table 15-1 are an incomplete

Table 15-2 *Form, Rate, Time and Method of NH_4NO_3 Application to* Artemisia tridentata *and* Atriplex confertifolia

Treatment no.	Application rate (kg/ha)		Method
	N	P	
Applied Oct 1973			
1	0.0	0.0	Broadcast
2	67	0.0	Broadcast
3	67	0.0	Injected
4	67	34	Injected
Applied Apr 1975			
5	0.0	0.0	Broadcast
6	34	0.0	Broadcast
7	67	0.0	Broadcast
8	134	0.0	Broadcast
9	134	0.0	Injected
10	134	34	Injected

Note: Phosphorus treatments are given for experimental design information only. Growth responses to phosphorus were not detected and therefore are not discussed here.

factorial combination of nitrogen and phosphorus fertilizer rates. The phosphorus treatments are given for experimental design information only and will not be further discussed since no phosphorus fertilizer effects on grass yield were observed. The bicarbonate-extractable phosphorus of this soil was 25.5 ppm (Evans and Jurinak 1976), which is well above the plant growth limiting level for this element in irrigated semiarid soils.

The shank or drill method involved injection of the dry fertilizer material to 10-cm depth with 30-cm row spacing. Fertilizer was metered into the shanks using a belt-type small plot fertilizer applicator. The shanks cut through some crested wheatgrass plant crowns, but apparently had no adverse effect on these plants.

Treatments 15 through 18 (Table 15-1) compared nitrate and ammonium N forms. These were applied as surface broadcast treatments in the fall. These treatments were also repeated in the spring before growth started, as represented by treatments 19 through 22. The residual effects of all 1973-74 treatments on herbage yields and nitrogen content were evaluated through 1975. Treatments 23 and 24 provided a "current season" control for the residual treatments. All treatments of Table 15-1 were applied in a randomized complete block design in eight replications.

Grass yield samples were collected by hand clipping a 3-m^2 area in each plot. These samples were taken on 2 July 1974 and 4 August 1975. At harvest the grass was mature but no seed had shattered in either year. The harvested material was dried and weighed. The results are expressed as kg oven-dry weight/ha. Subsamples of grass were ground in a Wiley mill and analyzed for total nitrogen by the Kjeldahl procedure. These results are expressed in terms of percent nitrogen.

Native Species

The area of the natural vegetation study was a stand of mixed *Artemisia* and *Atriplex* with only occasional plants of other, lower-statured species. Specimen plants of *Artemisia* were selected that were 50-60 cm high. A circular plot 1 mha (10 m^2) in size was measured with the specimen plant at the center of the plot. The *Atriplex* plants which occurred in each of the *Artemisia* plot areas were used as the specimen plants for this species. Fertilizer treatments summarized in Table 15-2 were applied in a completely randomized design in 10 replications. These plots for the 1974 and 1975 treatments were adjacent, but they were mutually exclusive and were analyzed as separate experiments.

The fertilizer injection treatments on the native vegetation plots were applied using a soil sampling tube. The fertilizer was applied to 20-25 random points within each plot to a depth of 10 cm. The soil plug was replaced after each injection.

Biomass production for both *Artemisia* and *Atriplex* was estimated in 1974 simply by measuring the length of vegetative growth of the current year. In 1975 the current year's shoot was again measured and then clipped, dried, weighed and analyzed for total nitrogen. Twenty vegetative shoots from each specimen sagebrush plant were selected for the biomass estimates by the method outlined by Smith and Doell (1968). For the *Atriplex*, five shoots from each of four plants in the plot area were selected at random

Soil Moisture

Soil samples were collected to the 75-cm depth in crested wheatgrass plots and analyzed for moisture content (dried overnight at 105 C). This was done at the beginning and at the end of the annual growth cycle. The data from the Desert Biome monthly climatological reports were used for precipitation inputs.

RESULTS

Available Moisture

Soil moisture results for both 1974 and 1975 are summarized in Figure 15-1. Using approximate figures for bulk density, the soil water depletion for 1974 was 8.7 cm and for 1975 it was 8.0 cm. Using climatological data in connection with the soil water depletion showed 13 cm of water lost in 1974 and 21 cm in 1975. It was estimated that 23 percent of the 1974 water loss and 37 percent of the 1975 water loss were used by the crop. The greater level of 1975 water availability should be borne in mind while evaluating the biomass yield results.

Crested Wheatgrass

Figure 15-2 gives the herbage yield and nitrogen content results for 1974. The results for yield were somewhat erratic for the middle nitrogen fertilizer rates, but a large difference in yield was associated with the high nitrogen treatment. The

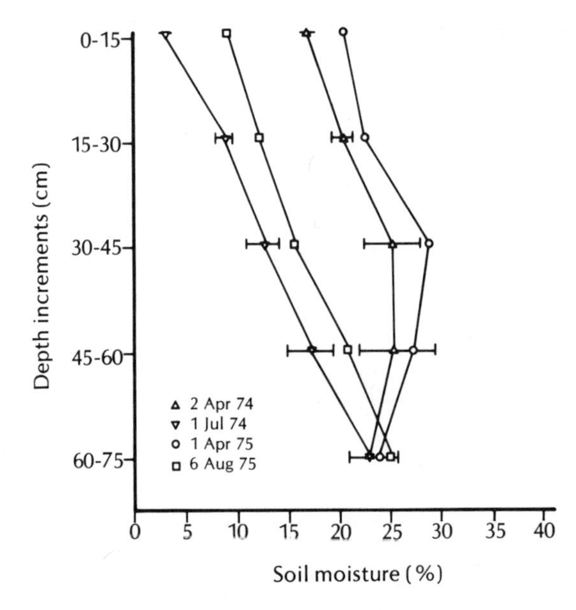

Figure 15-1 *Moisture distribution in the soil profile of the Curlew Valley* Agropyron *site sampled 2 April 1974, 1 July 1974, 1 April 1975 and 6 August 1975. Horizontal line segments are standard deviations for 1974 data only. Deviations in the 1975 results were similar.*

percent nitrogen responses indicated that nitrogen uptake increased regularly from zero to 112 kg N/ha of fertilizer applications. There was no difference between 112 and 224 kg N/ha on herbage nitrogen concentration. Between zero and 224 kg N/ha applications there was a net increase of 557 kg/ha of dry matter and 12.7 kg/ha of nitrogen in the grass.

Figures 15-3 and 15-4 show the 1975 yield results for both current and residual fertilizer treatments. As indicated, yield increased quite regularly for both the October 1973 and October 1974 treatments. There was a relatively smaller response to the residual treatments, indicating perhaps a decrease in nitrogen availability from earlier applications. But the herbage nitrogen content (Fig. 15-4) indicated that comparable nitrogen fertilizer rates gave essentially the same results for the two application times.

The maximum net increases measured in 1975 dry matter yields and nitrogen concentration from the October 1974 treatments were 1,080 and 22.0 kg/ha, respectively. The current year nitrogen treatments gave a maximum net increase of 913 kg dry weight and 13.4 kg N/ha. The maximum yield levels were of course not directly comparable between the two seasons since the highest rates of applied nitrogen were quite different.

The results from the supplemental nitrogen treatments applied in October 1973 together with the residual effects from fertilization in October 1974 are given in

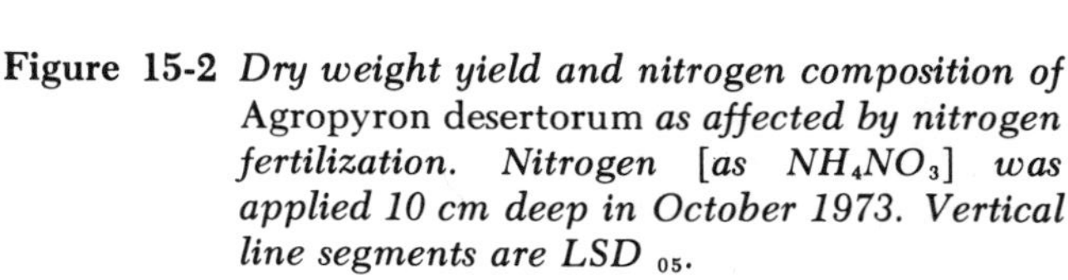

Figure 15-2 *Dry weight yield and nitrogen composition of* Agropyron desertorum *as affected by nitrogen fertilization. Nitrogen [as* NH_4NO_3*] was applied 10 cm deep in October 1973. Vertical line segments are LSD* $_{05}$.

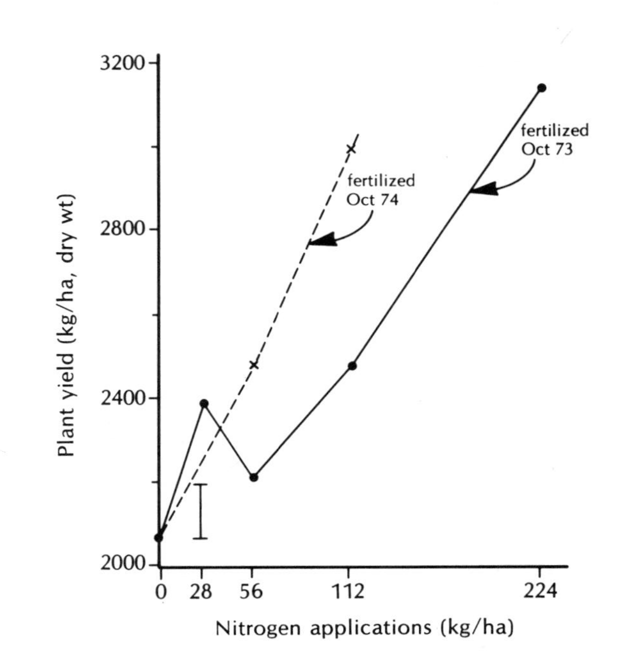

Figure 15-3 *Dry Weight yield of* Agropyron desertorum *harvested in 1975 as affected by the nitrogen injected October 1973 or surface applied April 1974. Vertical line segments are LSD* $_{05}$.

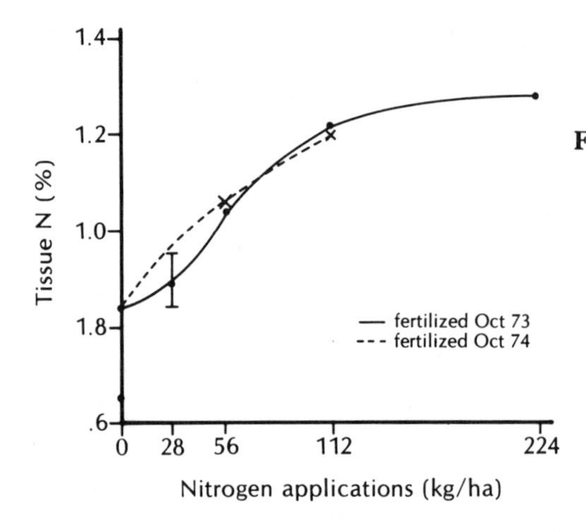

Figure 15-4 *Nitrogen concentration in 1975-harvested* Agropyron desertorum *[fertilizer injected 10 cm deep in October 1973 and surface applied in October 1974]. Vertical line segment is LSD* $_{05}$.

Table 15-3. In 1974 none of the treatment effects was significantly different within the group in Table 15-3 or between the group and the shank treatments (compare with Figs. 15-2 and 15-3). In 1975 differences were still not significant but the 56 kg N/ha treatment results tended to be higher than for the 28 kg N/ha treatment. No particular tendencies were noted for the nitrate N vs. ammonium N fertilizer forms nor for the fall vs. spring application times. With regard to percent nitrogen effects (Table 15-3), the supplemental treatments gave significant differences which favored 56 kg N/ha in both years, but this parameter did not segregate on the nitrogen fertilizer form or time of application.

Native Species

Figure 15-5 shows the 1974 and 1975 shoot growth results from the 1974 nitrogen treatments on *Artemisia* and *Atriplex.* Significant nitrogen responses

Table 15-3 Agropyron desertorum *Yields in 1974 and 1975, as Related to Time and Form of Nitrogen Application*

Treatment no.	kg N/ha	Time	Form	1974		1975	
				Yield (kg/ha)	N (%)	Yield (kg/ha)	N (%)
15	28	Oct 1973	NO_3	690	1.53	1847	0.96
17	56	Oct 1973	NO_3	723	1.86	2200	1.04
19	28	Apr 1974	NO_3	613	1.20	2070	0.90
21	56	Apr 1974	NO_3	753	1.32	2143	1.04
16	28	Oct 1973	NH_4	633	1.53	2203	0.92
18	56	Oct 1973	NH_4	740	1.56	2213	1.01
20	28	Apr 1974	NH_4	587	1.30	2203	1.03
22	56	Apr 1974	NH_4	547	1.37	2533	1.09

occurred in *Artemisia* in 1974 and in *Atriplex* in 1975. Nonsignificant trends occurred otherwise. Figure 15-6 gives the shoot weight and percent nitrogen in 1975 residual effects. In the *Artemisia* only the percent nitrogen was affected by nitrogen fertilization while in *Atriplex* both shoot weight and percent nitrogen were increased. Figures 15-5 and 15-6 do not show a general trend that would indicate a difference in nitrogen availability between the surface applied and the injected nitrogen treatments.

The results for the 1975 current nitrogen fertilization effects are shown in Figures 15-7 to 15-9. Differences in shoot weight were not significant for either species (Fig. 15-7), but there was a considerable overall contrast between the two. Both shoot length (Fig. 15-8) and nitrogen concentration (Fig. 15-9) were affected by the fertilizer treatments. The results for soil injected nitrogen (results not shown; cf. Table 15-2) in 1975 did not differ from the results for surface applied nitrogen.

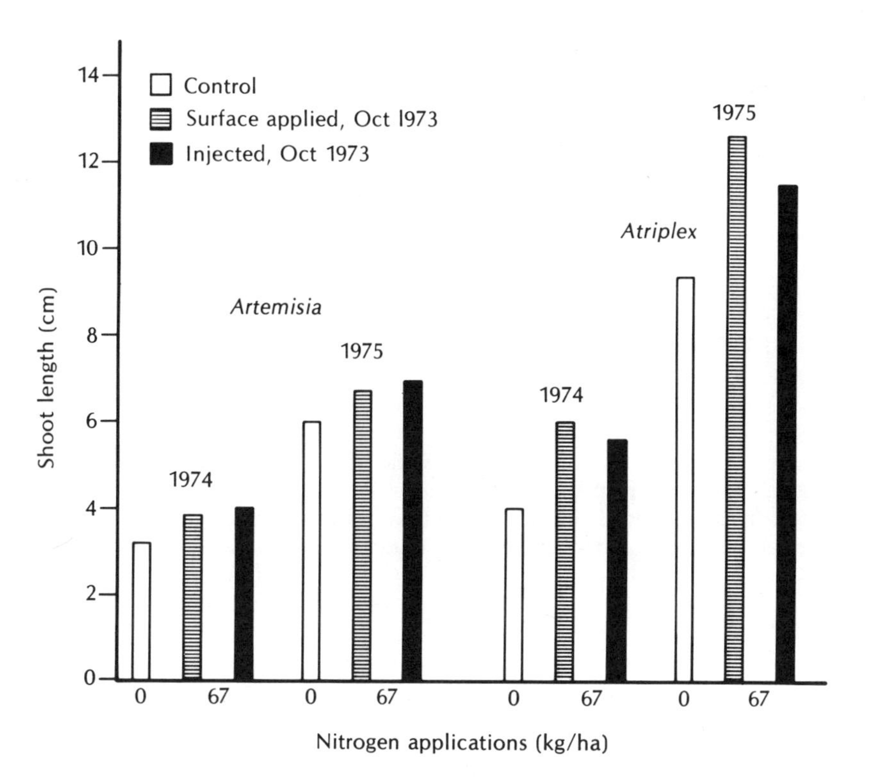

Figure 15-5 *Shoot length responses in 1974- and 1975-harvested* Artemisia tridentata *and* Atriplex confertifolia *to nitrogen fertilization either surface broadcast or injected October 1973. LSD* $_{05}$ *values were:* Artemisia *1974, 0.83;* Artemisia *1975, NS;* Atriplex *1974, NS;* Atriplex *1974, NS;* Atriplex *1975, 1.51.*

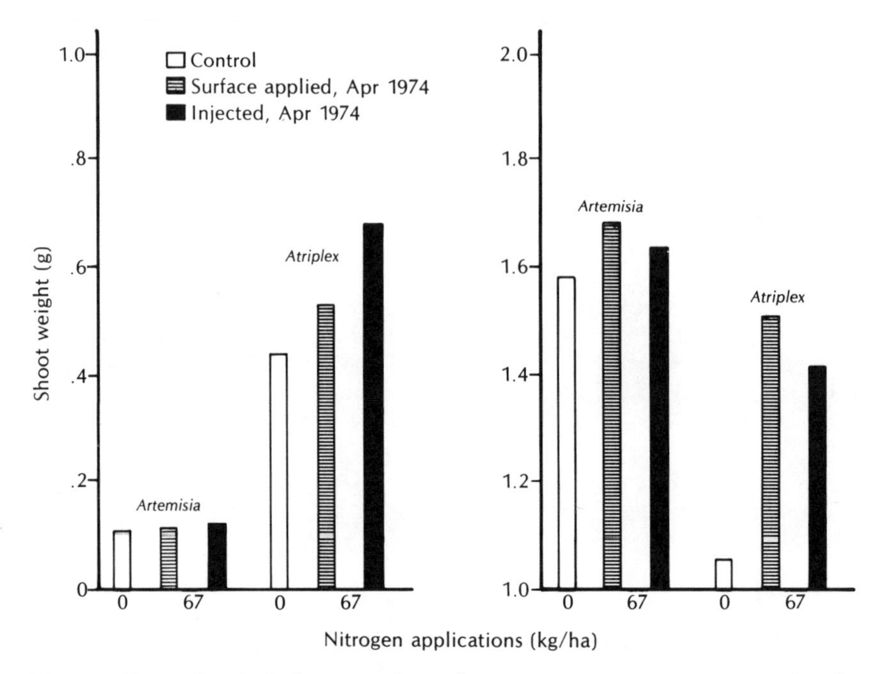

Figure 15-6 *The 1975 shoot weight and percent nitrogen responses to October 1973 nitrogen fertilization for* Artemisia *and* Atriplex. *The LSD values were: shoot weight for* Artemisia, *NS, and* Atriplex, *0.31; percent nitrogen for* Artemisia, *0.10, and for* Atriplex, *0.20.*

DISCUSSION

Crested Wheatgrass Responses

Under the conditions of this field work *Agropyron desertorum* was highly responsive to nitrogen fertilization, in terms of both quantity and quality of forage produced. It is apparent from the results that indigenous soil nitrogen availability was far below those levels required for maximum or even moderate levels of biomass production. No attempt was made here to evaluate the optimum nitrogen fertilizer rates in economic terms. There was no apparent loss of nitrogen efficiency from broadcast application as compared to fertilizer injected into the soil nor was there any apparent difference among sources of nitrogen. In a practical situation, range soil fertility management would involve the cheapest form and easiest method of fertilization. A large response was demonstrated here for 1-year-old nitrogen treatments. It is assumed that biomass responses would occur in the third season, although relative yield responses would probably diminish regularly with time. Therefore, the cost of nitrogen fertilization could be amortized over several years, adding to the feasibility of fertilizing this rangeland.

Soil moisture availability was very different between the 2 years of field observations. The overall results provide an indirect comparison of the soil nitrogen and soil water interaction in plant growth. In 1974, for example, without

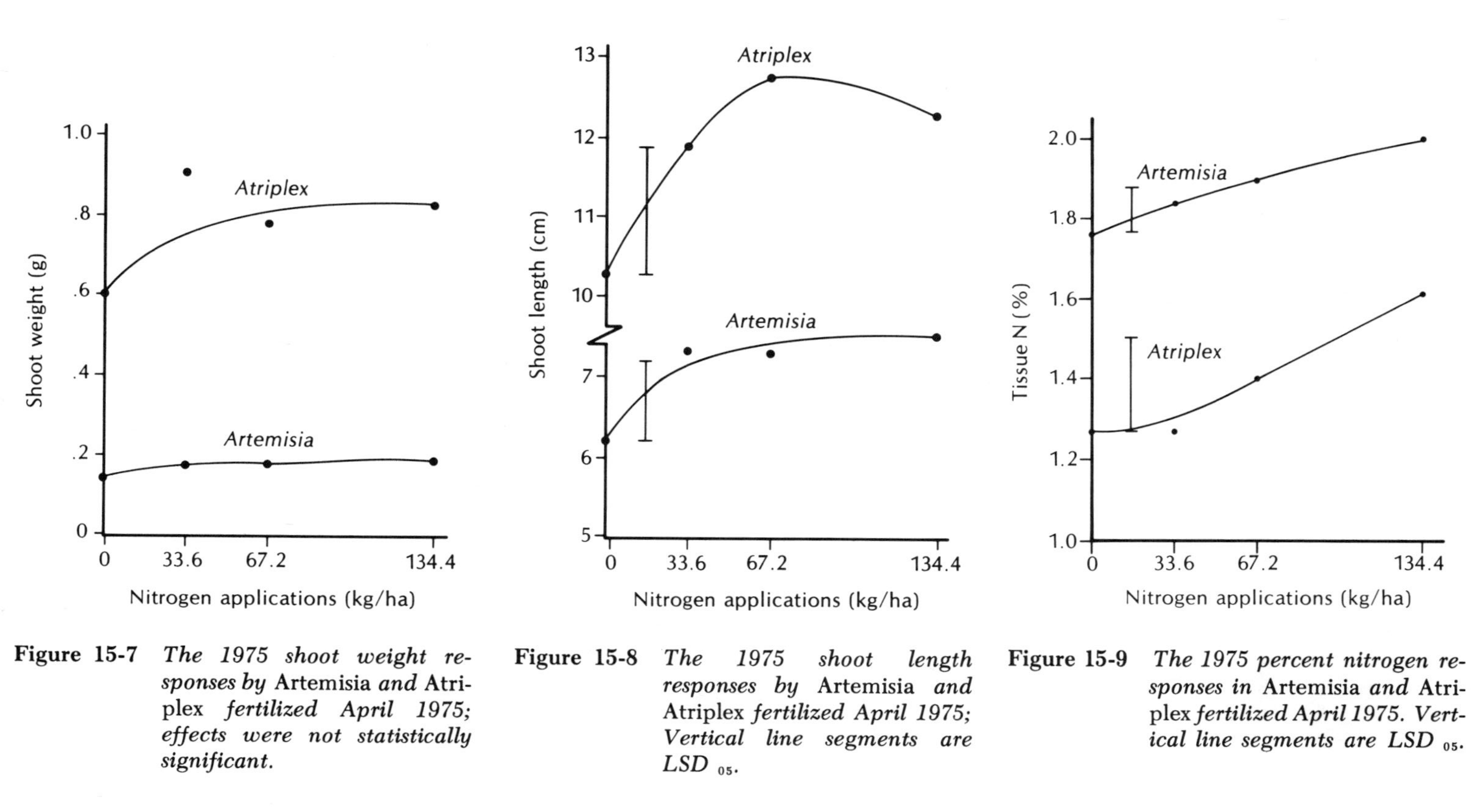

Figure 15-7 *The 1975 shoot weight responses by* Artemisia *and* Atriplex *fertilized April 1975; effects were not statistically significant.*

Figure 15-8 *The 1975 shoot length responses by* Artemisia *and* Atriplex *fertilized April 1975; Vertical line segments are LSD* $_{05}$.

Figure 15-9 *The 1975 percent nitrogen responses in* Artemisia *and* Atriplex *fertilized April 1975. Vertical line segments are LSD* $_{05}$.

fertilization the ratio of dry matter produced per unit of water consumed was 270 kg/ha per cm of water; at the high rate of fertilization the ratio was 541 kg/ha per cm of water. The 1975 residual effects of the 1974 treatments gave a ratio of 321 kg/cm for the control and 404 kg/ha for the 224 kg N/ha treatment. Assuming that nitrogen availability was the same both years in the control plots, the difference attributed to water availability between years (810 vs. 2,500 kg dry matter) indicated the change in indigenous nitrogen use efficiency as affected by moisture availability. Similar comparisons could be made for nitrogen concentration in the grass. The net result encourages nitrogen use in increasing forage production on these kinds of rangeland. Even in years of exceptionally low precipitation, fertilization will increase the efficiency of resource utilization. Also, in years of low precipitation, nitrogen consumption is reduced and residual nitrogen carryover is enhanced. Thus, the risk of poor return on the fertilizer investment is reduced or eliminated. The foregoing statements on water use efficiency are based on the assumption that residual seasonal soil moisture will be a function of weather and not of consumptive use by plants. The rooting depth was limited (40 cm) because of accumulated salts that restrict root extension to deeper soil layers at this research site.

The data presented here on nitrogen uptake as a function of nitrogen fertilizer placement (surface vs. subsurface), though somewhat qualitative, support the conclusions of Klubek et al. (Chapter 8 of this volume) that ammonia volatilization from the soil surface is a minor factor in biological soil nitrogen use efficiency. On the other hand, denitrification in desert soils appears to have a considerable potential for depressing the efficiency of soil nitrogen utilization (Wullstein and Gilmour 1964; Klubek et al., Chapter 8 of this volume; Westermann and Tucker, Chapter 7 of this volume). From the data in Figures 15-2 to 15-4 the rate of change of nitrogen removal in the grass was computed. The results for 1974 were: .054 kg N removed in the harvested material per kg fertilizer N applied; 1975, two-season total including the residual 1974 treatments: .156 kg N removed/kg N applied; and 1975 current season treatments: .164 kg N removed/kg N applied. The coefficient of determination was .99 for all these comparisons of nitrogen uptake per unit of nitrogen applied. These results indicating a low fertilizer use efficiency did not, of course, account for nitrogen stored in the root and the crown system of the plants nor for residual soil mineral nitrogen. Based on the results of Cook (1965) and Sneva (1973b), significant biomass responses to fertilization would not be expected beyond the third year, and within that time the projected nitrogen removal by the crop would not exceed 50 or 60 percent of the nitrogen applied. It would be worthwhile to determine what portion of the unaccounted-for nitrogen was still in the system as soil mineral fixed ammonia and organic nitrogen and what portion was denitrified. This could have an effect on long-term range soil fertility management.

Native Species Responses

There was a clear fertilizer effect on plant nitrogen concentration in both *Artemisia* and *Atriplex* in the 1 year in which analyses were made. However, results for actual biomass production were mixed. It is evident that a better or more thorough sampling procedure would be needed to establish the true effects of fertilization on these plants. Based on the data that were collected, it would appear that *Atriplex* was relatively more responsive to nitrogen fertilization in terms of both yield and quality than was *Artemisia*. The data were analyzed to determine the association between shoot length and shoot weight in the 1975 residual and current nitrogen plots. The calculated relationships were: *Artemisia* shoot weight (g) = —.058 + .032 shoot length (cm), $r^2 = .78$, $n = 10$; *Atriplex* shoot weight (g) = —.30 + .092 shoot length (cm), $r^2 = .64$, $n = 10$. Accordingly, shoot elongation may be regarded as a good first approximation of current seasonal biomass production for *Artemisia* and *Atriplex*.

CONCLUSIONS

Large responses to nitrogen fertilizer were observed in *Agropyron*. Smaller responses were observed in *Artemisia* and *Atriplex*. The results support the following conclusion.

1. Indigenous soil nitrogen was severely limiting to the growth of *Agropyron* and, to a lesser extent, to *Artemisia* and *Atriplex*.
2. Nitrogen fertilizer application gave increased yield and forage quality even in the year that soil moisture was exceptionally low.
3. Highly beneficial residual effects from nitrogen fertilizer can be expected for 2 or 3 years. The residual effect will be related to total moisture availability in any given year.
4. Efficiency of use of available soil water could be greatly enhanced by nitrogen fertilization of Great Basin Desert rangelands seeded with crested wheatgrass.
5. The first- and second-year growth results indicated that fertilizer nitrogen use efficiency was low. The ultimate level of efficiency will depend on what portion of the fertilizer nitrogen was stored in the soil as mineral fixed ammonium and organic nitrogen and what portion was lost by denitrification.
6. Economic worth of nitrogen fertilization will depend on the cost of the nitrogen applied vs. the value of the increased production.

16

PLANT RESPONSE TO NITROGEN FERTILIZATION IN THE NORTHERN MOHAVE DESERT AND ITS RELATIONSHIP TO WATER MANIPULATION

E. M. ROMNEY, A. WALLACE and R. B. HUNTER

INTRODUCTION

Water is generally considered to be the most limiting factor to primary productivity under desert conditions. If it is, then a logical conclusion is that additional water would increase productivity. This conclusion may not necessarily be correct if other limiting factors begin to operate immediately when more water is applied. Oxygen tension on roots is one possible example (Lunt et al. 1973). Nitrogen is generally considered to be the next limiting factor after water, but many subtle effects of both nitrogen and water must be considered in the interaction of limiting factors in determining whether or not response to either will be obtained.

EXPERIMENTAL STUDIES

Field experiments were conducted in the northern Mohave Desert ecosystem at the Nevada Test Site in order to determine the interaction from application of supplemental nitrogen and water. In one experiment, 100 x 100 m field plots were established in Rock Valley on soils with and without an underlying restrictive hardpan. Commercial fertilizer (NH_4NO_3) was broadcast onto the soil surface in an amount equivalent to 100 kg N/ha in April 1967. Observations were made of treatment response each year thereafter, and plant samples were collected for analysis during the peak of the growth season in 1969, 1970 and 1973. For most Mohave Desert vegetation the peak of growth occurs during April and May, but any given species will vary considerably from year to year, depending upon the early season moisture and temperature conditions.

A second experiment was conducted in Mercury Valley where advantage could be taken of a source of reclaimed sewage water from the town of Mercury, Nevada.

Table 16-1 *Amounts of Precipitation and Irrigation [cm] at Rock Valley and Mercury Valley, 1968-73*

Year	Rock Valley	Mercury Valley	
	Precipitation (cm)	Precipitation (cm)	Irrigation (cm)
1968	18.1	17.9	23
1969	25.0	22.1	30
1970	9.9	7.4	34
1971	10.3	7.8	--
1972	14.8	9.3	--
1973	26.1	22.0	--

Paired plots, 30.5 m in diameter, were treated with supplemental nitrogen at levels of 0, 100 and 200 kg N/ha (NH_4NO_3) broadcast uniformly onto the soil surface in March 1968 and again in October 1970. Supplemental moisture was applied by overhead sprinkler irrigation, starting in April 1968, in amounts sufficient to maintain the soil moistuere level above 5 percent by weight. Because of slow moisture penetration into the soil, it was necessary to irrigate for short periods, frequently, in order to avoid runoff from the plots. This became a factor in our decision to conduct a third experiment using the trickle-irrigation method of applying moisture. Our concern about keeping foliage wet for days at a time was another factor in making this decision, even though the effect observed was primarily a discoloration of foliage rather than growth inhibition. Inasmuch as the seasonal precipitation occurred mainly during late autumn and winter months, the sprinkler irrigation was applied during the period of naturally depleted soil moisture from mid-April through October. The dry plots received annual precipitation measured by rain gauge for the period of 1 September to 31 August in the amounts listed in Table 16-1. The watered plots received natural precipitation plus supplemental sprinkler irrigation. Due to evaporation loss, we estimate that only about one-half of this irrigation penetrated into the root zone during the hot summer months. We did, however, achieve our aim of maintaining the soil moisture level above 5 percent by weight on the watered plots during that period.

The irrigation water available from the sewage processing system was of marginal quality as indicated by the analyses in Table 16-2. It served the purpose, however, for the initial feasibility and response studies with Mohave Desert vegetation. Later experiments using the local municipal water supply gave results which confirmed that the high sodium adsorption ratio of the sewage water was not inhibitive to plant growth on the calcareous soil of the study area. Soil samples analyzed from the sprinkle-irrigation plots contained lime contents ranging from 15 to 20 percent (Romney et al. 1973).

The total nitrogen content of the sewage water, including that contributed by algae and other organic forms, varied from 5 to 7 μg/ml. During the 3-year period of sprinkler irrigation, the watered plots could have received additional nitrogen equivalent to 40-60 kg/ha. We do not know how rapidly this source of nitrogen became available as fertilizer, but presume that a normal mineralization of the organic forms occurred through microbial activity.

Table 16-2 *Chemical Composition of Irrigation Water*

Element	1967	1969	1973
Ca (μg/ml)	18.4	12.5	11.0
Mg "	3.1	2.8	2.7
K "	16.0	16.5	14.0
Na "	263.0	266.3	270.0
Cl^- "	132.0	108.4	---
$SO_4^=$ "	8.3	7.4	---
Na absorption ratio	21.1	23.6	26.56

Our third field experiment established in February 1974 involved a series of 30-m^2 circular plots in which trickle-irrigation treatments, using filtered municipal water, were established in combination with nitrogen treatments at levels equivalent to 0, 25 and 100 kg N/ha as NH_4NO_3 applied in March 1974. Water was applied to saturation of the soil during the periods of 15-19 April, 13-20 May, 16 June to 7 July, 30 August to 3 September and 23-26 October. A total of 20 cm water was applied in 1974 and again in 1975 (equivalent to twice the average natural rainfall). Detailed results from these field experiments have been given in US/IBP Desert Biome research memoranda (Romney et al. 1974; Hunter et al. 1975a; Hunter et al. 1976a).

RESULTS

Some examples of data given below indicate the kinds of responses we have observed in Mohave Desert vegetation, resulting from soil surface applications of supplemental nitrogen and water. Our results generally agree with findings reported for grasslands (Rogler and Lorenz 1957; Klipple and Retzer 1959; Dahl 1963;

Table 16-3 *Annual Plant Response to Surface Application of Nitrogen Fertilizer under Natural Rainfall Conditions in Rock Valley, Nevada*

Species	Plant yield (g/m^2)		Nitrogen content (%)	
	Control	100 kg N/ha[a]	Control	100 kg N/ha
Amsinckia tessellata	0.18	1.29	0.87	1.35
Chaenactis fremontii	7.00	9.08	0.87	0.81
Cryptantha nevadensis	0.30	2.13	0.75	0.85
C. pterocarya	0.72	2.26	0.72	0.72
Malacothrix glabrata	0.59	2.73	0.74	1.00
Mentzelia obscura	1.79	2.60	0.98	1.32
Phacelia vallis-mortae	0.24	0.65	0.78	0.98
Streptanthella longirostris	2.16	4.07	0.43	0.80

[a]Data listed are for samples collected 9 May 1970 from Rock Valley plots which had NH_4NO_3 broadcast onto the soil surface equivalent to 100 kg N/ha in Apr 1967.

Stroehlein et al. 1968; Owensby et al. 1970) in that surface applications of nitrogen generally were ineffective for increasing primary productivity when soil moisture was limiting. When followed by natural precipitation, broadcast fertilization of nitrogen was most immediately beneficial to the annual plant species, increasing both herbage yield and nitrogen content of plant tissue (Table 16-3). Residual effects of nitrogen treatments were observed on several annual species for as long as 5 years after fertilizer was applied. Shrubs responded in like manner after supplemental moisture had moved the nitrogen down into the root zone (Table 16-4). Plants having the highest natural grazing potential (viz., *Ambrosia dumosa, Ceratoides lanata* and *Lycium andersonii*) consistently showed favorable response. There was marked establishment of new seedlings of *Acamptopappus shockleyi, Ambrosia dumosa, Ceratoides lanata, Lepidium fremontii* and *Sphaeralcea ambigua* from sustained sprinkler-irrigation treatments (Table 16-5). One of the main effects of

Table 16-4 *Response of Shrubs to Supplemental Nitrogen Fertilizer and Sprinkler Irrigation in Mercury Valley, Nevada*

Species	New shoot growth (g dry wt)[a]		Nitrogen content (% dry leaf)	
	Control	200 kg N/ha[b]	Control	200 kg N/ha
		Natural desert conditions		
Acamptopappus shockleyi	1.87	1.26	1.17	1.11
Ambrosia dumosa	3.31	3.49	1.96	2.67
Ceratoides lanata	4.76	4.96	2.23	2.82
Grayia spinosa	7.20	7.62	2.15	2.37
Krameria parvifolia	1.08	1.32	2.03	2.92
Larrea tridentata	3.48	2.16	1.83	2.08
Lycium andersonii	2.78	3.90	4.85	6.00
Total	24.48	24.71		
		Supplemental moisture added[c]		
Acamptopappus shockleyi	4.90	6.80	1.53	1.46
Ambrosia dumosa	8.82	11.00	2.08	2.08
Ceratoides lanata	9.73	9.55	2.90	3.13
Grayia spinosa	11.75	10.27	2.31	2.57
Krameria parvifolia	4.76	3.52	1.87	2.50
Larrea tridentata	3.85	3.93	2.00	2.04
Lycium andersonii	4.84	8.50	3.53	4.52
Total	48.65	53.57		

[a]Weight of 50 new shoots picked at random from shrub species in each plot 1 May 1969.

[b]NH_3NO_4 broadcast on soil surface equivalent to 200 kg N/ha, Mar 1968.

[c]Sprinkler irrigation of 23 cm was applied in 1968 after 18 cm natural rainfall and winter rainfall in the following increments: 26 Apr to 17 May (7.3 cm); 27 Jun to 5 Jul (5.0 cm); 6-9 Aug (3.1 cm); 10-19 Sep (4.6 cm); 31 Oct to 4 Nov (3.0 cm). Natural rainfall of 20 cm occurred from Nov to Apr before samples were harvested and irrigation was resumed in May 1969.

Table 16-5 *Establishment of New Shrub Seedlings During and After 3 Years of Supplemental Sprinkler Irrigation in Mercury Valley, Nevada*

Species	Original plot population, (1968)	New seedlings surviving during irrigation, 1968-70	Seedlings surviving without irrigation, 1971-74
		Supplemental moisture[a] *(Plot 4)*	
Acamptopappus shockleyi	265	86	54
Ambrosia dumosa	148	17	13
Ceratoides lanata	138	64	36
Grayia spinosa	17	5	7
Krameria parvifolia	78	0	0
Larrea tridentata	32	1	0
Lepidium fremontii	0	8	25
Sphaeralcea ambigua	90	50	51
		Natural desert conditions (Plot 8)	
Acamptopappus shockleyi	76	15	9
Ambrosia dumosa	103	2	1
Ceratoides lanata	328	13	0
Grayia spinosa	30	12	12
Krameria parvifolia	29	1	1
Larrea tridentata	50	3	3
Lepidium fremontii	2	0	0
Sphaeralcea ambigua	49	2	0

[a]Water was applied periodically by sprinkler irrigation in 1968 (23 cm), 1969 (20 cm) and 1970 (34 cm) from mid-Apr through Oct.

supplemental irrigation was to lengthen the growing season of shrubs. Added moisture in late spring and early summer could not, however, overcome high temperature dormancy in the Mohave Desert vegetation. There appeared to be a growth-response plateau for most shrub species above which further additions of either supplement were not effective. Two species, *Ephedra funerea* and *Larrea tridentata*, showed some evidence of shoot discoloration from prolonged sprinkler irrigation. Growth of annual species on bare soil between shrub clumps was inhibited in sprinkler-irrigated plots, presumably from mechanical impact damage and sunscald. Inasmuch as soil moisture is one of the most limiting factors affecting primary productivity in the Mohave Desert, the addition of irrigation water generally masked any growth response that might have been attributable solely to supplemental nitrogen. Additions of supplemental moisture also markedly increased the productivity of native grasses (Wallace and Romney 1972a; Romney et al. 1974), indicating that sustained treatment should increase the livestock grazing potential of Mohave Desert areas. Residual effects of sprinkler irrigation applied to Mohave Desert shrub communities in 1968, 1969 and 1970 could still be observed in a few shrub species (viz., *Acamptopappus shockleyi*, *Ceratoides lanata* and *Grayia spinosa*) during the 1974 growth season. Two species, *Ephedra funera* and *Yucca schidigera*, showed virtually no positive growth response to sprinkler-irrigation treatments.

The results obtained from our trickle-irrigation experiment, using municipal drinking water, confirmed findings from sprinkler irrigation that several of the endemic species responded strongly to water, while a few species did not (Tables 16-6 to 16-8). Again, response to nitrogen fertilizer generally was not effective during the ensuing growth season unless accompanied by supplemental moisture. Invasion by short-lived winter annuals began during the first growth season after irrigation started in April 1974. As many as 30 different annual species appeared on the watered plots in 1975, but only a few contributed significant numbers and biomass (Table 16-9). After an initial surge of increased biomass during the 1974 growth season, it was observed that the vegetative production by shrubs shifted more from new stem growth toward proportionally greater production of deciduous structures in 1975 and 1976.

Table 16-6 *Examples of Shrub Species Growth Response to Nitrogen Fertilization on Trickle-Irrigated Plots Using Filtered Municipal Water in Mercury Valley, Nevada*

Species	Sampling date	Control		100 kg N/ha[a]	
		Biomass (kg/ha)	Percent increase	Biomass (kg/ha)	Percent increase
Acamptopappus shockleyi	Feb 1974	59		84	
	Oct 1974	155	163	197	136
	Oct 1975	200	29	234	19
Ambrosia dumosa	Feb 1974	150		80	
	Oct 1974	939	525	867	978
	Oct 1975	1,108	18	1,128	30
Atriplex confertifolia	Feb 1974	1,126		1,018	
	Oct 1974	1,572	40	1,563	54
	Oct 1975	1,871	19	2,568	64
Ephedra funera	Feb 1974	142		242	
	Oct 1974	222	57	400	65
	Oct 1975	262	18	480	20
Larrea tridentata	Feb 1974	588		713	
	Oct 1974	802	36	989	39
	Oct 1975	890	11	1,098	11
Lepidium fremontii	Feb 1974	0		13	
	Oct 1974	21	∞	58	338
	Oct 1975	97	360	281	385
Menodora spinescens	Feb 1974	89		82	
	Oct 1974	286	222	323	293
	Oct 1975	472	65	665	106

[a]Treatment level is kg/ha equivalent applied as NH_4NO_3 in Mar 1974. Water was applied to saturation of soil during the periods 15-19 Apr, 13-20 May, 16 Jun to 7 Jul, 30 Aug to 3 Sep, 23-26 Oct. A total of 20 cm water was applied in 1974 and again in 1975 (equivalent to twice the average natural rainfall). Biomass changes were measured by dimensional analysis methods (Wallace and Romney 1972a).

Table 16-7 *Change in Total Shrub Biomass [kg/ha] on Trickle-Irrigated and Nitrogen-Treated Plots in Mercury Valley, Nevada [as estimated by dimensional analysis methods]*

Sampling date	Natural rainfall		Supplemental irrigation	
	Control	100 kg N/ha[a]	Control	100 kg N/ha[a]
Feb 1974	3,133	2,484	2,171	2,253
Oct 1974	2,864	2,466	4,266	5,253
Oct 1975	3,599	2,860	5,203	6,475
Ratio (Oct 1975:Feb 1974)	1.15	1.15	2.40	2.87

[a]Nitrogen treatment level as kg/ha equivalent applied as NH_4NO_3 in Mar 1974 followed by periodic trickle irrigations at times indicated in the footnote of Table 16-6b.

As one should expect from broadcast fertilizer applications, the soil nitrate levels were elevated during the growth season of 1974 and 1975 by treatments up to 100 kg N/ha applied as NH_4NO_3 in March 1974. Ammonium concentrations in the root zone quickly decreased to control levels on irrigated plots, but remained high after fertilization of dry plots during the dry summer months (Table 16-10). Some volatilization loss of ammonia may have occurred, but the amount involved was not measured. Nitrogen content of plant tissue was little affected by nitrogen

Table 16-8 *Plant Species Response to Supplemental Trickle Irrigation in 1974 and 1975 Ranked According to Increase in Numbers and Biomass*

1974		1975	
Species	% increase in numbers	Species	% increase in numbers
Lepidium fremontii	7,982	*Sphaeralcea ambigua*	3,473
Sphaeralcea ambigua	1,008	*Ambrosia dumosa*	1,798
Ambrosia dumosa	364	*Lepidium fremontii*	1,777
Acamptopappus shockleyi	290	*Acamptopappus shockleyi*	942
Atriplex confertifolia	205	*Atriplex confertifolia*	402
Ephedra funerea	107	*Menodora spinescens*	229
Larrea tridentata	58	*Ephedra funerea*	84
Menodora spinescens	23	*Larrea tridentata*	11

Note: Data are from plots without supplemental nitrogen given irrigation treatments as listed in footnote of Table 16-6.

Table 16-9 *Dominant Winter Annuals Measured May 1975 on Trickle-Irrigated and on Dry Plots in Mercury Valley, Nevada*

Species	Number (10^3/ha)			Weight per plant (mg)		
	Irrigated	Dry	Ratio	Irrigated	Dry	Ratio
Caulanthus lasiophyllus	53.8	0.0	∞	53.7	0.0	∞
Schismus arabicus	106.1	1.7	62.4	90.9	400.0	0.2
Descurainia pinnata	140.3	3.6	39.0	87.9	30.6	2.9
Linanthus demissus	40.5	5.4	7.5	9.4	1.9	5.1
Langloisia setosissima	40.6	7.9	5.1	31.5	15.2	2.1
Gilia cana	93.7	18.6	5.0	22.0	45.7	0.5
Bromus rubens	2,037.0	814.0	2.5	129.0	126.2	1.0

Note: Trickle-irrigation treatments are given in footnote of Table 16-6.

fertilization until either natural rainfall or supplemental water from irrigation had leached the fertilizer amendment down into the root zone. The primary phenological effect of trickle-irrigation treatment was to extend the period of growth of stems and leaves of those shrubs not inhibited by soil and air temperatures. Another major effect of added water was increased fruit production in virtually all plant species growing in study plots. Nitrogen treatments, on the other hand, decreased fruit production in some species and increased it in others (Hunter et al. 1976a).

Table 16-10 *Examples of Soil Nitrogen Levels Affected by Trickle-Irrigation and Nitrogen Fertilizer Amendment on Mercury Valley Soil*

Treatment[a]	Nitrate N (ppm)		Ammonium N (ppm)	
	11 Jun[b]	30 Jul[c]	11 Jun	30 Jul
Wet, no N	17 ± 0.1	67 ± 39	0.7 ± 0.1	0.7 ± 0.1
Wet, 25 kg N	24 ± 0.1	89 ± 25	1.0 ± 0.1	0.9 ± 0.3
Wet, 100 kg N	113 ± 44	180 ± 63	1.3 ± 0.5	1.1 ± 0.4
Dry, no N	8 ± 0.0	6 ± 1	1.0 ± 0.1	1.1 ± 0.4
Dry, 25 kg N	42 ± 0.1	20 ± 1	3.4 ± 0.4	4.4 ± 0.1
Dry, 100 kg N	119 ± 0.4	90 ± 23	8 ± 3	21 ± 4

[a]Nitrogen applied as 0, 25 and 100 kg/ha equivalent NH_4NO_3 in Mar 1974. Date represents samples from 0-3 cm depth of soil profile collected in Jun and Jul 1974 after water had been applied as given in footnote of Table 16-6.

[b]90 days since treatment.

[c]140 days since treatment.

DISCUSSION

Among the more obvious aspects of water supplementation is timing. Water can be of relatively little value to plants if it is supplied at a time when it will be lost by evaporation, without plant use, or during a phenological stage of plant development when no response can be expected. Another aspect is the point at which water becomes limiting. In most of the spring seasons in the northern Mohave Desert, there is a period of time when water is not limiting to any of the plants. Soil moisture from winter rainfall is present and the efficient, extensive root systems are capable of extracting adequate moisture even if the soil appears dry on the surface. How long this time period lasts depends upon the recharge supply and the temperature conditions which determine how fast the plants use the available water.

Response to irrigation under such conditions, then, is not always a simple matter. If rainfall has been sparse prior to the onset of the spring growth season, the response can be dramatic when irrigation is superimposed. If, however, the soils already contain soil moisture recharge levels in the range for which plants develop and grow following phenological triggers, supplemental water does nothing more than extend the growth period until the soil eventually dries out. This extension of growth period usually is the most important effect of water addition (Hunter et al. 1975a, 1975b, 1976a). During this period, the biomass production can be greatly increased (several fold), depending upon the moisture supply.

There is a limit to increased growth response of shrubs to supplemental moisture which is imposed by the genetic nature of the plants (Noy-Meir 1973). If additional water is to give further yield increase, the species in the community must be changed to those which respond more to water. This is why extra water tends to change deserts to grasslands (Wallace and Romney 1972a). The annual plants are different in that additional water can result in more germination of annuals and in more and larger plants (Wallace and Romney 1972a). Furthermore, the soil moisture may at times be limiting for shallow-rooted annuals even when it is not limiting for the perennial plant species. The volume of soil available for each plant and the level of moisture present are critical and determine this relationship. This is a point for which more research information must be obtained.

The nitrogen cycle of the northern Mohave Desert operates in such a way that sufficient mineral nitrogen becomes available to supply plants with enough nitrogen for whatever growth potential is possible with the natural water supply (Wallace et al., Chapters 9 and 14 of this volume). The cycle seems able to adjust for increases or decreases of seasonal rainfall. Mineralization of nitrogen in the soil organic pool, like plant growth, seems to be related to soil moisture levels. These relationships are for natural conditions and may not hold when the growing season is prolonged with additional water. Further interaction, of course, would occur if so much additional water were added that much of the available nitrogen leached below the root zone. In that case, response to additional nitrogen would be certain. The root zone of annual plants is generally more shallow than that of perennials, so annuals would become nitrogen deficient first when soluble nitrogen is leached deep into the soil.

Nitrogen concentration of the perennial vegetation of the northern Mohave Desert is relatively high (Romney et al. 1973). The leaves of perennials often contain 3-4 percent nitrogen, dry weight. Such high levels would not suggest response to nitrogen fertilization. The perennial plants grow in groups spatially organized, mostly in so-called "fertile islands," in which soil organic matter contents are high (2 percent or more) in comparison to adjoining soil among the islands (0.5 percent or less; Romney et al. 1973). These islands have developed as the result of activities of plants and animals for decades, if not centuries (Roberts 1950; Paulsen 1953; Charley and Cowling 1968; Rickard 1965a, b; Garcia-Moya and McKell 1970; Charley 1972; Wallace and Romney 1972a; Sharma and Tongway 1973; Tiedemann and Klemmedson 1973a; Charley and West 1975). Rainfall is not sufficient to develop that kind of soil over all the desert, so the processes operate to utilize the water and energy available to develop the fertile islands on a small fraction of the soil area. The islands are as fertile as the entire area of more humid ecosystems. The existence of these islands makes it possible for the plants growing in them to develop to whatever limit the natural rainfall permits, without nitrogen deficiency (Romney et al., in press). Plants, particularly annuals, growing outside the fertile islands can be subject to nitrogen deficiency even without supplemental irrigation (Romney et al. 1974).

The fertile islands are a remarkable development in desert ecosystems. If, instead, they did not exist and the fertility in them were spread over the entire soil area, then nitrogen deficiency would be present everywhere for plants not having a biological nitrogen-fixation relationship (Garcia-Moya and McKell 1970; Wallace and Romney 1975; Charley and West 1975). Also, soil physical conditions would be very unfavorable for the establishment of new seedlings. Under those conditions, the productivity of the northern Mohave Desert would be much less than it is now. Land reclamation procedures are difficult in disturbed sites in the desert, and a major reason is that disturbance causes the fertile islands to disappear and then revegetation must occur in a more hostile environment (Wallace and Romney 1975).

If the addition of supplemental soil moisture results in the development of new plants in the bare areas between the fertile shrub-clump islands, then nitrogen fertilization would be necessary to achieve maximum productivity with those plants. In very heavy rainfall years when annuals do grow in the bare spaces, there is a size gradient according to distance from the shrub clump because of the nitrogen gradient in the soil (Wallace and Romney 1972a; Romney et al. 1974).

A given amount of water will sustain a given amount of primary productivity. If the amount of water were increased to a new level and maintained year after year, the system would adjust to the new level with a new productivity plateau. Water could be the new limiting factor, but it also could be nitrogen or some other nutrient. It could even be genetic, so that no additional growth would be possible even with input of more water or nitrogen. In fact, the genetic limit is soon reached with desert vegetation. This barrier may be overcome with increased density of the native species. More water would permit them to grow in the bare areas between the shrub clumps. This would require nitrogen fertilization to achieve added productivity, but eventually another plateau would be reached. Still another plateau could be reached if the system were changed to other species more responsive to water (grassland, for example), as shown in the hypothetical illustration in Figure 16-1. Supplemental nitrogen may not be necessary until the second and third limiting zones are reached. But, with time, we expect an equilibrium could occur where nitrogen fixation dependent upon species in the system could supply new nitrogen needed to maintain either of the two higher plateaus, provided the nitrogen cycled within the system (Day et al. 1975). The response to supplemental nitrogen application then would be only temporary.

If a native desert system is irrigated, it may take more than 1 year to reach the new productivity limit of the old plants imposed by the new soil moisture system,

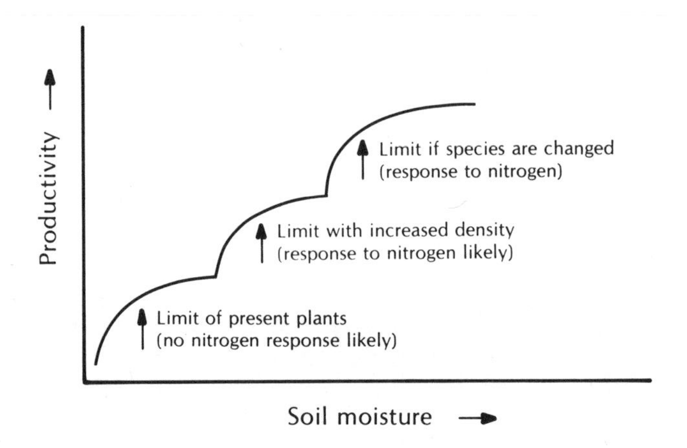

Figure 16-1 *Hypothetical response of desert vegetation to supplement moisture.*

even with adequate nitrogen. This is because larger plants will result which can support greater productivity, and 2 or 3 years may be needed to reach the new equilibrium size.

SUMMARY

Studies made of interactions between the most common factors limiting productivity of desert ecosystems (levels of soil moisture and available nitrogen) resulted in several salient conclusions:

1. Considerable water can be applied to Mohave Desert ecosystems with little if any, response in primary productivity if the timing of application is such that the moisture recharge level in soil is already high. If rainfall has been sparse and the recharge soil moisture level is low prior to the onset of the spring growth season, the response can be dramatic when irrigation is superimposed.
2. Water can be limiting for some plant species (especially annuals) and at the same time not be limiting to other plants (especially perennials) growing side by side when moisture near the soil surface is depleted.
3. The most common effect of supplemental moisture is that of extending the growth season of plants. Spring and summer irrigation, however, will not overcome high temperature dormancy of many Mohave Desert plants.
4. The nitrogen cycle in the northern Mohave Desert operates in a manner which will supply adequate nitrogen even in high rainfall years, but nitrogen stress can develop with supplemental irrigation which increases plant density and/or changes species composition.
5. The natural fertility of the northern Mohave Desert is concentrated in "fertile islands" produced underneath shrub clumps. Nitrogen stress occurs outside these islands when high rainfall years or supplemental moisture result in plants growing in the open area.

6. Many native desert species will not respond indefinitely to added moisture, but eventually will be replaced by other species (usually grasses and forbs) that use additional soil moisture more effectively.
7. Air and soil temperatures are important in determining responses to soil moisture and nitrogen.
8. Broadcast applications of nitrogen fertilizer generally are effective only after water from rainfall or irrigation has moved the nitrogen down into the root zones. Under conditions of soil moisture stress, addition of supplemental moisture generally will mask growth responses from supplemental nitrogen.

17

SUMMARY, CONCLUSIONS AND SUGGESTIONS FOR FURTHER RESEARCH

N. E. WEST and J. SKUJIŅŠ

DISTRIBUTION

Nitrogen in Ecosystem Structure

Relative importance and absolute importance of each process in the nitrogen cycle of desert ecosystems differ considerably from those given in general textbooks. Nearly all prior data contributing to understanding of terrestrial nitrogen cycles have come from either farmland or forest contexts. Overall, the nitrogen cycle in arid environments is quantitatively similar to that in more mesic environments only in terms of the available atmospheric nitrogen. Because of less biological activity and thus less biomass than in forests, grasslands and some tundras, deserts have comparatively little accumulated nitrogen in biomass and soils per unit area. What is present is highly concentrated spatially and temporally. Nitrogen is concentrated in vegetation; most of the nitrogen is concentrated as protein in new growth and shows higher concentrations than seen in the plants of other biomes. Since the plants are scattered, nitrogen is located in loci where plants take up and return nitrogen and otherwise modify the microenvironment. Nitrogen is also generally concentrated in the surface soil. Plant, animal and microbial activity are highly correlated with short periods of favorable environmental conditions, mainly when soil moisture is available. Changes in the form of nitrogen occur rapidly during these periods. Fixation and other metabolic processes occur at very high rates when these favorable periods "pulse" the system. This rapid, intense activity followed by low storage contrasts sharply with the low, slow fixation and rates of other processes in a high storage system such as a forest ecosystem. These differences, plus the organization of desert ecosystems into "fertile islands," are probably the long-term result of rainfall not being sufficient to develop rich soil over the whole desert landscape. Thus, processes operate to utilize water and energy over short periods of time in cells of concentrated activity covering a small fraction of the total surface area. These islands are often as fertile and as humid as the entire area of more mesic ecosystems. The existence of these islands makes it possible for the plants and animals growing there to develop to whatever limit the natural rainfall will permit, with the available nitrogen becoming limiting only at high moisture levels. The majority of the landscape is nitrogen limited most of the time, as are desert biological processes when expressed as functions per unit area of randomly-sampled surface.

The spatial and temporal distribution of nitrogen in desert ecosystems is closely correlated with biological activities in soils. The decomposition of organic matter is most intense in the upper few centimeters of the soil profile. Soil respiration in the surface 3-cm layer is, on the average, two to three times higher than in the 5-10 cm layer, and it decreases considerably with depth. Similar regularities apply also to dehydrogenase activity, characterizing total biological activity, and to proteolytic activity. The horizon of maximal biological activity does not coincide with the layer containing the maximal number of bacteria and streptomycetes (5-20 cm depth). The surface of desert soils, especially in the Great Basin, is dominated by blue-green and green algae and by lichens.

Both carbon and nitrogen have maximal accumulation in the surface layer of soils, with nitrogen reaching around 0.3 percent but commonly falling to 0.04 percent deeper in the profile. Nitrogen released during organic matter decomposition is in a cationic form (NH_4^+) and has limited mobility in the clay-containing soils. As the clay content decreases and sand content increases, the downward mobility of ammonium increases. Nitrification, however, may be responsible for downward movement with the gravitational flow of water during rainy periods.

As exemplified by the Great Basin Desert, the maximal nitrogen content in soil is present during the wet periods in spring and fall, coinciding with seasons of higher plant growth. Towards the end of the wet season the added nitrogen in the soil is lost by denitrification or volatilization. Organic compounds constitute more than 99 percent of the total soil nitrogen at all times. About 25 percent of this nitrogen is subjected to seasonal fluctuations; the remaining 75 percent presumably being part of slowly decomposing plant residue and continuously forming soil humic matter.

One practical implication of the pronounced spatial and temporal distribution patterns is the need for careful design of sampling schemes and consideration of spatial variation in model building. Another is the susceptibility of ecosystem productivity to the effects of disturbances. Erosion and dispersal of the nutrients stored in the islands of fertility may result in fewer favorable microsites for plant establishment and loci for animal and decomposer activity.

Although many descriptive data of nitrogen distribution exist, relatively little information on nitrogen dynamics in natural and perturbed desert systems is available. Some inferences on the relative importance of various components and rates of cycling processes may be obtained from structural data; however, it is preferable to study the processes themselves for a more accurate understanding.

PROCESSES

Fixation

Agronomic experience would lead the casual observer to suspect symbiotic microbial associations with legumes as the most likely natural mode of nitrogen fixation in deserts. In many arid to semiarid environments legumes are rare to absent; where they are dominant, they may lack observable nodules. There are indications that nodules on such legumes may be formed only when the right combination of temperature, moisture and illumination causes the necessary photosynthate to move into the roots. Indeed, a few studies designed on the premise of leguminous fixation by desert shrubs produced negative results. Here the problem might be fitting the plant with the right strain of *Rhizobium*. Relatively few "hard" data on this topic exist.

Some recent discoveries have implicated as nitrogen fixers a wide set of nonlegumes dominant in many desert environments. Numerous genera of shrubs, several grasses and at least two succulents have given positive results to modern tests of nitrogen-fixing abilities. These responses are highly seasonal, being restricted to cooler, wetter periods at the beginning of the growing season in the Great Basin and Mohave deserts. Since nodules are not always visible, there exists the possibility that fixation is really due to free-living bacteria, algae or actinomycetes in the rhizosphere or even phyllosphere. Although some symbiotic, nodule-forming nonleguminous plant-microbial associations in desert plants might exist, the associative symbiotic processes in the rhizosphere are more likely to prove important. More definitive research is now needed to pinpoint these sources of fixation within the soil profile. Such research could have practical implications for revegetating disturbed lands in arid to semiarid regions, since applications of artificial fertilizers to aid establishment could well become prohibitively expensive.

Generally, the root systems of plants in arid zones act as channels of biological activities. Nitrogen fixation and other processes of the nitrogen cycle are more pronounced in the rhizosphere around live or dead roots where organic matter is available, as compared to the generally inactive organic-matter-poor, nonrhizosphere soil. It is expected that the bulk of the nitrogen dynamics takes place in the root-influenced microenvironment.

Several definitive studies have shown the great importance of fixation by soil cryptogamic crusts in desert regions. Our present understanding indicates that this is the major site of nitrogen input to the soil of these systems, particularly of the cold-winter Great Basin Desert. The phycobionts of blue-green algae associated with several genera of lichen are involved. A few free-fixing algae are also commonly present. The cryptogamic crust formation is closely correlated with the clay content in arid soils, apparently because clays retain more water and nutrients and bind soil particles together. Once formed, the crusts counteract erosion. Since the cryptogamic crusts can be considerably altered by man's activities, future attention should be given to their destruction by animals, machines and pollutants.

Interesting leads on the role of allelochemics from higher plants in inhibiting nitrogen fixation and nitrification should be pursued. Such an inhibition contributes to nitrogen loss by denitrification and other processes of the nitrogen cycle that may also be altered by these plant-produced chemicals. Successional patterns may be at least partially controlled by these mechanisms.

Nitrogen fixation by heterotrophic, free-living microorganisms is generally a very minor process in deserts. The lack of major inputs of carbon is probably the major cause. Heterotrophic fixation occurs routinely only in carbon-rich microsites like the rhizosphere or under decaying cryptogamic soil crusts. The pH, temperature and moisture regimes of deserts are also not favorable to free-fixing organisms. Direct absorption of atmospheric ammonia by plants and soils is a mode of input found to be sometimes important in more mesic environments, but it has yet to be studied in desert conditions.

Litter Fall and Decomposition

Although litter and standing dead plant material may be sparse in deserts as compared to other biomes, nitrogen input from plant litter is a vital part of the cycle. Very little research has been done on litter fall and litter decomposition in deserts. What little has been done shows that soluble nitrogen is fairly rapidly leached from dead leaves; the remaining nitrogen is only slowly mineralized. The low carbon:nitrogen ratio, and especially the high proportion of lignin-type compounds in plant litter, is probably important in delaying general breakdown and nitrogen release. The seasonality of litter fall and decomposition varies considerably between hot and cool deserts. Coincidence of litter decomposition and wet periods is striking. Since the wet period is in the colder part of the year in the Great Basin, temperature often becomes limiting to decomposition processes. Because the wet period in the Chihuahuan Desert is during the summer, temperature limitations are not noted and comparatively more rapid decomposition ensues. It is evident that lower soil animals are comparatively more important in burying detritus and thus enhancing decomposition of litter in desert soils than in other biomes.

Proteolysis is the initial process of mineralizing the relatively small inputs of nitrogen in plant and animal detritus. The phenology of this process is apparently concomitant with decomposition and the occurence of the same appropriate environmental conditions. Ammonification of amino N is the final obligatory step of the mineralization process. It inherently contributes to nitrogen losses from desert soils by making possible subsequent volatilization, nitrification and denitrification processes. These impressions are, however, derived mostly from using artificial substrates and in vitro environments. Additional monitoring of proteolysis and ammonification of natural substrates in undisturbed field environments needs to be done.

Nitrification

Because of the possibility that nitrites may accumulate to toxic levels when oxidation to nitrate is impeded in alkaline soils, nitrification is a process of special interest in desert environments. Recent research sponsored by the IBP has shown that although there is a considerable potential for nitrite and nitrate formation, the accumulation of these products does not apparently happen in cold-winter deserts. There is evidence of buildup of nitrates in hot desert soils. The phenology and magnitude of nitrification is greatly affected by soil moisture, temperature, pH, other aspects of soil chemistry and their modifications by numerous physical properties. Allelopathic substances released by desert plants may considerably inhibit nitrite oxidation to nitrate. Regrettably, these results are derived only from studies of nitrification in the Great Basin Desert. Comparative work in the other three regional deserts is now needed to broaden the generalities.

Denitrification

Field and laboratory studies with ^{15}N have shown denitrification to be the major pathway for nitrogen loss from desert soils. Denitrification rates increased with increased moisture, temperature, organic carbon and nitrogen-containing amendments. The highest losses of ^{15}N occurred from native soils in horizons with highest organic content. These organic contents are comparatively low, so it is not surprising that amendments of straw along with ^{15}N-tagged fertilizer reduced the rate of loss, presumably by enhancing immobilization. It has been demonstrated that 99 percent of blue-green algae-lichen crust-fixed nitrogen may become denitrified during the same season. The highest decompositional activity of organic matter is in the soil surface layer during the wet seasons when nitrogen fixation also is at the highest level. Moisture during this period lowers the oxygen tension in crust microsites. The nitrifiers have a high affinity for oxygen, and conversion of ammonium to nitrite or nitrate may take place. At the same sites, however, the heterotrophic denitrifiers may use the easily accessible carbon sources from algae and in turn reduce the oxidized nitrogen. Denitrification occurs after or simultaneously with nitrification. Some comparison between soils from the regional deserts hints that denitrification pathway is as important in southern as in northern deserts. Much more comparative work needs to be done, however, because the practical implications would be great if this loss could be avoided and the nitrogen passed along into increased plant and animal productivity.

Volatilization of Ammonia

Although only Great Basin Desert soils have been covered, volatilization of ammonia apparently contributes much less than denitrification to the gaseous loss of nitrogen from desert soil. Volatilization is more significant in the case of decomposing higher plant residue. Volatilization losses from exogenously supplied $(NH_4)_2SO_4$ may be higher than endogenously produced ammonium N. While this would not be important under natural conditions, it may be important when nitrogen fertilizer is applied. It also might become significant in situations where the nitrification of algal crust-fixed N_2 is impeded.

The mosaic pattern of bare interspaces and scattered shrubs is important in this process as well. The greatest amount of plant residue accumulates beneath the plants from direct deposit or eolian transport. As the litter decomposes, much nitrogen is mineralized and may be utilized by the plants. Although ammonia volatilization is higher under the plant canopies than in the interspaces, the relative amounts of ammonium fixed on the soil exchange complex are greater in the former habitat. The net retention is apparently due to the comparatively low degradation rates of the plant litter (attributed to their high content of lignin-type compounds) and to more ammonium being retained by the clay complex before being nitrified and subsequently denitrified. In the interspace microsites the easily metabolizable amino N and photosynthate become available to microbial metabolism in comparatively large bursts of activity. The nitrogen becomes denitrified before the environmental conditions become suitable for ammonia volatilization. Allelochemics from the Great Basin Desert shrubs and their litter enhance the reactions that result in the greatest gaseous losses. More studies of these processes in other desert contexts are necessary to test their generality.

It is interesting to note that deserts lose, via gaseous losses, about the same percentage of fixed nitrogen that is lost via anion forms of nitrogen in the waters leaching through and draining from eastern deciduous forest watersheds. In the forests the carbon supply is excessive and nitrogen is limiting, but for different reasons than in our deserts.

Nitrogen Uptake by Plants

Nitrogen uptake and assimilation by higher plants are key steps in the nitrogen cycle of any ecosystem. Information on these processes is generally lacking for desert to semidesert plants and soils. The forms and amounts of nitrogen available on a seasonal basis have not been satisfactorily documented. A few initial studies indicate that average nitrate concentrations are obviously low, but they are at least seasonally high enough to meet much, if not most, of the short-term needs of the plants. In clay-containing desert soils with better nitrogen retention there appear to be sufficient quantities of ammonium to satisfy plant needs year-round.

Desert plants commonly respond to the addition of nitrogen fertilizer, especially when additional soil moisture is added. Mass flow of soluble nitrogen in water moved to roots could bring substantial quantities of nitrate to root surfaces even though its concentration in soil at any one time is low.

Recycling of nitrogen within the plant is a conservation measure which decreases the need for new supplies of nitrogen from the external media. Mass flow and recycling are perhaps the most important ways of meeting the needs of desert plants. The highly patterned concentrations of nitrogen in the "fertile islands" of desert ecosystems is an additional way of overcoming the problems of minimal supplies.

The presence of nitrate in the leaves of some plants implies that nitrate was absorbed. Since nitrate reductase is an induced enzyme, its presence in field-grown plants also implies that nitrate is being utilized. However, there is much more to be learned about the status and implication of nitrate reductase in desert plants.

Ammonium and nitrate uptake has been studied for several species of desert plants under glasshouse conditions. Results indicate that each source of nitrogen can be taken up almost uniformly. In some cases, however, there seemed to be a lag in the rate of nitrate uptake since the ratio of ammonium:nitrate uptake increased with time. No critical studies have yet been made of the kinetics of nitrogen uptake by desert plants to indicate the degree of affinity which such plants might have for particular ionic forms of nitrogen. The ionic balance in some field-grown desert plants for which cation-anion values were estimated implies that nitrate is the major form of nitrogen taken up by desert plants.

Of basic interest would be studies of whether microphylly, sclerophylly and other morphological and longevity factors associated with desert plants are related to nutrient limitations. Perhaps the high root:shoot ratios of Great Basin Desert shrubs will be found to be adaptations to facilitate uptake of limiting amounts of nitrogen from a larger portion of the soil profile.

Role of Animals

Although few direct data exist on the role of animals in desert nitrogen cycles, we can extrapolate from published data derived from the laboratory and other natural systems and reason that animals are probably more important than has been recognized. Their main role is to accelerate the flow of organic material from plants to detritus and decomposers. On an ecosystem scale they utilize little nitrogen in building their own body structure. Gaseous losses from urine are high. Feces are more readily attacked by decomposers following the external rumen concept.

Livestock on semidesert shrublands are thought to accelerate the nitrogen cycle and perhaps improve ecosystem productivity, at least over the short term. Livestock use, especially for sheep, is usually to maintain breeding animals; thus little net export of nitrogen results, although no one has done any balance studies in the western U.S. deserts accounting for the animal component.

Of considerable importance to the nitrogen cycle of deserts are the activities of lower animals, especially harvester ants and termites. Ants concentrate nitrogen in their mounds through their food gathering. Termites greatly accelerate the breakdown of the most resistant kinds of plant detritus. They also possess the unique ability to fix nitrogen through symbiotic microbes in their gut tracts. Further work may show this is important on an ecosystem basis. Many of the chewing insects concentrate their activity around shrubs, dropping nitrogen-rich and more easily decomposible frass under the plants. This is another means whereby nitrogen becomes concentrated into ecosystem "cells." It is evident that soil insects, such as cicadas, and other lower soil animals participate in direct, initial decomposition of below-ground litter and mix it into the soil.

Much more research on the role of all animals in desert nutrient cycles needs to be done. It is likely that in the future we will find that animals are much more influential in controlling the overall rates of energy flow and nutrient cycles in desert ecosystems than presently believed.

Physical Inputs

Few data are available on the magnitude of the various modes of physical inputs of nitrogen to desert ecosystems. The existing data indicate that total nitrogen in rainfall amounts to 4-6 kg/ha per year, with the values being significantly lower than for other systems. Ammonia N appears to be more abundant in the spring rainfall when warm temperatures stimulate ammonification. Nitrate N is comparatively more important in the cooler and/or drier months. Nitrite N is rare in rainfall. A high proportion of the total nitrogen input from the atmosphere is associated with particulate matter, either as dust or rainout. Deserts probably produce more dust than they receive. If so, then this would result in a net export of nitrogen.

Leaching from plants is an apparently minor redistribution process for nitrogen, although data from only one study exist. Because of comparatively high rates of nitrogen loss by erosion and gaseous transformations, the precipitation and dry fallout inputs become relatively important in maintaining the nitrogen cycle in deserts. Rising levels of nitrogen-rich pollutants may well add to these inputs.

Physical Transfers and Losses

Because nitrogen in deserts is concentrated near the soil surface and around the larger, longer-lived plants, it is very susceptible to being "lost" by erosional processes. (Lost is put in quotation marks, because whether an input, output or transfer is involved depends on one's definition of the extent of the system.)

It is well recognized that water and wind are major agents moving nitrogen-containing materials across desert landscapes. Very few direct data on these processes exist, however. Extrapolation of erosional formulae from agricultural contexts is made in this volume in an attempt to estimate the magnitude of these processes over western U.S. landscapes as a whole. Erosional losses in the Sonoran Desert are apparently several times more rapid than those in the northern Great Basin, where considerable surface stabilization by cryptogamic crust occurs. Data are now needed to verify these models and to check whether erosional dispersal of nitrogen will result in lowered ecosystem productivities, as has apparently happened in Australia. Knowledge of the rate of recovery of ecosystem nitrogen structure would also be interesting and of much practical importance.

SYNTHESIS

All studies of nitrogen distribution and processes are done with the implicit hope of eventually understanding the total nitrogen cycle. This holistic understanding requires some conscious synthesis of information on the parts (structure) and processes (functions). The most conventional approach to synthesis has been graphical. Figures or tables usually involve balance sheet-type estimates over a given period, usually a year. Shorter-term dynamics can be depicted by graphics or equations. Predictions of response to changes in the system require projections of the net effect of the interacting structure and processes. Some of the most dramatic changes could be predicted by intuition; however, a more modern approach to handling complex systems is through computer simulation of mathematically or statistically expressed relationships.

Great Basin Model

Only one of the Desert Biome sites (Curlew Valley) had enough data on all compartments and processes to justify an attempt to characterize the nitrogen cycle with the assistance of computer simulation. That attempt proved to be an efficient way of organizing existing information, recognizing deficiencies and encouraging research to fill the gaps. Simulations of normal responses seemed to give a realistic portrayal of the system. Attempts to simulate radical perturbations to the system gave counterintuitive results, indicating that either our intuition is unreliable or the modeling approach is too oversimplified to allow reasonable projections. Only observation of actual ecosystem perturbation will give us further insight into which problem is greater. Meanwhile, the model could be made more sophisticated as more directly relevant data are accumulated.

Mohave Desert Model

Understanding of various aspects of the nitrogen cycle in the Mohave Desert has developed to the point that a schematic model has been made and implications derived by intuition.

Considerable differences in the magnitude of certain processes contrast the Mohave Desert and Great Basin Desert models. For instance, the amount of cryptogamic crust and its importance in nitrogen fixation are much less in the former. Nitrates appear to accumulate in Mohave Desert soils, but not in the ones examined so far in the Great Basin. Due to the limited nitrogen input, the nitrogen processes do not appear to be carbon limited in Mohave Desert soils as they do in the Great Basin. The differences are mostly due to contrasting climate and soil physical and chemical properties.

Manipulation Studies

Although more direct research on the processes would help to elucidate differences between deserts, we can also learn from the application of amendments to these systems. Since fertilization and irrigation of native communities were, even before the "energy crunch," patently uneconomical, little work on the fertilization of drier rangelands has been done. Two studies of nitrogen additions to desert vegetation have given us tests of the predictive powers of our increasing fund of basic information and pointed to further potential applications.

Great Basin Desert

Discovery of the major leakage of nitrogen in a gaseous form via the process of denitrification at Great Basin Desert sites led us to believe that any supplemental nitrogen would also be largely lost to the atmosphere. This is apparently not always the case where pastures of crested wheatgrasses (*Agropyron desertorum*) have been established in the Great Basin Desert. The positive plant growth responses reported herein may be due to the method of application—mechanically inserting the amendments below the soil surface. Application of fertilizers to soil surfaces in even wetter environments has often resulted in little or no plant response. Only in wetter than average years would some of the nitrogen limitations be alleviated by nitrogen amendments. Many of the desert soils are too rocky for subsurface application and native vegetation is easily altered by mechanical treatments, with the invasion of undesirable annual plants being fostered by any tillage. The experimental treatments used would therefore likely be of little practical value in improving primary production for the bulk of the Great Basin region. Addition of carbon to enhance immobilization of naturally fixed nitrogen and introduction of nitrification inhibitors seem to be more likely to be effective and perhaps more economical alternatives.

Mohave Desert

Annual plants of the Mohave Desert respond to additional nitrogen when soil moisture is adequate. Native grasses respond more to increased water than to nitrogen. Native shrubs show only slight responses; they have high amounts of nitrogen in their tissues normally. They grow in the localized islands of fertility that probably have adequate levels of nitrogen maintained in the more favorable microenvironment, and thus all processes proceed much more favorably than in the bare interspaces. As with Great Basin shrubs, there appear to be little if any evolutionary preadaptations to greater supplies of nitrogen.

General Observations

Research on nitrogen in the western deserts of the United States has shown that two distinct local environments exist; one under the plant canopies, the other in the interspaces. For the understanding of the total ecosystem, the biological and nitrogen processes in both kinds of sites should be integrated. Destruction of this mosaic and dilution of the nitrogen pool to form a more homogeneous nutrient landscape will probably require man's addition of nitrogen to repair the damage that natural sucession only very slowly re-establishes.

Some interesting leads to further research in both fundamental and applied directions are given. We can see how nitrogen is only one thread sewn through the ecosystem. We will have to weave this thread in with those of energy flow, the carbon cycle, phosphorus cycle, allelochemics, soil properties and many others in order to understand the pattern of the whole cloth—the ecosystem.

LITERATURE CITED

ABD-EL-MALEK, Y. 1971. Free-living nitrogen-fixing bacteria in Egyptian soils and their possible contribution to soil fertility. Pages 377-391 *in* T. A. Lie and E. G. Mulder, eds. Biological nitrogen fixation in natural and agricultural habitats. Martinus Nijhoff, The Hague, Netherlands.

ACQUAYE, D. K., and R. K. CUNNINGHAM. 1965. Losses of nitrogen by ammonia volatilization from surface fertilized tropical forest soils. Trop. Agric. (London) 42:281-292.

AHMADJIAN, V. 1967. The lichen symbiosis. Blaisdell Publ. Co., Waltham, Mass. 152 pp.

ALEEM, M. I. H. 1970. Oxidation of inorganic nitrogen compounds. Annu. Rev. Plant Physiol. 21:67-90.

ALEXANDER, M. 1961a. Denitrification. Pages 292-308 *in* M. Alexander, ed. Introduction to soil microbiology. John Wiley & Sons, Inc., New York.

ALEXANDER, M. 1961b. Introduction to soil microbiology. John Wiley & Sons, Inc., New York. 471 pp.

ALEXANDER, M. 1965. Denitrifying bacteria. Pages 1484-1486 *in* Methods of soil analysis, Part II. Am. Soc. Agron. Monogr. No. 9. Madison, Wis.

ALLEN, E. K., and O. N. ALLEN. 1961. The scope of nodulation in the Leguminosae. Rec. Adv. Bot. 2:585-588.

ALLEN, M. B., and C. B. VAN NEIL. 1952. Experiments on bacterial denitrification. J. Bacteriol. 64:397-412.

ALLEN, O. N. 1957. Experiments in soil bacteriology, 3rd rev. ed. Burgess Publ. Co., Minneapolis, Minn. 177 pp.

ALLISON, F. E., J. N. CARTER, and L. D. STERLING. 1960. The effect of partial pressure of oxygen on denitrification in soil. Soil Sci. Soc. Am. Proc. 24:283-285.

ALLISON, F. E., and J. DOETSCH. 1950. Nitrogen gas production by the reaction of nitrites with amino acids in slightly acidic media. Soil Sci. Soc. Am. Proc. 15:163-166.

ALLISON, F. E., J. H. DOETSCH, and L. D. STERLING. 1952. Nitrogen gas formation by interaction of nitrites and amino acids. Soil Sci. Soc. Am. Proc. 74:311-314.

ANDERSON, O. E., and R. C. BOSWELL. 1964. Influence of low temperature and various concentrations of ammonium nitrate on nitrification in acid soils. Soil Sci. Soc. Am. Proc. 28:525-529.

ARDAKANI, M. S., L. W. BELSER, and A. D. MCLAREN. 1975. Reduction in nitrate in a soil column during continuous flow. Soil Sci. Soc. Am. Proc. 39:290-294.

ARDAKANI, M. S., J. T. REHBOCK and A. D. MCLAREN.1973. Oxidation of nitrite to nitrate in a soil column. Soil Sci. Soc. Am. Proc. 37:53-56.

ARDAKANI, M. S., R. K. SCHULZ, and A. D. MCLAREN. 1974. A kinetic study of ammonium and nitrite oxidation in a soil field plot. Soil Sci. Soc. Am. Proc. 38:273-277.

ARNOLD, J. F., and H. G. REYNOLDS. 1943. Droppings of Arizona and antelope jackrabbits and the pellet census. J. Wildl. Manage. 7:322-327.

ASHCROFT, R. T., A. WALLACE, and A. M. ABOU-ZAMZAM. 1972. Nitrogen pretreatments *vs* nitrate treatments after detopping on xylem exudation in tobacco. Plant Soil 36:407-416.

BALIGAR, V. C., and S. V. PATIL. 1968. Volatile losses of NH_3 as influenced by rate and methods of urea application to soils. Indian J. Agron. 13:230-233.

BALPH, D. F., coordinator, et al. 1973. Curlew Valley Validation Site report. US/IBP Desert Biome Res. Memo. 73-1. Utah State Univ., Logan. 336 pp.

BARBER, S. A. 1971. A kinetic approach to the evaluation of the soil nutrient potential. Pages 259-273 *in* R. M. Samish, ed. Recent advances in plant nutrition Vol. 1. Gordon and Breach Science Publ., New York.

BARICA, J., and F. A. J. ARMSTRONG. 1971. Contribution by snow to the nutrient budget of some small northwest Ontario lakes. Limnol. Oceanogr. 16:891-899.

BARROW, N. J. 1967. Some aspects of the effects of grazing on the nutrition of pastures. J. Aust. Inst. Agric. Sci. 33:254.

BAYOUMI, M. A. 1975. Response to fertilization of big game winter range vegetation. Ph.D. Dissertation. Utah State Univ., Logan. 104 pp.

BEADLE, N. C. W. 1959. Some aspects of ecological research in semi-arid Australia. Pages 452-460 *in* A. T. Keast, ed. Biogeography and ecology in Australia. Dr. W. Junk Publ., The Hague, Netherlands.

BEADLE, N. C. W., and Y. T. TCHAN. 1955. Nitrogen economy in semi-arid plant communities. I. Environment and general considerations. Proc. Linn. Soc. N.S.W. 80:62-70.

BEATLEY, J. C. 1976. Vascular plants of the Nevada Test Site: ecologic and geographic distributions. Office of Information Services, TID-26881. Springfield, Va. 308 pp.

BECKING, J. H. 1956. On the mechanism of ammonium ion uptake by maize roots. Acta Bot. Neerl. (Netherlands) 5:1-9.

BENEMANN, J. R. 1973. Nitrogen fixation in termites. Science 181:164-165.

BENNET, J. P. 1974. Concepts of mathematical modeling of sediment yield. Water Resour. Res. 10:485-492.

BIRCH, H. F. 1958. The effect of oil drying on humus decomposition and nitrogen availability. Plant Soil 10:9-31.

BIRCH, H. F. 1959. Further observations on humus decomposition and nitrification. Plant Soil 11:262-286.

BIRCH, H. F. 1960. Nitrification in soils after different periods of dryness. Plant Soil 12:81-96.

BJERREGAARD, R. S., 1971. The nitrogen budget of two salt desert shrub plant communities of western Utah. Ph.D. Dissertation. Utah State Univ., Logan. 110 pp. University Microfilms, Ann Arbor, Mich. (Diss. Abstr. 73-873).

BLACKBURN, W. H., R. E. ECKERT, and P. TUELLER. 1969. Vegetation and soils of Nevada watersheds. Univ. of Nevada Agric. Exp. Sta., Reno.

BLASCO, M. J., and A. H. CORNFIELD. 1966. Volatilization of nitrogen as ammonia from acid soils. Nature 212:1279-1280.

BLEAK, A. T. 1970. Disappearance of plant material under a winter snow cover. Ecology 51:915-917.

BLOCKER, H. D. 1969. The impact of insects as herbivores in grassland ecosystems. Pages 290-299 *in* R. L. Dix and R. G. Beidleman, eds. The grassland ecosystem. Range Sci. Dep. Sci. Ser. No. 2. Colorado State Univ., Ft. Collins.

BLOWER, J. G. 1955. Millipedes and centipedes as soil animals. Pages 138-151 *in* D. K. McE. Kevan, ed. Soil zoology. Academic Press, New York.

BOBRITSKAYA. M. A. 1962. Nitrogen intake in soil from atmospheric precipitation in various zones of European USSR [in Russian]. Pochvovedenie 1962 (12):1363-1368.

BOCOCK, K. L., and O. J. W. GILBERT. 1957. The disappearance of leaf litter under different woodland conditions. Plant Soil 9:179-195.

BOCOCK, K. L., O. J. W. GILBERT, C. K. CAPSTICK, D. C. TWINN, J. S. WAID, and M. J. WOODMAN. 1960. Changes in leaf litter when placed on the surface of soils with contrasting humus types: I. Losses in dry weight of oak and ash leaf litter. J. Soil Sci. 11:1-9.

BOHN, H. L., D. B. FENN, and W. J. MOORE. 1969. Electrode potentials of nitrogen and sulfur half-reactions. Soil Sci. 108:95-101.

BOND, G. 1967. Fixation of nitrogen by higher plants other than legumes. Annu. Rev. Plant Physiol. 18:107-126.

BOND, G. 1971. Root nodule formation in non-leguminous angiosperms. Pages 316-324 *in* T. A. Lie and E. G. Mulder, eds. Biological nitrogen fixation in natural and agricultural habitats. Martinus Nijhoff, The Hague, Netherlands.

BOOTH, W. E. 1941. Algae as pioneers in plant succession and their importance in erosion control. Ecology 22:38-46.

BORMANN, F. H. 1969. A holistic approach to nutrient cycling problems in plant communities. Pages 149-165 *in* K. N. H. Greenidge, ed. Essays in plant georaphy and ecology. Nova Scotia Museum, Halifax.

BORNEBUSCH, C. H. 1930. The fauna of the forest floor [in Danish, English summary]. Forstl. Forsoegsvaes. Dan. 11:1-224.

BOWEN, H. J. M. 1966. Trace elements in biochemistry. Academic Press, New York. 241 pp.

BOWMAN, R. A., and D. D. FOCHT. 1974. The influence of glucose and nitrate concentrations upon denitrification rates in sandy soils. Soil Biol. Biochem. 6:297-301.

BRANSON, F. A., G. F. GIFFORD, and J. R. OWEN. 1972. Rangeland hydrology. Range Sci. Ser. No. 1, Soc. Range Manage., Denver, Colo. 84 pp.

BRAY, H. G., and K. WHITE. 1966. Kinetics and thermodynamics in biochemistry. Academic Press, New York.

BREMNER, J. M. 1965a. Organic nitrogen in soils. Pages 93-149 *in* W. V. Bartholomew and F. E. Clark, eds. Soil nitrogen. Agronomy 10. Am. Soc. Agron., Madison, Wis.

BREMNER, J. M. 1965b. Total nitrogen. Pages 1149-1178 *in* C. A. Black, ed. Methods of soil analysis Part 2. Am. Soc. Agron., Madison, Wis.

BREMNER, J. M. 1965c. Inorganic forms of nitrogen. Pages 1179-1237 *in* C. A. Black, ed. Methods of soil analysis Part 2. Am. Soc. Agron., Madison, Wis.

BREMNER, J. M. 1965d. Isotope-ratio analysis of nitrogen in nitrogen-15 tracer investigations. Pages 1256-1286 *in* C. A. Black, ed. Methods of soil analysis Part 2. Am. Soc. Agron., Madison, Wis.

BREMNER, J. M., and K. SHAW. 1958a. Denitrification in soil: I. Methods of investigation. J. Agric. Sci. 51:22-39.

BREMNER, J. M., and K. SHAW. 1958b. Denitrification in soil: II. Factors affecting denitrification. J. Agric. Sci. 51:40-52.

BREZNAK, J. A., W. J. BRILL, J. W. MERTINS, and H. C. COPPEL. 1973. Nitrogen fixation in termites. Nature 244:577-580.

BROADBENT, F. E., and F. E. CLARK. 1951. Denitrification in some California soils. Soil Sci. 72:129-137.

BROADBENT, F. E., and F. E. CLARK. 1965. Denitrification. Pages 344-359 *in* W. V. Bartholomew and F. E. Clark, eds. Soil nitrogen. Agronomy 10. Am. Soc. Agron., Madison, Wis.

BROADBENT, F. E., and B. F. STOJANOVIC. 1952. The effect of partial pressure of oxygen on some soil nitrogen transformations. Soil Sci. Soc. Am. Proc. 16:359-363.

BROADBENT, F. E., and M. E. TUSNEEM. 1971. Losses of nitrogen from some flooded soils in tracer experiments. Soil Sci. Soc. Am. Proc. 35:922-926.

BROADBENT, F. E., K. B. TYLER, and G. N. HILL. 1957. Nitrification of ammoniacal fertilizers in some California soils. Hilgardia 27:247-267.

BROADFOOT, W. M., and W. H. PIERRE. 1939. Forest soil studies: I. Relation of rate of decomposition of tree leaves to their acid-base balance and other chemical properties. Soil Sci. 48:329-348.

BROWN, G. W., JR. 1970. Nitrogen metabolism in birds. Pages 711-793 *in* J. W. Campbell, ed. Comparative biochemistry of nitrogen metabolism. Academic Press, New York.

BURRIS, R. H. 1975. The acetylene-reduction technique. Pages 249-269 *in* W. D. P. Stewart, ed. Nitrogen fixation by free-living microorganisms. Cambridge Univ. Press, London.

BURSELL, E. 1967. The excretion of nitrogen in insects. Pages 33-67 *in* J. W. L. Begment, J. E. Treherne, and V. B. Wigglesworth, eds. Advances in insect physiology Vol. 4. Academic Press, New York.

CADY, F. B., and W. V. BARTHOLOMEW. 1960. Sequential products of anaerobic denitrification in Norfolk soil material. Soil Sci. Soc. Am. Proc. 24:477-482.

CADY, F. B., and W. V. BARTHOLOMEW. 1961. Influence of low pO_2 on denitrification processes and products. Soil Sci. Soc. Am. Proc. 25:362-365.

CADY, F. B., and W. V. BARTHOLOMEW. 1963. Investigations of nitric oxide reactions in soils. Soil Sci. Soc. Am. Proc. 27:546-549.

CALDWELL, M. M. 1975. Primary production of grazing lands. Pages 41-75 *in* J. P. Cooper, ed. Photosynthesis and productivity in different environments. IBP Vol. 3. Cambridge Univ. Press, London.

CALDWELL, M. M., and L. B. CAMP. 1974. Belowground productivity of two cool desert communities. Oecologia (Berlin) 17:123-130.

CAMERON, R. E. 1962. Species of *Nostoc* Vaucher occurring in the Sonoran Desert in Arizona. Trans. Am. Microsc. Soc. 81:379-384.

CAMERON, R. E. 1963. Algae of southern Arizona. Rev. Agrol. 4:232-318.

CAMERON, R. E., and G. B. BLANK. 1966. Soil studies—desert microflora. XI. Desert soil algae survival in extremely low temperatures. Space Programs Summary No. 37-37. 4:174-181. Jet Propulsion Laboratory, Calif. Inst. Technol., Pasadena.

CAMERON, R. E., and W. H. FULLER. 1960. Nitrogen fixation by some soil algae in Arizona soils. Soil Sci. Soc. Am. Proc. 24:353-356.

CAMPBELL, J. W., R. B. DROTMAN, J. A. MCDONALD, and P. R. TRAMELL. 1972. Nitrogen metabolism in terrestrial invertebrates. Pages 1-54 *in* J. W. Campbell and L. Goldstein, eds. Nitrogen metabolism and the environment. Academic Press, New York.

CAMPBELL, N. E. R., and H. LEES. 1967. The nitrogen cycle. Pages 194-215 *in* A. B. McLaren and G. H. Peterson, eds. Soil biochemistry, Vol. III. Marcel Dekker, New York.

CARPENTER, L. H. 1972. Middle Park deer study: range fertilization. Game Research Report. Part II. Colorado Div. of Wildlife, Denver.

CARTER, J. N., and F. E. ALLISON. 1960. Investigations on denitrification in well-aerated soils. Soil Sci. 90:173-179.

CHAO, T. T., and W. KROONTJE. 1964. Relationships between ammonia volatilization, ammonia concentration, and water evaporation. Soil Sci. Soc. Am. Proc. 28:393-395.

CHAPIN, J. D., and P. D. UTTORMARK. 1972. Atmospheric contribution of nitrogen and phosphorus. Water Resources Center Tech. Rep. WIS-WRC 73-2. Univ. of Wisconsin, Madison.

CHAPMAN, H. D., and F. G. LIEBIG, JR. 1952. Field and laboratory studies of nitrite accumulation in soils. Soil Sci. Soc. Am. Proc. 16:276-282.

CHAPMAN, R. F. 1969. The insects: structure and function. The English Univ. Press Ltd, London. 869 pp.

CHARLEY, J. L. 1972. The role of shrubs in nutrient cycling. Pages 182-203 *in* C. M. McKell, J. P. Blaisdell, and J. R. Goodin, eds. Wildland shrubs—their biology and utilization. U.S. Dep. Agric. For. Serv. Gen. Tech. Rep. INT-1. 500 pp.

CHARLEY, J. L., and S. W. COWLING. 1968. Changes in soil nutrient status resulting from overgrazing and their consequences in plant communities of semi-arid zones. Proc. Ecol. Soc. Aust. 3:28-38.

CHARLEY, J. L., and J. W. MCGARITY. 1964. High soil nitrate levels in patterned saltbush communities. Nature 201:1351-1352.

CHARLEY, J. L., and N. E. WEST. 1975. Plant-induced soil chemical patterns in some shrub-dominated semi-desert ecosystems in Utah. J. Ecol. 63:945-963.

CHARLEY, J., L., and N. E. WEST. 1978. Micropatterns of nitrogen mineralization activity in soils of some shrub-dominated semi-desert ecosystems in Utah. Soil Biol. Biochem. 9:357-365.

CHARREAU, C. 1974. Soils of tropical dry and dry-wet climatic areas of West Africa: their use and management. Agronomy Mimeo 74-76. Dep. Agron., Cornell Univ., Ithaca, New York.

CHASE, F. E., C. T. CORKE, and J. B. ROBINSON. 1967. Nitrifying bacteria in soil. Pages 593-611 *in* T. R. G. Gray and C. Parkinson, eds. Ecology of soil bacteria. Liverpool Univ. Press, Liverpool, England.

CHEN, A. W-C., and D. M. GRIFFIN. 1966. Soil physical factors and the ecology of fungi: V. Further studies in relatively dry soils. Trans. Br. Mycol. Soc. 49:419-426.

CHEPIL, W. S. 1945a. Dynamics of wind erosion: I. Nature of movement of soil by wind. Soil Sci. 60:305-320.

CHEPIL, W. S. 1945b. Dynamics of wind erosion: II. Initiation of soil movement. Soil Sci. 60:397-411.

CHEPIL, W. S. 1945c. Dynamics of wind erosion: III. The transport capacity of the wind. Soil Sci. 60:475-480.

CHEPIL, W. S. 1946a. Dynamics of wind erosion: V. Cumulative intensity of soil drifting across eroding fields. Soil Sci. 61:257-263.

CHEPIL, W. S. 1946b. Dynamics of wind erosion: VI. Sorting of soil material by the wind. Soil Sci. 61:331-340.
CHEPIL, W. S. 1953. Factors that influence clod structure and erodibility of soil by wind: I. Soil texture. Soil Sci. 75:473-483.
CHEPIL, W. S. 1954. Factors that influence clod structure and erodibility of soil by wind: II. Water-stable structure. Soil Sci. 76:389-399.
CHEPIL, W. S. 1955a. Factors that influence clod structure and erodibility of soil by wind: III. Calcium carbonate and decomposed organic matter. Soil Sci. 77:473-480.
CHEPIL, W. S. 1955b. Factors that influence clod structure and erodibility of soil by wind: IV. Sand, silt, and clay. Soil Sci. 80:155-162.
CHEPIL, W. S. 1955c. Factors that influence clod structure and erodibility of soil by wind. V. Organic matter at various stages of decomposition. Soil Sci. 80:413-421.
CHEPIL, W. S. 1957a. Sedimentary characteristics of dust storms: I. Sorting of wind-eroded soil material. Am. J. Sci. 255:12-22.
CHEPIL, W. S. 1957b. Sedimentary characteristics of dust storms: III. Composition of suspended dust. Am. J. Sci. 255:206-213.
CHEPIL, W. S. 1957c. Width of field strips to control wind erosion. Kansas Agric. Exp. Sta. Tech. Bull. 92. 16 pp.
CHEPIL, W. S. 1958. Soil conditions that influence wind erosion. U.S. Dep. Agric. Tech. Bull. 1185. 40 pp.
CHEPIL, W. S., and N. P. WOODRUFF. 1957. Sedimentary characteristics of dust storms: II. Visibility and dust concentration. Am. J. Sci. 255:104-114.
CHEPIL, W. S., N. P. WOODRUFF, and A. W. ZINGG. 1955. Field study of wind erosion in western Texas. U.S. Dep. Agric. Soil Conserv. Serv. TP-125.
CHEW, R. M., and A. E. CHEW. 1970. Energy relationships of the mammals of a desert shrub (*Larrea tridentata*) community. Ecol. Monogr. 40:1-21.
CHILD, J. J. 1976. New developments in nitrogen fixation research. BioScience 26:614-617.
CHO, C. M. 1971. Convective transport of ammonium with nitrification in soil. Can. J. Soil Sci. 51:339-350.
CLARK, F. E., and W. E. BEARD. 1960. Influence of organic matter on volatile loss of nitrogen from soil. Trans. 7th Int. Congr. Soil Sci. 2:501-508.
CLAYTON, J. L. 1976. Nutrient gains to adjacent ecosystems during a forest fire: An evaluation. For. Sci. 22:162-166.
CLOUDSLEY-THOMPSON, J. L. 1971. The temperature and water relations of reptiles. Merrow Publ. Co. Ltd, Watford Herts., England. 159 pp.
COLE, L. C. 1946. A study of the cryptozoa of an Illinois woodland. Ecol. Monogr. 16:50-86.
COLLINS, F. M., and C. M. SIMS. 1956. A compact soil perfusion apparatus. Nature 178:1073-1074.
COMANOR, P. L., and D. C. PRUSSO. 1974. Decomposition and mineralization in an *Artemisia tridentata* community in northern Nevada. US/IBP Desert Biome Res. Memo. 74-40. Utah State Univ., Logan. 10 pp.
COOK, C. W. 1965. Plant and livestock responses to fertilized rangelands. Utah Agric. Exp. Sta. Bull. 455.
COOK, C. W. 1971. Effects of season and intensity of use on desert vegetation. Utah Agric. Exp. Sta. Bull. 483.

Cook, C. W., and L. A. Stoddart. 1953. Effects of grazing intensity upon the nutritive value of range forage. J. Range Manage. 6:51-54.

Cook, C. W., L. A. Stoddart, and L. E. Harris. 1954. The nutritive value of winter range plants in the Great Basin. Utah Agric. Exp. Sta. Bull. 372.

Cooper, G. S., and R. L. Smith. 1963. Sequence of products formed during denitrification in some diverse western soils. Soil Sci. Soc. Am. Proc. 27:659-662.

Cornforth, I. S., and H. A. D. Chesney. 1971. Nitrification inhibitors and ammonia volatilization. Plant Soil 34:497-501.

Coulson, R. A., and T. Hernandez. 1970. Nitrogen metabolism in the living reptile. Pages 639-710 *in* J. W. Campbell, ed. Comparative biochemistry of nitrogen metabolism. Vol. 2. Academic Press, New York.

Cowling, S. W. 1969. A study of vegetation activity patterns in a semi-arid environment. Ph.D. Dissertation. Univ. of New England, Armidale, New South Wales, Australia. 286 pp.

Cowling, S. W. 1977. Effects of herbivores on nutrient cycling and distribution in rangeland ecosystems. pp. 277-298 *in* The impact of herbivores on arid to semi-arid rangelands. Proc. 2nd US/Australian Rangeland Panel. Adelaide, 1972. Australian Rangeland Soc. Cottesloe, West. Aust.

Craig, R. 1960. The physiology of excretion in the insect. Pages 53-68 *in* E. A. Steinhaus and R. F. Smith, eds. Annual review of entomology. Vol. 5. Annual Reviews, Inc., Palo Alto, Calif.

Crawford, C. S., and R. F. Harwood. 1964. Bionomics and control of insects affecting Washington grass seed fields. Washington Agric. Exp. Sta. Tech. Bull. 44.

Crossley, D. A., Jr. 1970. Roles of microflora and microfauna in soil systems. Pages 30-35 *in* Pesticides in the soil: Ecology, degradation and movement. Int. Symp. on Pesticides in the Soil. Michigan State Univ., East Lansing.

Dahl, B. E. 1963. Soil moisture as a predictive index to forage yield for sandhills range type. J. Range Manage. 16:128-132.

Daubenmire, R. 1975. Ecology of *Artemisia tridentata* subsp. *tridentata* in the state of Washington. Northwest Sci. 49:24-35.

Day, J. M., D. Harris, P. J. Dart, and P. van Berkum. 1975. The Broadbalk experiment. An investigation of nitrogen gains from nonsymbiotic nitrogen fixation. Pages 71-84 *in* W. D. P. Stewart, ed. Nitrogen fixation by free-living microorganisms. Cambridge Univ. Press, London.

DeKock, P. C. 1970. The mineral nutrition of plants supplied with nitrate or ammonium nitrogen. Pages 39-44 *in* E. A. Kirkby, ed. Nitrogen nutrition of the plant. Univ. of Leeds, Agr. Chem. Symp., The Univ. of Leeds, England.

Delwiche, C. C. 1956. Denitrification. Pages 233-256 *in* W. D. McElroy and B. Glass, eds. Inorganic nitrogen metabolism. Johns Hopkins Univ. Press, Baltimore.

Delwiche, C. C. 1970. The nitrogen cycle. Sci. Am. 223:137-146.

Denmead, O. T., J. R. Simpson, and J. R. Freney. 1974. Ammonia flux into the atmosphere from a grazed pasture. Science 183:609-610.

Dhar, N. R., and A. K. Banerjee. 1966. Loss of nitrogen from alkaline soils under aerated conditions, when inorganic nitrogenous fertilizers are incorporated and its retardation by a grass and Bihar rock phosphate. Natl. Acad. Sci. India Annu. Number 31:312-316.

Dijkshoorn, W. 1971. Partition of ionic constituents between organs. Pages 447-476 *in* R. M. Samish, ed. Recent advances in plant nutrition. Vol. 2. Gordon and Breach Science Publ., New York.

DOAK, B. W. 1952. Some chemical changes in the nitrogen constituents of urea when voided on pasture. J. Agric. Sci. 42:162-171.

DOGAN, A. 1975. Some effects of microflora on surface runoff quality. Ph.D. Dissertation. Utah State Univ., Logan. 127 pp. Univ. Microfilms, Ann Arbor, Mich. (Diss. Abstr. 75-14,429).

DOMMERGUES, Y. 1960. Nitrogen mineralization at low mositure contents. Trans. 7th Int. Congr. Soil Sci. 2:672.

DONER, H. E., M. G. VOLZ, and A. D. MCLAREN. 1974. Column studies on denitrification in soil. Soil Biol. Biochem. 6:341-346.

DREGNE, H. E. 1968. Surface materials of desert environments. Pages 287-377 *in* W. W. McGinnies, B. J. Goldman, and P. Paylore, eds. Deserts of the world: an appraisal of research into their physical and biological environments. Univ. of Arizona Press, Tucson.

DRILHON, A., and F. MARCOUX. 1942. Biochemical studies of the blood and urine of a land turtle, *Testudo mauritanica* [in French, English summary]. Bull. Soc. Chem. Biol. 24:103-107.

DROVER, D. P., and I. P. BARRET-LENNARD. 1953. The amount of nitrite and ammonium ion in rainwater for six West Australia centres. Pages 3-5 *in* Aust. Conf. Soil Sci., 2, 6, 13. Adelaide.

DUBEY, H. D. 1968. Effect of soil moisture levels on nitrification. Can. J. Microbiol. 14:1348-1350.

DU PLESSIS, M. C. F., and W. KROONTJE. 1964. The relationship between pH and ammonia equilibrium in soil. Soil Sci. Soc. Am. Proc. 28:751-754.

DUTT, G. R., and G. M. MARION. 1974. Predicting nitrogen transformations and osmotic potentials in warm desert soils. US/IBP Desert Biome Res. Memo. 74-47. Utah State Univ., Logan. 20 pp.

EDWARDS, C. H., D. E. REICHLE, and D. A. CROSSLEY, JR. 1970. The role of invertebrates in turnover of organic matter and nutrients. Pages 147-172 *in* D. E. Reichle, ed. Analysis of temperate forest ecosystems. Springer-Verlag, New York.

EGUNJOBI. J. K. 1969. Dry matter and nitrogen accumulation in secondary successions involving gorse (*Ulex europaeus* L.) and associated shrubs and trees. N. Z. J. Sci. 12:175-193.

ELEUSENOVA, N. G., and I. A. SELIVANOV. 1975. Seasonal changes in mycorrhizae of desert plants. Mikol Fitopatol. 9:473-476.

EPPLEY, R. W., J. N. ROGERS, and J. J. MCCARTHY. 1969. Half-saturation constants for uptake of nitrate and ammonium by marine phytoplankton. Limnol. Oceanogr. 14:912-920.

EPPLEY, R. W., and W. H. THOMAS. 1969. Comparison of half-saturation constants for growth and nitrate uptake of marine phytoplankton. J. Phycol. 5:375-379.

EPSTEIN, E. 1972. Mineral nutrition of plants: principles and perspectives. John Wiley & Sons, New York. 412 pp.

ERIKSSON, E. 1952. Composition of atmospheric precipitation. I. Nitrogen compounds. Tellus 4:215-232.

EVANS, R. L., and J. J. JURINAK. 1976. Kinetics of phosphate release from a desert soil. Soil Sci. 121:205-211.

FALCONER, J. G., J. W. WRIGHT, and H. W. BEALL. 1933. The decomposition of certain types of forest litter under field conditions. Am. J. Bot. 20:196-203.

FANELLI, C., and S. G. ALBONETTI. 1972. Nitrogen fixation linked with mycorrhizas. Ann. Bot. (Rome) 31:175-186.

FARNSWORTH, R. B. 1975. Nodulation and nitrogen fixation in shrubs. Pages 32-71 *in* H. C. Stutz, ed. Proc. Symp. and Workshop on Wildland Shrubs. Brigham Young Univ. Press, Provo, Utah.

FARNSWORTH, R. B., E. M. ROMNEY, and A. WALLACE. 1976. Implications of symbiotic nitrogen fixation by desert plants. Great Basin Nat. 36:65-80.

FAURIE, G., A. JOSSERAND, and R. BARDIN. 1975. Influence des colloides argileux sur la retention d'ammonium et la nitrification [in French, English summary]. Rev. Ecol. Biol. Sol. (Paris) 12:201-210.

FENN, L. B. 1975. Ammonia volatilization from surface applications of ammonium compounds on calcareous soils: III. Effects of mixing low and high loss ammonium compounds. Soil Sci. Soc. Am. Proc. 39:366-368.

FENN, L. B., and D. E. KISSEL. 1973. Ammonia volatilization from surface applications of ammonium compounds on calcareous soils: I. General theory. Soil Sci. Soc. Am. Proc. 37:855-859.

FENN, L. B., and D. E. KISSEL. 1974. Ammonia volatilization from surface applications of ammonium compounds on calcareous soils: II. Effects of temperature and rate of ammonium nitrogen application. Soil Sci. Soc. Am. Proc. 38:606-610.

FENN, L. B., and D. E. KISSEL. 1976. The influence of cation exchange capacity and depth of incorporation on ammonia volatilization from ammonium compounds applied to calcareous soils. Soil Sci. Soc. Am. Proc. 40:395-398.

FERNANDEZ, O. A. 1974. The dynamics of root growth and the partitioning of photosynthates in cool desert shrubs. Ph.D. Dissertation. Utah State Univ., Logan. 132 pp. (Diss. Abst. 75-14,431).

FIREMAN, M., and H. E. HAYWARD. 1952. Indicator significance of some shrubs in the Escalante Desert, Utah. Bot. Gaz. 114:143-155.

FLETCHER, J. E. 1961. The effect of plant growth on infiltration in the Southwest. Pages 51-63 *in* Proc. Am. Asso. Adv. Sci., Southw. and Rock Mtn. Div., Alpine, Texas.

FLETCHER, J. E., and E. L. BEUTNER. 1941. Erodibility investigations on some soils of upper Gila watershed, Arizona. U.S. Dep. Agric. Tech. Bull. 794. Washington, D.C. 32 pp.

FLETCHER, J. E., and W. P. MARTIN. 1948. Some effects of algae and molds in the rain-crust of desert soils. Ecology 29:95-100.

FOCHT, D. D. 1974. The effect of temperature, pH, and aeration on the production of nitrous oxide and gaseous nitrogen—a zero-order kinetic model. Soil Sci. 118:173-179.

FOGG, G. E., W. D. P. STEWART, P. FAY, and A. E. WALSBY. 1973. The blue-green algae. Academic Press, New York. 459 pp.

FOWLER, G. J., and Y. N. KOTWAL. 1924. Chemical factors in denitrification. J. Indian Instit. Sci. 7:29-37.

FRANCIS, C. W., and M. W. CALLAHAN. 1975. Biological denitrification and its application in treatment of high-nitrate waste water. J. Environ. Qual. Saf. 4:153-163.

FRANZ, C. E., O. J. REICHMAN, and K. M. VAN DE GRAAFF. 1973. Diets, food preferences and reproductive cycles of some desert rodents. US/IBP Desert Biome Res. Memo. 73-24. Utah State Univ., Logan. 128 pp.

FRENCH, J. R. J., G. L. TURNER, and J. F. BRADBURY. 1976. Nitrogen fixation by bacteria from the hindgut of termites. J. Gen. Microbiol. 95:202-206.

French, N. R. 1959. Iodine metabolism in wild jackrabbits. Pages 113-121 *in* Oklahoma conference—radioisotopes in agriculture. U.S. Atomic Energy Comm. Rep. TID-7578. Office of Information Services, Springfield, Va.

Friedmann, E. I., Y. Lipkin, and R. Ocampo-Paus. 1967. Desert algae of the Negev (Israel). Phycologia 6:185-200.

Friedmann, E. I., and R. Ocampo-Paus. 1976. Endolithic blue-green algae in the dry valleys: primary producers in the Antarctic desert ecosystem. Science 193: 1247-1249.

Fuller, W. H. 1974. Desert soils. Pages 32-101 *in* G. W. Brown, Jr., ed. Desert biology. Vol. 2. Academic Press, New York.

Fuller, W. H. 1975a. Management of southwestern desert soils. Univ. of Arizona Press, Tucson. 195 pp.

Fuller, W. H. 1975b. Soils of the desert Southwest. Univ. of Arizona Press, Tucson.

Fuller, W. H., R. E. Cameron, and N. Raica, Jr. 1960. Fixation of nitrogen in desert soils by algae. Trans. 7th Int. Congr. Soil Sci. 2:617-624.

Garbosky, A. J., and N. Giambiagi. 1962. The survival of nitrifying bacteria in the soil. Plant Soil 17:271-278.

Garcia-Moya, E., and C. M. McKell. 1970. Contribution of shrubs to the nitrogen economy of a desert-wash plant community. Ecology 51:81-88.

Gardner, W. R. 1965. Movement of nitrogen in soil. Pages 500-572 *in* W. Bartholomew and F. Clark, eds. Soil nitrogen. Agronomy 10. Am. Soc. Agron., Madison, Wis.

Gentry, J. B., and K. L. Sitritz. 1972. The role of the Florida harvester ant, *Pogonomyrmex badius*, in old field nutrient relationships. Environ. Entomol. 1:39-41.

Gerretsen, F. C., and H. de Hoop. 1957. Nitrogen losses during nitrification in solutions and in acid sandy soils. Can. J. Microbiol. 3:359-380.

Ghilarov, M. S. 1967. On the relationship between soil dwelling invertebrates and soil microorganisms. Pages 355-359 *in* A. Burgess and F. Raw, eds. Soil biology. Academic Press, New York.

Gifford, G. F., and F. E. Busby. 1973. Loss of particulate organic materials from semiarid watersheds as a result of extreme hydrologic events. Water Resou. Res. 9:1443-1449.

Gilbert, O., and K. L. Bocock. 1960. Changes in leaf litter when placed on the surface of soils with contrasting humus types: II. Changes in the nitrogen content of oak and ash leaf letter. J. Soil Sci. 11:10-19.

Gist, C. S. 1972. Analysis of mineral pathways in a cryptozoan food-web. Eastern Deciduous Forest Biome Rep. EDFB-IBP-72-23. Oak Ridge National Laboratory, Oak Ridge, Tenn.

Goodall, D. W. 1970. Studying the effects of environmental factors on ecosystems. Pages 19-28 *in* D. E. Reichle, ed. Ecological studies. Analysis and synthesis. Vol. I. Analysis of temperate forest ecosystems. Springer-Verlag, Berlin.

Goodman, P. J. 1973. Physiological and ecotypic adaptations of plants to salt desert conditions in Utah. J. Ecol. 61:473-494.

Goodman, P. J., and M. M. Caldwell. 1971. Shrub ecotypes in a salt desert. Nature 232:571-572.

Gross, J. E., L. C. Stoddart, and F. H. Wagner. 1974. Demographic analysis of a northern Utah jackrabbit population. Wildl. Monogr. 40. 68 pp.

HADDAD, S. G. 1972. Identification of a *Bacillus* sp. and the ecological significance of its protease. M.S. Thesis. New Mexico State Univ., Las Cruces.

HAGIN, J. 1955. Rates of nitrification in nature and "conditioner"-formed soil aggregates of various sizes. Bull. Res. Counc. Isr. 5B:98-104.

HALE, W. G. 1967. Collembola. Pages 397-411 *in* A. Burgess and F. Raw, eds. Soil biology. Academic Press, New York.

HALVORSON, A. R., and A. C. CALDWELL. 1948. Factors affecting nitrate producing power of Minnesota soil. Soil Sci. Soc. Am. Proc. 13:258-260.

HAMISSA, M. R. A., and M. Y. SHAWARBI. 1962. Volatilization of ammonia from ammoniacal fertilizers applied to soils. Agric. Res. Rev. (Cairo) 40:147-159.

HANKS, R. J., and N. M. Nimah. 1973. Model for estimating soil water and atmospheric interrelations: II. Field test of the model. Soil Sci. Soc. Am. Proc. 37:528-532.

HARDING, E. L. 1975. Aerobic sporeforming bacteria in northern Chihuahuan Desert soils. Ph.D. Dissertation. New Mexico State Univ., Las Cruces. 61 pp. University Microfilms, Ann Arbor, Mich. (Diss. Abstr. 76-19,685.)

HARDY, R. W. F., R. C. BURNS, R. R. HEBERT, R. D. HOLSTEN, and E. K. JACKSON. 1971. Biological nitrogen fixation: a key to world protein. Pages 561-590 *in* Biological fixation in natural and agricultural habitats. Martinus Nijhoff, The Hague, Netherlands.

HARDY, R. W. F., R. C. BURNS, AND R. D. HOLSTEN. 1973. Applications of the acetylene-ethylene assay for measurement of nitrogen fixation. Soil Biol. Biochem. 5:47-81.

HARLEY, J. L. 1970. The biology of mycorrhiza, 2nd ed. Leonard Hill, London. 233 pp.

HARMSEN, G. W., and G. J. KOLENBRANDER. 1965. Soil inorganic nitrogen. Pages 43-92 *in* W. V. Bartholomew and F. E. Clark, eds. Soil nitrogen. Agronomy 10. Am. Soc. Agron., Madison, Wis.

HASTINGS, J. R., AND R. M. TURNER. 1965. The changing mile: an ecological study of vegetation change with time in the lower mile of an arid and semi-arid region. Univ. of Arizona Press, Tucson. 317 pp.

HAUCK, R. D., and S. W. MELSTED. 1956. Some aspects of the problem of evaluating denitrification in soils. Soil Sci. Soc. Am. Proc. 29:361-364.

HAYES, A. J. 1965. Studies on the decomposition of coniferous leaf litter: I. Physical and chemical changes. J. Soil Sci. 16:121-140.

HENRIKSSON, E., B. ENGLUND, M. HEDEN, and I. WAS. 1972. Nitrogen fixation in Swedish soils by blue-green algae. *In* T. V. Desikachary, ed. Taxonomy and biology of blue-green algae. 1st Int. Symp., Madras, Spain. 591 pp.

HERMANN, F. A., and E. GORHAM. 1957. Total mineral material, acidity, sulphur and nitrogen in rain and snow at Kentville, Nova Scotia. Tellus 9:180-183.

HETHENER, P. 1967. Activité microbiologique des sols à *Cupressus Dupreziana* A. Camus au Tassili N'Ajjer (Sahara central). Bull. Soc. Hist. Nat. Afr. Nord. (Algeria) 58:39-100.

HIDY, G. M. 1970. Theory of diffusive and impactive scavenging. Pages 355-371 *in* R. J. Englemann and W. G. N. Shinn, eds. Precipitation scavenging. U.S. Atomic Energy Comm. Symp. Ser. 22, Conf.-700601. Division of Technical Information, Oak Ridge, Tenn.

HOLMGREN, R. C., and S. F. BREWSTER, JR. 1972. Distribution of organic matter reserve in a desert shrub community. U.S. Dep. Agric. For. Serv. Res. Paper

INT-30.

HUMPHREY, R. R. 1974. Fire in deserts and desert grasslands of North America. Pages 366-400 *in* T. T. Kozlowski and C. E. Ahlgren, eds. Fire and ecosystems. Academic Press, New York.

HUNGERFORD, C. R., C. H. LOWE, and J. P. GRAY. 1972. Population studies of desert cottontail (*Sylvilagus auduboni*), black-tailed jackrabbit (*Lepus californicus*) and Allen's jackrabbit (*Lepus alleni*) in the Sonoran Desert. US/IBP Desert Biome Res. Memo. 72-27. Utah State Univ., Logan 8 pp.

HUNGERFORD, C. R., C. H. LOWE, and R. L. MADSEN. 1973. Population studies of the desert cottontail (*Sylvilagus auduboni*) and black-tailed jackrabbit (*Lepus californicus*) in the Sonoran Desert. US/IBP Desert Biome Res. Memo. 73-20. Utah State Univ., Logan. 15 pp.

HUNTER, R. B., E. M. ROMNEY, J. D. CHILDRESS, and J. E. KINNEAR. 1975a. Responses and interactions in desert plants as influenced by irrigation and nitrogen applications. US/IBP Desert Biome Res. Memo. 75-13. Utah State Univ., Logan. 14 pp.

HUNTER, R. B., A. WALLACE, E. M. ROMNEY, and P. A. T. WIELAND. 1975b. Nitrogen transformations in Rock Valley and adjacent areas of the Mohave Desert. US/IBP Desert Biome Res. Memo. 75-35. Utah State Univ., Logan. 8 pp.

HUNTER, R. B., E. M. ROMNEY, A. WALLACE, H. O. HILL, T. A. ACKERMAN, and J. E. KINNEAR. 1976a. Responses and interactions in desert plants as influenced by irrigation and nitrogen applications. US/IBP Desert Biome Res. Memo. 76-14. Utah State Univ., Logan. 7 pp.

HUNTER, R. B., A. WALLACE, and E. M. ROMNEY. 1976b. Nitrogen transformations in Rock Valley and adjacent areas of the Mohave Desert. US/IBP Desert Biome Res. Memo. 76-28. Utah State Univ., Logan. 10 pp.

HUTCHINSON, G. L., R. J. MILLINGTON, and D. B. PETERS. 1972. Atmospheric ammonia: absorption by plant leaves. Science 175:771-772.

INGHAM, G. 1950. The mineral content of air and rain and its importance to agriculture. J. Agric. Sci. 40:55-61.

JACKSON, M. L. 1958. Soil chemical analysis. Prentice Hall, New Jersey. 498 pp.

JACKSON, W. A., D. FLESHER, and R. H. HAGEMAN. 1973. Nitrate uptake by dark-grown corn seedlings. Plant Physiol. 51:120-127.

JANSSON, S. L., and F. E. CLARK. 1952. Losses of nitrogen during decomposition of plant material in presence of inorganic nitrogen. Soil Sci. Soc. Am. Proc. 16:330-334.

JENNY, H. 1930. A study of the influence of climate upon the nitrogen and organic matter content of the soil. Missouri Agric. Exp. Sta. Res. Bull. 152:1-66.

JENNY, H., S. P. GESSEL, and F. T. BINGHAM. 1949. Comparative study of decomposition ratio of organic matter in temperate and tropical regions. Soil Sci. 68:419-432.

JENNY, H. A., A. D. AYERS, and J. S. HASKINGS. 1945. Comparative behavior of ammonia and ammonium salts in soils. Hilgardia 16:429-457.

JENSEN, V. 1974. Decomposition of angiosperm tree leaf litter. Pages 69-104 *in* C. H. Dickinson and G. J. F. Pugh, eds. Biology of plant litter decomposition. Vol. 1. Academic Press, New York.

JOHNSON, D. D., and W. D. GUENZI. 1963. Influence of salts on ammonium oxidation and CO_2 evolution from soil. Soil Sci. Soc. Am. Proc. 27:663-665.

JOHNSON, S. 1965. An ecological study of the chuckwalla, *Sauromalus obesus* Baird, in the western Mojave Desert. Am. Midl. Nat. 73:1-29.

JONES, E. J. 1951. Loss of elemental nitrogen from soils under anaerobic conditions. Soil Sci. 71:193-196.

JUNGE, C. E. 1958. The distribution of ammonia and nitrate in rainwater over the United Sates. Trans. Am. Geophys. Union 39:241-248.

JUSTICE, J. K., and R. L. SMITH. 1962. Nitrification of ammonium sulfate in a calcareous soil as influenced by combinations of moisture, temperature, and levels of added nitrogen. Soil Sci. Soc. Am. Proc. 26:246-250.

KAI, H., and T. HARADA. 1969. Studies on the environmental conditions controlling nitrification in soil. II. Effect of soil clay minerals on the rate of nitrification. Soil Sci. Plant Nutr. 15:1-10.

KAPUSTKA, L. A., and E. RICE. 1976. Acetylene reduction (N_2-fixation) in soil and old field succession in central Oklahoma. Soil Biol. Biochem. 8: 497-503.

KEFAUVER, M., and F. E. ALLISON. 1957. Nitrite reduction by *Bacterium denitrificans* in relation to oxidation-reduction potential and oxygen tension. J. Bacteriol. 73:8-14.

KELLY, G. D., and W. W. MIDDLEKAUFF. 1961. Biological studies of *Dissosteira spurcata* Saussure with distributional notes on related California species (Orthoptera-Acrididae). Hilgardia 30:395-424.

KEVAN, D. K. MCE. 1962. Soil animals. H. F. & G. Witherby Ltd London. 237 pp.

KHALIL, F. , and G. HAGGAG. 1955. Ureotelism and uricotelism in tortoises. J. Exp. Zool. 130:423-432.

KHAN, A. G. 1974. The occurrence of mycorrhizas in halophytes, hydrophytes and xerophytes, and of Endogone spores in adjacent soils. J. Gen. Microbiol. 81:7-14.

KHUDAIRI, A. K. 1969. Mycorrhiza in desert soils. BioScience 19:598,599.

KING, H. G. C., and C . W. HEATH. 1967. The chemical analysis of small samples of leaf material and the relationship between disappearance and composition of leaves. Pedobiologia 7:192-197.

KIRDA, C., J. L. STARR, C. MISRA, J. W. BIGGAR, and D. R. NIELSEN. 1974. Nitrification and denitrification during miscible displacement in unsaturated soil. Soil Sci. Soc. Am. Proc. 38:772-776.

KIRKBY, E. A. 1969. Ion uptake and ionic balance in plants in relation to the form of nitrogen nutrition. Pages 215-235 *in* I. H. Rorison, ed. Ecological aspects of the mineral nutrition of plants. Blackwell Scientific Publ., Oxford.

KIRKBY, E. A., ed. 1970. Nitrogen nutrition of the plant. The Univ. of Leeds, England.

KIRKHAM, D., and W. L. POWERS. 1972. Advanced soil physics. Wiley Interscience, New York.

KITAZAWA, Y. 1967. Community metabolism of soil invertebrates in forest ecosystems in Japan. Pages 649-661 *in* K. Petrusewicz, ed. Secondary productivity of terrestrial ecosystems. Polish Acad. Sci., Warsaw.

KLEMMEDSON, J. O. 1975. Nitrogen and carbon regimes in an ecosystem of young dense ponderosa pine in Arizona. For. Sci. 21:163-168.

KLEMMEDSON, J. O., and R. C. BARTH, 1975. Distribution and balance of biomass and nutrients in desert shrub ecosystems. US/IBP Desert Biome Res. Memo. 75-5. Utah State Univ., Logan, 18 pp.

KLINGEBIEL, A. A. 1961. Soil factor and soil loss tolerance. Pages 18-19 *in* W. H. Wischmeier, ed. Soil loss prediction in North Dakota, South Dakota, Nebraska and Kansas. U.S. Dep. Agric. Soil Conserv. Serv., Lincoln, Neb.

KLIPPLE, G. E., and J. L. RETZER. 1959. Response of native vegetation of the central Great Plains to application of corral manure and commercial fertilizer. J. Range Manage. 12:239-243.

KOMAREK, E. V., SR. 1970. Controlled burning and air pollution: an ecological review. Proc. Tall Timbers Fire Ecol. Conf. 10:141-173.

KONONOVA, M. M. 1961. Soil organic matter, its nature, its role in soil formation and in soil fertility. (Transl. from Russian.) Pergamon Press, New York. 450 pp.

KORSAKOVA, M. P. 1941. Effect of aeration on the process of bacterial nitrate reduction [in Russian, English summary]. Mikrobiol. (Moscow) 10:163-178.

KOWAL, N. E. 1969. Effect of leaching on pine litter decomposition rate. Ecology 50:739-740.

KOZLAVA, A. V. 1951. Nitrate accumulation in Turkmenian termite mounds [in Russian, English summary]. Pochvovedenie 1951:626-631.

KREBILL, R. G., and J. M. MUIR. 1974. Morphological characterization of *Frankia purshiae*, the endophyte in root nodules of bitterbrush, *Purshia tridentata*. Northwest Sci. 48:266-268.

KUCERA, C. L. 1959. Weathering characteristics of deciduous leaf litter. Ecology 40: 485-487.

KUNHELT, W. 1955. An introduction to the study of soil animals. Pages 3-19 *in* D. K. McE. Kevan, ed. Soil zoology. Academic Press, New York.

LARSEN, S., and D. GUNARY. 1962. Ammonia loss from ammoniacal fertilizers applied to calcareous soils. J. Sci. Food Agric. 13:566-572.

LEE, K., and T. G. WOOD. 1971. Termites and soils. Academic Press, New York. 251 pp.

LEES, H., and J. H. QUASTEL. 1946. Biochemistry of nitrification in soil. II. The site of soil nitrification. Biochem. J. 40:815-823.

LEMÉE, G., and N. BICHAUT. 1973. Resérches sur les écosystemès des réserves biologiques de la forêt de Fontainebleau. II. Décomposition de la litière de feuilles des arbres et libération des bioéléments. Oecol. Plant (Paris). 8:153-174.

LIEBIG, J. 1855. Principles of agricultural chemistry with special reference to late researches made in England. Taylor and Walton, London. (Reprinted in L. R. Pomeroy, ed. 1974. Cycles of essential elements. Dowden, Hutchinson and Ross Inc., Stroudsburg, Penn.)

LIKENS, G. E., F. H. BORMANN, N. M. JOHNSON, D. W. FISHER, and R. S. PIERCE. 1970. Effects of forest cutting and herbicide treatment on nutrient budgets in the Hubbard Brook watershed-ecosystem. Ecol. Monogr. 40:23-47.

LIPS, S. H., A. BEN-ZIONI, and Y. VAADIA. 1971. K^+ recirculation in plants and its importance for adequate nitrate nutrition. Pages 207-215 *in* R. M. Samish, ed. Recent advances in plant nutrition. Vol. 1. Gordon and Breach Science Publ., New York.

LODGE, J. P., JR., J. B. PATE, W. BASBERGELL, G. S. SWANSON, K. C. HILL, E. LORANGE, and A. L. LAZARUS. 1968. Chemistry of United States precipitation. Final Report on the National Precipitation Sampling Network. National Center for Atmospheric Research. Boulder, Colo. 66 pp.

LOEWENSTEIN, H., L. E. ENGELBERT, O. J. ATTOE, and O. N. ALLEN. 1957. Nitrogen loss in gaseous form from soils as influenced by fertilizers and management. Soil

Sci. Soc. Am. Proc. 21:397-400.

LOFTIS, J. R., and C. E. SCARSBROOK. 1969. Ammonia volatilization from ratios of formamide and urea solutions in soils. Agron. J. 61:725-727.

LOOPE, W. L., and G. F. GIFFORD. 1972. Influence of a soil microbial crust on select properties of soils under pinyon-juniper in southeastern Utah. J. Soil Water Conserv. 27:164-167.

LOWRIE, D. C. 1942. The ecology of the spiders of xeric dunelands in the Chicago area. Bull. Chic. Acad. Sci. 6:161-189.

LUNT, O. R., J. LETEY, JR., and S. B. CLARK. 1973. Oxygen requirements for root growth in three desert species of desert shrubs. Ecology 54:1356-1362.

LYDA, S. D., and G. D. ROBINSON. 1969. Soil respiratory activity and organic matter depletion in an arid Nevada soil. Soil Sci. Soc. Am. Proc. 33:92-94.

LYNN, R. I., and R. E. CAMERON. 1971. Role of algae in crust formation in desert soils. US/IBP Desert Biome Res. Memo. 71-15. Utah State Univ., Logan. 4 pp.

LYNN, R. I., and R. E. CAMERON. 1972. The role of algae in crust formation and nitrogen cycling in desert soils. US/IBP Desert Biome Res. Memo. 72-43. Utah State Univ., Logan. 15 pp.

LYNN, R. I., and R. E. CAMERON. 1973. The role of algae in crust formation and nitrogen cycling in desert soils. US/IBP Desert Biome Res. Memo. 73-40. Utah State Univ., Logan. 26 pp.

LYNN, R. I., and M. C. VOGELSBERG. 1974. The role of algae in crust formation and nitrogen cycling in desert soils. US/IBP Desert Biome Res. Memo. 74-41. Utah State univ., Logan. 10 pp.

MACGREGOR., A. N. 1972. Gaseous losses of nitrogen from freshly wetted desert soils. Soil Sci. Soc. Am. Proc. 36:594-596.

MACGREGOR, A. N., and D. E. JOHNSON. 1971. Capacity of desert algal crusts to fix atmospheric nitrogen. Soil Sci. Soc. Am. Proc. 35:843-844.

MACK, R. N., 1971. Mineral cycling in *Artemisia tridentata*. Ph.D. Dissertation. Washington State Univ., Pullman. 106 pp. Univ. Microfilms, Ann Arbor, Mich. (Diss. Abstr. 72-7661.)

MACRAE, I. C., and R. ANCAJAS. 1970. Volatilization of ammonia from submerged tropical soils. Plant Soil 33:97-103.

MAHENDRAPPA, M. K., and R. L. SMITH. 1967. Some effects of moisture on denitrification in acid and alkaline soils. Soil Sci. Soc. Am. Proc. 31:212-215.

MAHENDRAPPA, M. K., R. L. SMITH, and A. T. CHRISTIANSEN. 1966. Nitrifiying organisms affected by climatic region in western United States. Soil Sci. Soc. Am. Proc. 30:60-63.

MAHMOUD, S. A. Z., M. ABOU EL-FADL, and M. K. ELMOFTY. 1964. Studies on the rhizosphere microflora of a desert plant. Folia Microbiol. 9:1-8.

MALECHEK, J. C., and B. M. SMITH. 1972. Energy consumption by cattle. Pages 182-188 *in* D. F. Balph, coordinator, et al., Curlew Valley Validation Site report. US/IBP Desert Biome Res. Memo. 72-1. Utah State Univ., Logan. 188 pp.

MARSHALL, J. K. 1973. Drought, land use and soil erosion. Pages 55-77 *in* J. V. Lovett, ed. Drought. Angus and Robertson, Sydney, Australia.

MARSHALL, R. O., H. J. DISHBURGER, R. MACVICAR, and G. D. HALLMARK. 1953. Studies on the effect of aeration on nitrate reduction by *Pseudomonas* species using ^{15}N. J. Bacteriol. 66:254-258.

MARTIN, J. P., and H. D. CHAPMAN. 1951. Volatilization of ammonia from surface-fertilized soils. Soil Sci. 71:25-34.

MARTIN, W. P., T. F. BUEHRER, and A. B. CASTER. 1942. Threshold pH value for the nitrification of ammonia in desert soils. Soil Sci. Soc. Am. Proc. 7:223-228.

MARTIN, W. P., and J. E. FLETCHER. 1943. Vertical zonation of great soil groups on Mt. Graham, Arizona, as correlated with climate, vegetation, and profile characteristics. Arizona Agric. Exp. Sta. Tech. Bull. 99. Univ. of Arizona, Tucson. 153 pp.

MASON, W. H., and E. P. ODUM. 1969. The effect of coprophagy on retention and bioelimination of radionuclides by detritus feeding animals. Pages 721-724 *in* D. J. Nelson and F. C. Evans, eds. Symposium on Radioecology. U.S. Atomic Energy Commission Conf.-67053. Washington, D.C.

MAYLAND, H. F. 1969. Rangeland fertilization for balanced forage and cattle nutrition. Pages 192-199 *in* Proc. 20th Annu. Pac. Northw. Fertilizer Conf., Spokane, Wash.

MAYLAND, H. F., and T. H. MCINTOSH. 1966. Availability of biologically fixed atmospheric nitrogen-15 to higher plants. Nature 209:421-422.

MAYLAND, H. F., T. H. MCINTOSH, and W. H. FULLER. 1966. Fixation of isotopic nitrogen on a semiarid soil by algal crust organisms. Soil Sci. Soc. Am. Proc. 30:56-60.

MAYNARD, E. A. 1951. A monograph of the Collembola or springtailed insects of New York. Comstock Publ. Co., Ithaca, New York. 339 pp.

MCBRAYER, J. F., D. E. REICHLE, and M. WHITKAMP. 1973. Energy flow and nutrient cycling in a cryptozoan food-web. Eastern Deciduous Forest Biome Rep. EDFB-IBP-73-8. Oak Ridge National Laboratory, Oak Ridge, Tenn.

MCCORMICK, P. W., and J. P. WORKMAN. 1975. Early range readiness with nitrogen fertilizer: An economic analysis. J. Range Manage. 28:181-184.

MCGARITY, J. W. 1961. Denitrification studies on some south Australian soils. Plant Soil 14:1-21.

MCGARITY, J. W. 1962. Effect of freezing soil on denitrification. Nature 196: 1342-1343.

MCGARITY, J. W., and E. H. HOULT. 1971. The plant component as a factor in ammonia volatilization from pasture swards. J. Br. Grassl. Soc. 26:31-34.

MCGARITY, J. W., and R. J. K. MYERS. 1968. Denitrifying activity in solodized solonetz soils of eastern Australia. Soil Sci. Soc. Am. Proc. 32:812-817.

MCGINNIES, W. G., B. J. GOLDMAN, and P. PAYLORE, eds. 1970. Deserts of the world: an appraisal of research into their physical and biological environments. Univ. of Arizona Press, Tucson. 788 pp.

MCLAREN, A. D. 1969a. Steady state studies of nitrification in soil: theoretical considerations. Soil Sci. Soc. Am. Proc. 33:273-275.

MCLAREN, A. D. 1969b. Nitrification in soil: systems approaching a steady state. Soil Sci. Soc. Am. Proc. 33-551-556.

MCLAREN, A. D. 1970. Temporal and vectorial reactions of nitrogen in soil: a review. Can. J. Soil Sci. 50:97-109.

MCLAREN, A. D. 1971. Kinetics of nitrification in soil growth of nitrifiers. Soil Sci. Soc. Am. Proc. 35:91-95.

MEDWECKA-KORNAS, A. 1971. Plant litter. Pages 24-33 *in* Methods of study in quantitative soil ecology: population, production and energy flow. IBP Handbook No. 18. Blackwell Scientific Publ., Oxford.

MEEK, B. D., L. B. GRASS, and A. J. MACKENZIE. 1969. Applied nitrogen losses in relation to oxygen status of soils. Soil Sci. Soc. Am. Proc. 33:575-578.

MEEK, B. D., and A. J. MACKENZIE. 1965. The effect of nitrite and organic matter on aerobic gaseous losses of nitrogen from calcareous soil. Soil Sci. Soc. Am. Proc. 29:176-178.

MEIKLEJOHN, J. 1940. Aerobic denitrification. Ann. Appl. Biol. 27:558-573.

MELIN, E. 1930. Biological decomposition of some types of litter from North American forests. Ecology 11:72-101.

MICHAEL, G., P. MARTIN, and I. OWASSIA. 1970. The uptake of ammonium and nitrate from labelled ammonium nitrate in relation to the carbohydrate supply of the roots. Pages 22-29 *in* E. A. Kirkby, ed. Nitrogen nutrition of the plant. The University of Leeds Agric. Chem. Symp. The Univ. of Leeds, England.

MIKOLA, P. 1958. Studies on the decomposition of forest litter by basidiomycetes. Commun. Inst. Forest. Fenniae 48:1-22.

MILLBANK, J. W., and K. A. KERSHAW. 1969. Nitrogen metabolism in lichens. I. nitrogen fixation in the cephalodia of *Peltigera apthosa*. New Phytol. 68: 721-729.

MILLER, J. F., R. H. FREDERICK, and R. J. TRACEY. 1973. Precipitation frequency atlas of western United States. NOAA [Nat. Ocean. Atmos. Admin.], Nat. Weather Serv., Atlas No. 2. Silver Spring, Maryland. 545 pp.

MILLER, R. D., and D. D. JOHNSON. 1964. The effect of soil moisture tension on carbon dioxide evolution, nitrification and nitrogen mineralization. Soil Sci. Soc. Am. Proc. 28:644-647.

MINDERMAN, G. 1968. Addition, decomposition and accumulation of organic matter in forests. Ecology 56:355-362.

MISHUSTIN, E. N., and V. K. SHIL-NIKOVA. 1971. Biological fixation of atmospheric nitrogen, English ed. Pennsylvania State Univ. Press, University Park. 420 pp.

MISRA, C., D. R. NIELSEN, and J. W. BIGGAR. 1974a. Nitrogen transformations in soil during leaching: I. Theoretical considerations. Soil Sci. Am. Proc. 38:289-293.

MISRA, C., D. R. NIELSEN, and J. W. BIGGAR. 1974b. Nitrogen transformations in soil during leaching: II. Steady state nitrification and nitrate reduction. Soil Sci. Soc. Am. Proc. 38:294-299.

MISRA, C., D. R. NIELSEN, and J. W. BIGGAR. 1974c. Nitrogen transformations in soil during leaching: III. Nitrate reduction in soil columns. Soil Sci. Soc. Am. Proc. 38:300-304.

MISRA, S. G., and B. SINGH. 1970. Effect of different treatments on volatile losses of nitrogen. Bol. Inst. Nac. Invest. Agron. (Madrid) 30:91-98.

MITCHELL, J. E., J. E. WAIDE, and R. L. TODD. 1975. A preliminary compartment model of a nitrogen cycle in a deciduous forest ecosystem. Cycling in southwestern ecosystems. Pages 41-46 *in* USERDA Symposium Series Conf. 740513.

MITCHELL, J. E., N. E. WEST, and R. W. MILLER. 1966. Soil physical properties in relation to plant community patterns in the shadscale zone of northwestern Utah. Ecology 47:824-837.

MOORE, B. P. 1969. Biochemical studies in termites. Pages 407-432 *in* K. Krishna and F. M. Weesner, eds. Biology of termites. Vol. 1. Academic Press, New York.

MORGAN, M. A., R. J. VOLK, and W. A. JACKSON. 1973. Simultaneous influx and efflux of nitrate during uptake by perennial ryegrass. Plant Physiol. 51:267-272.

MOULDER, B. C., D. E. REICHLE, and S. I. AUERBACH. 1970. Significance of spider predation in the dynamics of forest floor arthropod communities. ORNL-2806. Oak Ridge National Laboratory, Oak Ridge, Tenn. 170 pp.

MULKERN, G. B., D. R. TOCZEK, and M. A. BRUSVEN. 1964. Biology and ecology of North Dakota grasshoppers. II. Food habits and preferences of grasshoppers associated with the sand hills prairie. North Dakota Agric. Exp. Sta. Res. Rep. No. 11.

MULLIN, D., and P. E. HUNTER. 1964. Mites associated with the Passalus beetle. II. Biological studies of *Cosmolaelaps passali*. Acarologia 6:421-431.

MUNRO, P. E. 1966. Inhibition of nitrifiers by grass root extracts. J. Appl. Ecol. 3:231-238.

MURRAY, R. B. 1975. Effect of *Artemisia tridentata* removal on mineral cycling. Ph.D. Dissertation. Washington State Univ., Pullman. 203 pp. University Microfilms, Ann Arbor, Mich. (Diss. Abstr. 76-322.)

MUSGRAVE, G. W. 1947. The quantitative evaluation of factors in water erosion—a first approximation. J. Soil Water Conserv. 2:133-138.

NASH, T. H., III, S. WHITE, and J. M. NASH. 1974. composition and biomass contribution of lichen and moss communities in the hot desert ecosystems. US/IBP Desert Biome Res. Memo. 74-19. Utah State Univ., Logan. 7 pp.

NEAL, J. L., JR. 1969. Inhibition of nitrifying bacteria by grass and forb root extracts. Can. J. Microbiol. 15:633-635.

NEMERYUK, G. E., V. P. PALTSEV, and V. F. SINEGLOZOVA. 1965. Factors affecting ammonia migration from soils [in Russian, English summary]. Pochvovedenie 1965(4):55-64.

NESS, J. C., R. C. DUGDALE, J. J. GOERING, and V. A. DUGDALE. 1963. Use of nitrogen-15 for measurement of rates in the nitrogen cycle. Pages 480-485 *in* V. Schultz and A. W. Klement, Jr. Radioecology. Reinhold Publ. Corp. New York.

NEWBOULD, P. J. 1967. Methods for estimating the primary production of forests. IBP Handbook No. 2. Blackwell Scientific Publ., Oxford. 62 pp.

NISHITA, N., and R. M. HAUG. 1973. Distribution of different forms of nitrogen in some desert soils. Soil Sci. 116:51-58.

NOMMIK, H. 1956. Investigations on denitrification in soil. Acta Agric. Scand. VI. 2:195-228.

NORRIS, K. 1953. The ecology of the desert iguana *Dipsosaurus dorsalis*. Ecology 34: 265-287.

NOY-MEIR, I. 1973. Desert ecosystems: environment and producers. Annu. Rev. Ecol. Syst. 4:25-51.

NUTMAN, P. S. 1965. Symbiotic nitrogen fixation. Pages 360-383 *in* W. V. Bartholomew and F. E. Clark, eds. Soil nitrogen. Agronomy 10. Am. Soc. Agron., Madison, Wis.

NUTTING, W. L., M. I. HAVERTY, and J. P. LAFAGE. 1973. Colony characters of termites as related to population density and habitat. US/IBP Desert Biome Res. Memo. 73-30. Utah State Univ., Logan. 17 pp.

NUTTING, W. L., M. I. HAVERTY, and J. P. LAFAGE. 1975. Demography of termite colonies as related to various environmental factors: Population dynamics and role in the detritus cycle. US/IBP Desert Biome Res. Memo. 75-31. Utah State Univ., Logan. 26 pp.

O'BRIEN, R. T. 1972. Proteolytic activity of soil microorganisms. US/IBP Desert Biome Res. Memo. 72-42. Utah State Univ., Logan. 8 pp.

O'BRIEN, R. T. 1973. Proteolytic activity of soil microorganisms. US/IBP Desert Biome Res. Memo. 73-38. Utah State Univ., Logan. 10 pp.

OLSON, T. C. 1961. Determining "K" values. Pages 20-26 *in* W. H. Wischmeier, ed. Soil loss prediction in North Dakota, South Dakota, Nebraska, and Kansas. U.S. Dep. Agric. Soil Conserv. Serv., Lincoln Neb. 62 pp.

OSBORNE, H. 1912. Leafhoppers affecting cereal and forage crops. U.S. Dep. Agric. Bur. Entomol. Bull. 108. 123 pp.

OSBORNE, H. 1939. Meadow and pasture insects. Educators Press, Columbus, Ohio. 288 pp.

OVERREIN, L. N. 1969. Lysimeter studies on tracer nitrogen in forest soil: 2. Comparative losses of nitrogen through leaching and volatilization after the addition of urea-, ammonium-, and nitrate-N^{15}. Soil Sci. 107:149-159.

OWENSBY, C. E., R. M. HYDE, and K. L. ANDERSON. 1970. Effect of clipping and supplemental nitrogen and water on loamy upland bluestem range. J. Range Manage. 23:341-346.

PARKINSON, D., T. R. G. GRAY, and S. T. WILLIAMS. 1971. Methods for studying the ecology of soil microorganisms. IBP Handbook No. 19. Blackwell Scientific Publ., Oxford. 116 pp.

PATRICK, W. H., JR. 1960. Nitrate reduction rates in a submerged soil as affected by redox-potential. Trans. 7th Int. Congr. Soil Sci. 2:494-500.

PAUL, E. A. 1970. Characterization and turnover rate of soil humic constituents. Pages 63-76 *in* S. Pawluck, ed. Pedology and quarternary research. Univ. of Alberta Press, Edmonton.

PAULSEN, H. A. 1953. A comparison of surface soil properties under mesquite and perennial grass. Ecology 34:727-732.

PAYNE, W. J. 1973. Reduction of nitrogenous oxides by microorganisms. Bacteriol. Rev. 37:409-452.

PEARSALL, W. H., and C. H. MORTIMER. 1939. Oxidation-reduction potentials in waterlogged soils, natural waters and muds. J. Ecol. 27:483-501.

PEARSON, L. C. 1972. Survey of lichens. Pages 55-60 *in* D. F. Balph, coordinator, et al. Curlew Valley Validation Site report. US/IBP Desert Biome Res. Memo. 72-1. Utah State Univ., Logan. 188 pp.

POCHON, J., DE BARJAC, and J. LAJUDIE. 1957. Recherches sur la microflore des sols sahariens. Ann. Inst. Pasteur (Paris) 92:833-836.

PORCELLA, D. B., J. E. FLETCHER, D. L. SORENSEN, G. C. PIDGE, and A. DOGAN. 1973. Nitrogen and carbon flux in a soil-vegetation complex in the desert biome. US/IBP Desert Biome Res. Memo. 73-36. Utah State Univ., Logan. 23 pp.

PROSSER, C. L., D. W. BISHOP, F. A. BROWN, JR., T. L. JAHN, and V. J. WULFF. 1952. Comparative animal physiology. W. B. Saunders Co., Philadelphia. 888 pp.

PUGH, G. J. F. 1974. Terrestrial fungi. Pages 303-336 *in* C. H. Dickinson and G. J. F. Pugh, eds. Biology of plant litter decomposition. Vol. 2. Academic Press, New York.

Puh, Y. -s. 1964. Effects of soil temperature, mineral type, cation ratio, and degree of base saturation on volatilization losses of ammonia. Ph.D. Dissertation. The Pennsylvania State Univ. 110 pp. Univ. Microfilms, Ann Arbor, Mich. (Diss. Abstr. 24:4898-4899.)

Quastel, J. H., and P. G. Scholefield. 1951. Biochemistry of nitrification in soil. Bacteriol. Rev. 15:1-53.

Rao, K. P., and T. Gopalakrishnareddy. 1962. Nitrogen excretion in arachnids. Comp. Biochem. Physiol. 7:175-178.

Raw, F. 1967. Arthropoda (except Acari and Collembola). Pages 323-362 *in* A. Burgess and F. Raw, eds. Soil biology. Academic Press, New York.

Reichle, D. E. 1975. Advances in ecosystem analysis. BioScience 25:257-264.

Renard, K. G., J. R. Simanton, and H. B. Osborn. 1974. Applicability of the universal soil loss equation to semi-arid rangeland conditions. Hydrology and water resources in Arizona and the Southwest. Proc. Hydrol. Sec., Am. Water Resour. Assoc., Arizona Acad. Sci., Flagstaff.

Rice, E. L. 1974. Allelopathy. Academic Press, New York. 353 pp.

Rickard, W. H. 1965a. The influence of greasewood on soil moisture and soil chemistry. Northwest. Sci. 39:36-42.

Rickard, W. H. 1965b. Sodium and potassium accumulation by greasewood and hopsage leaves. Bot. Gaz. 126:116-119.

Rickard, W. H., and J. F. Cline. 1970. Litter fall. Pages 57-62 *in* Plant specialists' meeting. US/IBP Desert Biome Res. Memo. 70-3. Utah State Univ., Logan. 144 pp.

Rixon, A. J. 1969. Cycling of nutrients in a grazed *Atriplex vesicaria* community. Pages 87-95 *in* R. Jones, ed. The biology of *Atriplex*. CSIRO, Canberra, Australia.

Rixon, A. J. 1971. Oxygen uptake and nitrification by soil within a grazed *Atriplex vesicaria* community in semi-arid rangeland. J. Range Manage. 24:435-439.

Roberts, R. C. 1950. Chemical effects of salt-tolerant shrubs on soils. Proc. 4th Int. Congr. Soil Sci. 1:404-406.

Robinson, J. B. 1957. The critical relationship between mositure content in the region on wilting point and the mineralization of natural soil nitrogen. J. Agric. Sci. 49:100.

Rodin, L. E., and N. I. Bazilevich. 1967. Production and mineral cycling in terrestrial vegetation. (Transl. from Russian.) Oliver and Boyd, London. 288 pp.

Rodin, L. E., and N. I. Bazilevich. 1968. World distribution of plant biomass. Pages 45-50 *in* F. C. Eckardt, ed. Functioning of terrestrial ecosystems of the primary production level. UNESCO, Paris.

Rogers, R. W., R. T. Lange, and D. J. D. Nicholas. 1966. Nitrogen fixation by lichens of arid soil crusts. Nature 209:96-97.

Rogler, G. A., and R. J. Lorenz. 1957. Nitrogen fertilization of northern Great Plains rangelands. J. Range Manage. 10:156-160.

Romney, E. M., V. Q. Hale, A. Wallace, O. R. Lunt, J. D. Childress, H. Kaaz, G. V. Alexander, J. E. Kinnear, and T. L. Ackerman. 1973. Some characteristics of soil and perennial vegetation in northern Mojave Desert areas of the Nevada Test Site. U.S. Atomic Energy Comm. Rep. UCLA 12-916. Office of Information Services, Springfield, Va. 340 pp.

ROMNEY, E. M., A. WALLACE, J. D. CHILDRESS, J. E. KINNEAR, H. KAAZ, P. A. T. WIELAND, M. LEE, and T. L. ACKERMAN. 1974. Responses and interactions in desert plants as influenced by irrigation and nitrogen applications. US/IBP Desert Biome Res. Memo. 74-17. Utah State Univ., Logan. 12 pp.

ROMNEY, E. M., A. WALLACE, H. KAAZ, V. Q. HALE, and J. D. CHILDRESS. Effect of shrubs on redistribution of mineral nutrients in zones near roots in the Mojave Desert. *In* J. K. Marshall, ed. Proc. Belowground Ecosystem Symp. Ft. Collins, Colo. (In press)

RYCHERT, R. C., and J. SKUJINS. 1973. Microbial activity in arid soils. Utah Sci. 34:96-98.

RYCHERT, R. C., and J. SKUJINS. 1974a. Inhibition of algae-lichen crust nitrogen fixation in desert shrub communities. 74th Annu. Meeting Am. Soc. Microbiol., Chicago, Ill. (Abstr.)

RYCHERT, R. C., and J. SKUJINS. 1974b. Nitrogen fixation by blue-green algae-lichen crusts in the Great Basin Desert. Soil Sci. Soc. Am. Proc. 38:768-771.

SABBE, W. E., and L. W. REED. 1964. Investigations concerning nitrogen loss through chemical reactions involving urea and nitrite. Soil Sci. Soc. Am. Proc. 28:478-481.

SABEY, B. R. 1969. Influence of soil moisture tension in nitrate accumulation in soils. Soil Sci. Soc. Am. Proc. 33:263-266.

SABEY, B. R., L. R. FREDERICK, and W. V. BARTHOLOMEW. 1959. The formation of nitrate from ammonium in soils: III. Influence of temperature and initial population on nitrifying organisms on the maximum rate and delay period. Soil Sci. Soc. Am. Proc. 23:462-465.

SACKS, L. E. 1948. Metabolic studies on denitrification by *Pseudomonas denitrificans*. Ph.D. Thesis. Univ. of California, Berkeley. 47 pp.

SACKS, L. E., and H. A. BARKER. 1949. The influence of oxygen on nitrate and nitrite reduction. J. Bacteriol. 58:11-22.

SCHAEFER, R. 1964. Observations on the effect of incubation temperature and particularly of repeated freezing on nitrifying activity and CO_2 evolution in calcic mull soils from series of hydromorphic soils [in French, English summary]. Ann. Inst. Pasteur (Paris) 107:534.

SCHEER, B. T. 1963. Animal physiology. John Wiley and Sons, Inc., New York. 409 pp.

SCHLESINGER, W. H., and W. A. REINERS. 1974. Deposition of water and cations on artificial foliar collectors in fir Krummholtz of New England mountains. Ecology 55:378-386.

SCHMITT, E. L. 1972. The traditional plate count technique among modern methods in microbial ecology. Bull. Ecol. Res. Comm. (Stockholm) 17:453-454.

SCHWARTZ, W. 1959. Bacterian and actinomycetan symbiosen. Encycl. Plant Physiol. 11:560-572.

SCHWARTZBECK, R. A., J. V. MACGREGOR, and E. L. SCHMIDT. 1961. Gaseous nitrogen losses from nitrogen fertilized soils measured with infrared and mass spectroscopy. Soil Sci. Soc. Am. Proc. 25:186-189.

SCHWITZBEGEL, R. B., and D. A. WILBUR. 1943. Diptera associated with ironweed, *Vernonia interior* Small in Kansas. J. Kansas Entomol. Soc. 16:4-13.

SEIFERT, J. 1966. Ecology of soil microbes. Acta Univ. Carol. Biol. (Prague). 1: 139-146.

SHANKARACHARYA, N. B., and B. V. MEHTA. 1969. Evaluation of loss of nitrogen by ammonia volatilization from soil fertilized with urea. J. Indian Soc. Soil Sci. 17:423-430.

SHANKS, R. E., and J. S. OLSON. 1961. First year breakdown of leaf litter in southern Appalachian forests. Science 134:194-195.

SHARMA, M. L., and D. J. TONGWAY. 1973. Plant induced soil salinity patterns in two saltbush (*Atriplex* spp.) communities. J. Range Manage. 26:121-125.

SHIELDS, L. M. 1957. Algal and lichen floras in relation to nitrogen content of certain volcanic and arid range soils. Ecology 38:661-663.

SHIELDS, L. M., and F. DROUET. 1962. Distribution of terrestrial algae within the Nevada Test Site. Am. J. Bot. 49:547-554.

SHIELDS, L. M., C. MITCHELL, and F. DROUET. 1957. Algae and lichens stabilized surface crusts as soil nitrogen sources. Am. J. Bot. 44:489-498.

SHINN, R. S. 1975. The influence of climate on biomass and mineralomass of a crested wheatgrass community in northern Utah. M.S. Thesis. Utah State Univ. Logan. 47 pp.

SHOEMAKER, V. H., K. A. NAGY, and W. R. COSTA. 1973. The consumption, utilization and modification of nutritional resources by the jackrabbit (*Lepus californicus*) in the Mojave Desert. US/IBP Desert Biome Res. Memo. 73-22. Utah State Univ., Logan. 13 pp.

SHREVE, F. 1942. The desert vegetation of North America. Bot. Rev. 8:195-246.

SILVER, W. S., J. M. CENTRIFANTO, and D. J. NICHOLAS. 1963. Nitrogen fixation by the leaf-nodule endophyte of *Psychotria bacteriophila.* Nature 199:396-397.

SIMPSON, J. R. 1968. Losses of urea nitrogen from the surface of pasture soils. Pages 459-465 *in* Trans. 9th Int. Congr. Soil Sci., Adelaide, Australia.

SIMS, C. M., and F. M. COLLINS. 1960. The numbers and distribution of ammonia-oxidizing bacteria in some northern territory and south Australian soils. Aust. J. Agric. Res. 11:505-512.

SINDHU, M. A., and A. H. CORNFIELD. 1967. Comparative effects of varying levels of chlorides and sulphates of sodium, potassium, calcium, and magnesium on ammonification and nitrification during incubation of soil. Plant Soil 27:468-472.

SKIDMORE, E. L., and N. P. WOODRUFF. 1968. Wind erosion forces in the United States and their use in predicting soil loss. U.S. Dep. Agric. Handbook 346. 42 pp.

SKUJINS, J. J. 1964. Spectrophotometric determination of nitrate with 4-methylumbelliferone. Anal. Chem. 36:240-241.

SKUJINS, J. 1972. Nitrogen dynamics in desert soils. I. Nitrification. US/IBP Desert Biome Res. Memo. 72-40. Utah State Univ., Logan. 33 pp.

SKUJINS, J. 1973. Dehydrogenase: an indicator of biological activities in arid soils. Bull. Ecol. Res. Comm. (Stockholm) 17:235-241.

SKUJINS, J. 1975. Nitrogen dynamics in stands dominated by some major cool desert shrubs. IV. Inhibition by plant litter and limiting factors of several processes. US/IBP Desert Biome Res. Memo. 75-33. Utah State Univ., Logan. 38 pp.

SKUJINS, J., and P. J. EBERHARDT. 1973. *In situ* ^{15}N studies of the nitrogen cycle in cold desert soils. Am. Soc. Agron. Abstr. 65th Annu. Meeting: 94.

SKUJINS, J., and A. D. MCLAREN. 1968. Persistence of enzymatic activities in stored and geologically preserved soils. Enzymologia 34:213-225.

SKUJINS, J. J., and N. E. WEST. 1973. Nitrogen dynamics in stands dominated by some major cold desert shrubs. US/IBP Desert Biome Res. Memo. 73-35. Utah State Univ., Logan. 91 pp.

SKUJINS, J. J., and N. E. WEST. 1974. Nitrogen dynamics in stands dominated by some major cool desert shrubs. US/IBP Desert Biome Res. Memo. 74-42. Utah State Univ., Logan. 56 pp.

SLATYER, R. O. 1960. Principles and problems of plant production in arid regions. Pages 7-1 to 7-14 *in* Proc. Arid Zone Tech. Conf. Vol. 2. Warburton, Victoria.

SMITH, A. D., and D. D. DOELL. 1968. Guides to allocating forage between cattle and wildlife on a big-game winter range. Utah Div. Fish Game Bull. 68:11.

SMITH, B. M. 1973. The relationship of climatic factors to grazing activities of cows on winter and spring ranges. M.S.Thesis. Utah State Univ., Logan. 90 pp.

SMITH, D. D. 1941. Interpretation of soil conservation data for field use. Agric. Eng. 22:173-175.

SMITH, D. D., and D. M. WHITT. 1947. Estimating soil loss from field areas of claypan soils. Soil Sci. Soc. Am. Proc. 11:485-490.

SMITH, D. H., and F. E. CLARK. 1960. Volatile losses of nitrogen from acid or neutral soils or solution containing nitrite and ammonium ions. Soil Sci. 90: 86-92.

SMITH, J. H. 1964. Relationships between cation exchange capacity and the toxicity of ammonia to the nitrification process. Soil Sci. Soc. Am. Proc. 28:640-643.

SMITH, S. J., and L. O. LEGG. 1971. Reflections on the *A* value concept of soil nutrient availability. Soil Sci. 112:373-375.

SNEVA, F. A. 1973a. Crested wheatgrass response to nitrogen and clipping. J. Range Manage. 26:47-50.

SNEVA, F. A. 1973b. Wheatgrass response to seasonal applications of two nitrogen sources. J. Range Manage. 26:137-139.

SNEVA, F. A. 1973c. Nitrogen and paraquat saves range forage for fall grazing. J. Range Manage. 26:294-295.

SNYDER, J. M., and L. H. WULLSTEIN. 1973a. Nitrogen fixation on granite outcrop pioneer ecosystems. Bryologist 76:196-199.

SNYDER, J. M., and L. H. WULLSTEIN. 1973b. The role of desert cryptogams in nitrogen fixation. Am. Midl. Nat. 90:257-265.

SORENSEN, D. L. 1975. *In situ* nitrogen fixation in cold desert soil-algae crust in northern Utah. M.S. Thesis. Utah State Univ., Logan. 65 pp.

SORENSEN, D. L., and D. B. PORCELLA. 1974. Nitrogen erosion and fixation in cool desert soil-algal crusts in northern Utah. US/IBP Desert Biome Res. Memo. 74-37. Utah State Univ., Logan. 18 pp.

SPECTOR, W. S., ed. 1956. Handbook of biological data. Wright Air Devel. Center, U.S. Air Force, Dayton, Ohio. 584 pp.

SPEEG, K. V., JR., and J. W. CAMPBELL. 1968. Formation and volatilization of ammonia gas by terrestrial snails. Am. J. Physiol. 214:1392-1402.

SPRENT, J. I. 1971. Effects of water stress on nitrogen fixation in root nodules. Pages 225-228 *in* T. A. Lie and E. G. Mulder, eds. Biological nitrogen fixation in natural and agricultural habitats. Martinus Nijhoff, The Hague, Netherlands.

STAFFELDT, E. E., and K. B. VOGT. 1974. Measurements of carbon and nitrogen changes in soil. US/IBP Desert Biome Res. Memo. 74-38. Utah State Univ., Logan. 16 pp.

STANFORD, G., R. A. VANDER POL, and S. DZIENIA. 1974. Denitrification rates in relation to total and extractable soil carbon. Soil Sci. Soc. Am. Proc. 39: 284-289.

STARK, N. 1973. Nutrient cycling in a Jeffrey pine ecosystem. Institute for Microbiology. Univ. of Montana, Missoula. 389 pp.

STARKS, K. J., and R. THURSTON. 1962. Silvertop of bluegrass. J. Econ. Entomol. 55:865-867.

STARR, J. L., F. E. BROADBENT, and D. R. NIELSEN. 1974. Nitrogen transformations during continuous leaching. Soil Sci. Soc. Am. Proc. 38:283-289.

STERNSEL, H. D., R. C. LOEHR, and A. W. LAWRENCE. 1973. Biological kinetics of suspended-growth denitrification. J. Water Pollut. Control Fed. 45:249-261.

STEVENSON, F. J. 1965. Origin and distribution of nitrogen in soil. Pages 1-42 *in* W. V. Bartholomew and F. E. Clark, eds. Soil nitrogen. Agronomy 10. Am. Soc. Agron., Madison, Wis.

STEVENSON, F. J., R. M. HARRISON, R. WETSELAAR, and R. A. LEEPER. 1970. Nitrosation of soil organic matter: III. Nature of gases produces by reaction of nitrite with lignins, humic substances, and phenolic constituents under neutral and slightly acidic conditions. Soil Sci. Soc. Am. Proc. 34:430-435.

STEVENSON, F. J., and R. J. SWABY. 1964. Nitrosation of soil organic matter: I. Nature of gases evolved during nitrous acid treatment of lignins and humic substances. Soil Sci. Soc. Am. Proc. 28:773-778.

STEVENSON, G. B. 1953. Bacterial symbiosis in some New Zealand plants. Ann. Bot. 17:343.

STEWART, W. D. P. 1970. Alagal fixation of atmospheric nitrogen. Plant Soil 32: 555-588.

STEWART, W. D. P. 1971. Physiological studies on nitrogen-fixing blue-green algae. Pages 377-391 *in* T. A. Lie and E. G. Mulder, eds. Biological nitrogen fixation in natural and agricultural habitats. Martinus Nijhoff, The Hague, Netherlands.

STEWART, W. D. P. 1973. Nitrogen fixation. Pages 260-278 *in* N. G. Carr and B. A. Whitton, eds. The biology of blue-green algae. Univ. of California Press, Berkeley.

STEYN, P. L., and C. C. DELWICHE. 1970. Nitrogen fixation by nonsymbiotic microorganisms in some California soils. Environ. Sci. Tech. 4:1122-1128.

STODDART, L. C. 1972. Population biology of the black-tailed jackrabbit (*Lepus californicus*) in northern Utah. Ph.D. Dissertation. Utah State Univ., Logan. 151 pp. (Diss. Abstr. 73-919).

STOJANOVIC, B. J., and M. ALEXANDER. 1958. Effect of inorganic nitrogen on nitrification. Soil Sci. 86:208-215.

STROEHLEIN, J. L., P. R. OGDEN, and B. BILLY. 1968. Time of fertilizer application on desert grasslands. J. Range Manage. 21:86-89.

STURGES, D. C. 1975. Hydrologic relationships on undisturbed and converted big sagebrush lands: The status of our knowledge. U.S. Dep. Agric. For. Serv. Res. Paper RM-140. Rocky Mtn. For. Range Exp. Sta., Ft. Collins, Colo. 23 pp.

STURKIE, P. D. 1965. Avian physiology, 2nd ed. Comstock Publ. Asso. Cornell Univ. Press, Ithaca, New York. 766 pp.

SULLIVAN, M. J., JR. 1942. The effect of field applications of organic matter on the properties of some Arizona soils. Ph.D. Dissertation. Univ. of Arizona, Tucson.

TCHAN, Y. T., and N. C. W. BEADLE. 1955. Nitrogen economy in semi-arid plant communities. II. The non-symbiotic nitrogen-fixing organisms. Proc. Linn. Soc. N.S.W. 80:97-104.

TERMAN, G. L., and C. M. HUNT. 1964. Volatilization losses of nitrogen from surface-applied fertilizers, as measured by crop response. Soil Sci. Soc. Am. Proc. 28:667-672.

THAMES, J. L., coordinator, et al. 1973. Tucson Basin Validation Sites report. US/IBP Desert Biome Res. Memo. 73-3. Utah State Univ., Logan. 101 pp.

THOMAS, W. 1968. Decomposition of loblolly pine needles with and without addition of dogwood leaves. Ecology 49:568-571.

TIEDEMANN, A. R., and J. O. KLEMMEDSON. 1973a. Effect of mesquite on physical and chemical properties of the soil. J. Range Manage. 26:27-29.

TIEDEMANN, A. R., and J. O. KLEMMEDSON. 1973b. Nutrient availablity in desert grassland soils under mesquite (*Prosopis juliflora*) trees and adjacent open areas. Soil Sci. Soc. Am. Proc. 37:107-111.

TOETZ, D. W., L. P. VARGA, and E. D. LOUGHRAN. 1973. Half-saturation constants for uptake of nitrate and ammonium by reservoir plankton. Ecology 54: 903-908.

TRUJILLO Y FULGHAM, P. T., B. KLUBEK, and J. SKUJINS. 1975. Differential inhibition of nitrification in desert shrub communities. 75th Annu. Meeting Am. Soc. Microbiol. (Abstr.)

TRUMBLE, H. C., and K. WOODROFFE. 1954. the influence of climatic factors on the reaction of desert shrubs to grazing by sheep. Pages 129-147 *in* J. L. Cloudsley-Thompson, ed. Biology of deserts. Inst. Biology, London.

TUCKER, T. C., and R. L. WESTERMAN. 1974. Gaseous losses of nitrogen from the soil of semi-arid regions. US/IBP Desert Biome Res. Memo. 74-39. Utah State Univ., Logan. 11 pp.

TUKEY, H. B., JR. 1970. Leaching of substances from plants. Annu. Rev. Plant Physiol. 21:305-324.

TURNER, F. B., coordinator, et al., 1973. Rock Valley Validation Site report. US/IBP Desert Biome Res. Memo. 73-2. Utah State Univ., Logan. 211 pp.

TURNER, F. B., and J. F. MCBRAYER, eds. 1974. Rock Valley Validation Site report. US/IBP Desert Biome Res. Memo. 74-2. Utah State Univ., Logan. 64 pp.

TYLER, K. B., and F. E. BROADBENT. 1960. Nitrite transformations in California soils. Soil Sci. Soc. Am. Proc. 24:279-282.

UMAN, M. A. 1971. Understanding lightning. Bek Technol. Publ., Inc., Carnegie, Penn. 166 pp.

U.S. DEPARTMENT OF AGRICULTURE. 1961. A universal equation for predicting rainfall-erosion losses. ARS 22-66. 11 pp.

VALLENTINE, J. F. 1971. Range development and improvements. Brigham Young Univ. Press, Provo, Utah. 516 pp.

VAN CLEVE, K., L. A. VIERECK, and R. L. SCHLENTNER. 1971. Accumulation of nitrogen in alder (*Alnus*) ecosystems near Fairbanks, Alaska. Arct. Alp. Res. 3: 101-114.

VAN DEN HONERT, T. H., and J. J. M. HOOYMANS. 1955. On the absorption of nitrate by maize in water culture. Acta Bot. Neerl. 4:376-384.

VAN SCHREVEN, D. A., and D. W. HARMSEN. 1968. Soil bacteria in relation to the development of polders in the region of the former Zuider Zee. Pages 474-

499 *in* T. R. G. Gray and D. Parkinson, eds. The ecology of soil bacteria. Univ. of Toronto Press, Toronto.

VERCOE, J. M. 1962. Some observations on the nitrogen and energy losses in the feces and urine of grazing sheep. Proc. Aust. Soc. Anim. Prod. N.S.W. Branch Bull. 4:160-162.

VINCENT, J. M. 1974. Root-nodule symbioses with *Rhizobium*. Pages 266-307 *in* A. Quispel, ed. The biology of nitrogen fixation. North-Holland Publ. Co., Amsterdam.

VITOUSEK, P. M., and W. A. REINERS. 1975. Ecosystem succession and nutrient retention: A hypothesis. BioScience 25:376-381.

VOJINOVIC, Z., and S. SESTIC. 1967. Losses of N by volatilization of NH_3 after applying urea to some soils. Zemljiste Biljka (Belgrade) 16A:43-51.

VOLK, G. M. 1970. Gaseous loss of ammonia from prilled urea applied to slash pine. Soil Sci. Soc. Am. Proc. 34:513-516.

WAGNER, G. H., and G. E. SMITH. 1958. Nitrogen losses from soils fertilized with different nitrogen carriers. Soil Sci. 85:125-229.

WAKSMAN, S. A., and F. C. GERRETSON. 1931. Influence of temperature and moisture upon the nature and extent of decomposition of plant residues by microorganisms. Ecology 12:33-60.

WAKSMAN, S. A., and F. G. TENNEY. 1927. The composition of natural organic materials and their decomposition in the soil: II. Influence of age of plants upon rapidity and nature of their decomposition. Soil Sci. 24:317-333.

WAKSMAN, S. A., and F. G. TENNEY. 1928. Composition of natural organic materials and their decomposition in the soil: III. The influence of nature of plant upon the rapidity of its decomposition. Soil Sci. 26:155-171.

WALLACE, A., and R. T. ASHCROFT. 1956. Preliminary comparisons of the effects of urea and other nitrogen sources on the mineral composition of rough lemon and bean plants. Proc. Am. Soc. Hortic. Sci. 68:227-233.

WALLACE, A., R. T. ASHCROFT, and O. R. LUNT. 1967. Day-night periodicity of exudation of detopped tobacco. Plant Physiol. 42:238-242.

WALLACE, A., and R. T. MUELLER. 1957. Ammonium and nitrate nitrogen absorption from sand culture by rough lemon cuttings. Proc. Am. Soc. Hortic. Sci. 69: 183-188.

WALLACE, A., and R. T. MUELLER. 1963. K^{42} vs $HC^{14}O_3$-uptake (less free space components) from $KHCO_3$ solutions. Pages 60-62 *in* A. Wallace, ed. Solute uptake by intact plants. A. Wallace, Los Angeles.

WALLACE, A., and E. M. ROMNEY. 1972a. Radioecology and ecophysiology of desert plants at the Nevada Test Site. U.S. Atomic Energy Comm. TID-26954. Office of Information Services, Springfield, Va. 439 pp.

WALLACE, A. and E. M. ROMNEY. 1972b. Estimate of primary productivity in the Rock Valley area of the Nevada Test Site for 1966-1968. Pages 230-246 *in* A. Wallace and E. M. Romney, eds. Radioecology and ecophysiology of desert plants at the Nevada Test Site. U.S. Atomic Energy Comm. TID-25954. Office of Information Services, Springfield, Va.

WALLACE, A., and E. M. ROMNEY. 1972c. Approximate age of shrub clumps at the Nevada Test Site. Pages 307-309 *in* A. Wallace and E. M. Romney, eds. Radioecology and ecophysiology of desert plants at the Nevada Test Site. U.S. Atomic Energy Comm. TID-25954. Office of Information Services, Springfield Va.

WALLACE, A., and E. M. ROMNEY. 1972d. Preliminary studies on nitrogen cycling. Pages 339-342 *in* A. Wallace and E. M. Romney, eds. Radioecology and ecophysiology of desert plants at the Nevada Test Site. U.S. Atomic Energy Comm. TID-25954. Office of Information Services, Springfield, Va.

WALLACE, A., and E. M. ROMNEY. 1972e. Response of Mojave Desert vegetation to nitrogen fertilizer and supplemental moisture. Pages 349-351 *in* A. Wallace and E. M. Romney, eds. Radioecology and ecophysiology of desert plants at the Nevada Test Site. U.S. Atomic Energy Comm. TID-25954. Office of Information Services, Springfield, Va.

WALLACE, A., and E. M. ROMNEY. 1972f. Symbiotic nitrogen fixation in desert vegetation and its relationship to nitrogen cycling. Pages 352-357 *in* A. Wallace and E. M. Romney, eds. Radioecology and ecophysiology of desert plants at the Nevada Test Site. U.S. Atomic Energy Comm. TID-25954. Office of Information Services, Springfield, Va.

WALLACE, A., and E. M. ROMNEY. 1975. Feasibility and alternate procedures for decontamination and post-treatment management of Pu-contaminated areas in Nevada. Pages 251-337 *in* The radioecology of plutonium and other transuranics in desert environments: Nevada Applied Ecology Group Progress Report. USERDA Rep. NVO-153. 90 pp.

WALLACE, A., E. M. ROMNEY, J. W. CHA, and S. M. SOUFI. 1974. Nitrogen transformations in Rock Valley and adjacent areas of the Mojave Desert. US/IBP Desert Biome Res. Memo. 74-36. Utah State Univ., Logan. 25 pp.

WALLACE, A., and R. L. SMITH. 1954. Nitrogen interchange during decomposition of orange and avocado tree residues in soil. Soil Sci. 78:231-242.

WALLWORK, J. A. 1962. Distribution patterns and population dynamics of the microarthropods of a desert soil in southern California. J. Am. Ecol. 41: 291-310.

WALTER, H. 1963. Climatic diagrams as a means to comprehend the various climatic types for ecological and agricultural purposes. Pages 3-9 *in* A. J. Rutter and F. H. Whitehead, eds. The water relations of plants. John Wiley & Sons, Inc., New York.

WALTER, H. 1971. Ecology of tropical and sub-tropical vegetation. (Transl. from German.) Oliver and Boyd, Edinburgh.

WALTER, H. 1973. Vegetation of the earth in relation to climate and the ecophysiological conditions. (Transl. from German.) Springer-Verlag, New York.

WATKINS, S. H., R. F. STRAND, D. S. DEBELL, and J. ESCH. 1972. Factors influencing ammonia losses from urea applied to northwestern forest soils. Soil Sci. Soc. Am. Proc. 36:354-357.

WATSON, C. W. 1930. Studies of decomposition of forest organic matter as related to its composition. Ph.D. Dissertation. Yale Univ., New Haven, Conn.

WATSON, E. R., and P. LAPINS. 1964. The influence of subterranean clover pastures on soil fertility. II. The effects of certain management systems. Aust. J. Agric. Res. 15:885-894.

WATSON, E. R., and P. LAPINS. 1969. Losses of nitrogen from urine on soils from southwestern Australia. Aust. J. Exp. Agric. Animal Husb. 9:85-91.

WATTS, J. G., and A. C. BELLOTTI. 1967. Some new and little-known insects of economic importance on range grasses. J. Econ. Entomol. 60:961-963.

WEISSMAN, G. S. 1950. Growth and nitrogen absorption of wheat seedlings as influenced by ammonium: Nitrate ratio and the hydrogen ion concentration.

Am. J. Bot. 37:725-738.

WELCH, L. F., and A. D. SCOTT. 1960. Nitrification of fixed ammonium in clay minerals as affected by added potassium. Soil Sci. 90:79-85.

WELCH, T. G., and J. O. KLEMMEDSON. 1975. Influence of the biotic factor and parent material on distribution of nitrogen and carbon in ponderosa pine ecosystems. Pages 159-178 *in* B. Bernier and C. H. Winget, eds. Forest soils and forest land management. Les Presses de l'Univ. Laval, Quebec.

WELLS, K. F. 1967. Aspects of shrub-herb productivity in an arid environment. M.S. Thesis. Univ. of California, Berkeley. 67 pp.

WENT, F. W., and N. STARK. 1968. Mycorrhiza. BioScience 18:1035-1039.

WENT, F. W., J. WHEELER, and G. C. WHEELER. 1972. Feeding and digestion in some ants (*Veromessor* and *Manica*). BioScience 22:82-88.

WEST, N. E. 1972. Biomass and nutrient dynamics of some major cold desert shrubs. US/IBP Desert Biome Res. Memo. 72-15. Utah State Univ., Logan. 24 pp.

WEST, N. E. Formation, distribution and function of plant litter in desert ecosystems. *In* Raj Kumar Gupta, ed. IBP Arid Land Synthesis. Cambridge Univ. Press, London. (In press a)

WEST, N. E. Nutrient cycling in desert ecosystems. *In* I. Noy-Meir, ed. Short-term dynamics of arid ecosystems. Cambridge Univ. Press, London (In press b)

WEST, N. E. Nutrient cycling in desert ecosystems. *In* R. A. Perry and D. W. Goodall, eds. Arid ecosystems. Cambridge Univ. Press, London. (In press c)

WEST, N. E., and M. FAREED. 1973. Shoot growth and litter fall processes as they bear on primary production of some cool desert shrubs. US/IBP Desert Biome Res. Memo. 73-9. Utah State Univ., Logan. 13 pp.

WEST, N. E., and G. F. GIFFORD. 1976. Rainfall interception by cool-desert shrubs. J. Range Manage. 29:171-172.

WEST, N. E., and C. GUNN. 1974. Phenology, productivity and nutrient dynamics of some cool desert shrubs. US/IBP Desert Biome Res. Memo. 74-7. Utah State Univ., Logan. 6 pp.

WEST, N. E., and J. SKUJINS. 1977. The nitrogen cycle in North American cold winter semi-desert ecosystems. Oecol. Plant. (Paris) 12:45-53.

WESTOBY, M. 1973. The impact of black-tailed jackrabbits (*Lepus californicus*) on vegetation in Curlew Valley, northern Utah. Ph.D. Dissertation, Utah State Univ., Logan. 159 pp.

WHISLER, F. O., J. C. LANCE, and R. S. LINEBARGER. 1974. Redox-potentials in soil columns intermittently flooded with sewage water. J. Environ. Qual. 3:74-78.

WHITFORD, W. G. 1973. Deomography and bioenergetics of herbivorous ants in a desert ecosystem as functions of vegetation, soil type and weather variables. US/IBP Desert Biome Res. Memo. 73-29. Utah State Univ., Logan. 63 pp.

WHITFORD, W. G., coordinator, et al. 1973. Jornada Validation Site report. US/IBP Desert Biome Res. Memo. 73-4. Utah State Univ., Logan. 332 pp.

WIELAND, P. A. T., E. F. FROLICH, and A. WALLACE. 1971. Vegetative propagation of woody shrub species from the northern Mojave and southern Great Basin Deserts. Madroño 21:149-152.

WIJLER, J., and C. C. DELWICHE. 1954. Investigations on the denitrifying process in soil. Plant Soil 5:155-169.

WILKIN, D. 1973. Special purpose models: model of response of a desert shrub community to sheep grazing. US/IBP Desert Biome Res. Memo. 73-59. Utah State Univ., Logan. 19 pp.

WILLIAMS, C. R. 1970. Pasture nitrogen in Australia. J. Aust. Inst. Agric. Sci. 36:199-205.

WILLIAMS, S. E., and E. F. ALDON. 1974. Growth of *Atriplex canescens* (Pursh) Nutt. improved by formation of vesicular-arbuscular mycorrhizae, *Endogone mosseae*. Soil Sci. Soc. Am. Proc. 38:962-965.

WILLIAMS, S. E., and E. F. ALDON. 1976. Endomycorrhizal (vesicular-arbuscular) associations of some arid zone shrubs. Southwest Nat. 20:437-444.

WILLIAMS, S. T., and T. R. G. GRAY. 1974. Decomposition of litter on the soil surface. Pages 611-632 *in* C. H. Dickinson and G. J. F. Pugh, eds. Biology of plant litter decomnposition. Vol. 2. Academic Press, New York.

WISCHMEIER, W. H. 1959. A rainfall erosion index for a universal erosion equation. Soil Sci. Soc. Am. Proc. 23:246-249.

WISCHMEIER, W. H. 1961. Soil loss prediction in North Dakota, South Dakota, Nebraska, and Kansas. U.S. Dep. Agric. Soil Conserv. Serv., Lincoln, Neb. 62 pp.

WISCHMEIER, W. H., and D. D. SMITH. 1965. Predicting rainfall erosion losses from croplands east of the Rocky Mountains. U.S. Dep. Agric. Handbook 282. U.S. Goverment Printing Office, Washington, D. C. 42 pp.

WISER, W., and G. SCHWEIZER. 1970. A re-examination of the excretion of nitrogen by terrestrial arthropods. J. Exp. Biol. 52:267-274.

WITKAMP, M. 1963. Microbial populations of leaf litter in relation to environmental conditions and decomposition. Ecology 44:370-377.

WITKAMP, M. 1966. Decomposition of leaf litter in relation to environment, microflora, and microbial respiration. Ecology 47:194-202.

WOLDENDROP, J. W. 1962. The quantitative influence of the rhizosphere on denitrification. Plant Soil 17:267-270.

WOOD, T. G. 1971. The distribution and abundance of *Folsomides deserticola* (Collembola: Isotomidae) and other micro-arthropods in arid and semi-arid soils in southern Australia, with a note on nematode populations. Pedobiology 11:446-468.

WOODMANSEE, R. G. Nitrogen budget for a shortgrass prairie—additions and losses. *In* D. C. Adriano and I. L. Brisbin, eds. Environmental Chemistry and Cycling Processes Symp. U.S. Energy and Resource Development Admin. (In press)

WOODRUFF, N. P., and F. H. SIDDOWAY. 1965. A wind erosion equation. Soil Sci. Soc. Am. Proc. 29:602-608.

WULLSTEIN, L. H., and C. M. GILMOUR. 1964. Gaseous nitrogen losses and range fertilization. J. Range Manage. 17:203.

YAALON, D. H. 1964. The concentration of ammonia and nitrate in rain water over Israel in relation to environmental factors. Tellus 16:200-204.

ZINGG, A. W. 1940. Degree and length of land slope as it affects soil loss in runoff. Agric. Eng. 21:59-64.

ZINGG, A. W. 1953. Wind tunnel studies of movement of sedimentary material. Pages 111-135 *in* Proc. 5th Hydraul Conf. Bull. 34. Iowa State Univ., Ames.

ZINKE, P. J. 1962. The pattern of influence of individual forest trees on soil properties. Ecology. 43:130-133.

ZINKE, P. J., and R. L. CROCKER. 1962. The influence of giant sequoia on soil properties. For. Sci. 8:2-11.

AUTHOR INDEX

TAXONOMIC INDEX

SUBJECT MATTER INDEX